T0228888

QUANTUM COHERENCE AND DECOHERENCE

North-Holland
Delta Series

ELSEVIER
Amsterdam – Lausanne – New York – Oxford – Shannon – Singapore – Tokyo

Quantum Coherence and Decoherence

Proceedings of the 6th International Symposium on Foundations of Quantum
Mechanics in the Light of New Technology (ISQM-Tokyo '98),
Advanced Research Laboratory, Hitachi, Ltd., Hatoyama, Saitama, Japan,
August 24-27, 1998

Edited by

Y.A. Ono

Advanced Research Laboratory
Hitachi, Ltd.
Saitama, Japan

K. Fujikawa

Department of Physics
The University of Tokyo
Tokyo, Japan

1999

ELSEVIER
Amsterdam – Lausanne – New York – Oxford – Shannon – Singapore – Tokyo

ELSEVIER SCIENCE B.V.
Sara Burgerhartstraat 25
P.O. Box 211, 1000 AE Amsterdam, The Netherlands

1999 Elsevier Science B.V. All rights reserved

This work is protected under copyright by Elsevier Science, and the following terms and conditions apply to its use:

Photocopying
Single photocopies of single chapters may be made for personal use as allowed by national copyright laws. Permission of the Publisher and payment of a fee is required for all other photocopying, including multiple or systematic copying, copying for advertising or promotional purposes, resale, and all forms of document delivery. Special rates are available for educational institutions that wish to make photocopies for non-profit educational, classroom use.

Permissions may be sought directly from Elsevier Science Rights & Permissions Department, PO Box 800, Oxford OX5 1DX, UK; phone: (+44) 1865 843830, fax: (+44) 1865 853333, e-mail: permissions@elsevier.co.uk. You may also contact Rights & Permissions directly through Elsevier's home page (http://www.elsevier.nl), selecting first 'Customer Support', then 'General Information', then 'Permissions Query Form'.

In the USA, users may clear permissions and make payments through the Copyright Clearance Center, Inc., 222 Rosewood Drive, Danvers, MA 01923, USA; phone: (978) 7508400, fax: (978) 7504744, and in the UK through the Copyright Licensing Agency Rapid Clearance Service (CLARCS), 90 Tottenham Court Road, London W1P 0LP, UK; phone: (+44) 171 631 5555; fax: (+44) 171 631 5500. Other countries may have a local reprographic rights agency for payments.

Derivative Works
Tables of contents may be reproduced for internal circulation, but permission of Elsevier Science is required for external resale or distribution of such material. Permission of the Publisher is required for all other derivative works, including compilations and translations.

Electronic Storage or Usage
Permission of the Publisher is required to store or use electronically any material contained in this work, including any chapter or part of a chapter.

Except as outlined above, no part of this book may be reproduced, stored in a retrieval system or transmitted in any form or by any means, electronic, mechanical, photocopying, recording or otherwise, without prior written permission of the Publisher. Address permissions requests to: Elsevier Science Rights & Permissions Department, at the mail, fax and e-mail addresses noted above.

Notice
No responsibility is assumed by the Publisher for any injury and/or damage to persons or property as a matter of products liability, negligence or otherwise, or from any use or operation of any methods, products, instructions or ideas contained in the material herein. Because of rapid advances in the medical sciences, in particular, independent verification of diagnoses and drug dosages should be made.

First edition 1999

Library of Congress Cataloging in Publication Data
International Symposium Foundations of Quantum Mechanics in the Light of New Technology (6th : 1998 : Hatoyama-machi, Japan)
    Quantum coherence and decoherence : proceedings of the 6th International Symposium of Quantum Mechanics in the Light of New Technology (ISQM-Tokyo'98), Advanced Research Laboratory, Hitachi Ltd., Hatoyatama, Saitama, Japan, August 24-27, 1998 / edited by Y.A. Ono, K. Fujikawa.-- 1st ed.
        p. cm. -- (North-Holland delta series)
    Includes bibliographical references and index.
    ISBN 0-444-50091-X (alk. paper)
    1. Coherence (Nuclear physics)--Congresses. 2. Quantum theory--Congresses. I. Ono, Yoshimasa A. II. Fujikawa, K. III. Title. IV. Series.

QC794.6.C58 I55 1998
539--dc21
                                                          99-056387

ISBN: 0 444 50091 X

♾ The paper used in this publication meets the requirements of ANSI/NISO Z39.48-1992 (Permanence of Paper).

Printed and bound by Antony Rowe Ltd, Eastbourne
Transferred to digital printing 2006

PREFACE

The Sixth International Symposium on Foundations of Quantum Mechanics in the Light of New Technology (ISQM-Tokyo '98) was held on August 24-27, 1998 at the Advanced Research Laboratory, Hitachi, Ltd. in Hatoyama, Saitama, Japan. The symposium was organized by its own Scientific Committee under the auspices of the Physical Society of Japan, the Japan Society of Applied Physics, and the Advanced Research Laboratory, Hitachi, Ltd. A total of 135 participants (36 from abroad) attended the symposium, and 27 invited oral papers, 19 contributed oral papers, and 34 poster papers were presented.

Just as in the previous five symposia, the aim of this symposium was to link the recent advances in technology with fundamental problems in quantum mechanics. It provided a unique interdisciplinary forum where scientists with different backgrounds, who would otherwise never meet, were given the opportunity to discuss basic problems of common interest in quantum science and technology from various aspects and "in the light of new technology."

The theme of the fifth symposium of this series (ISQM-Tokyo '95), *Quantum Coherence and Decoherence,* was again chosen as the main theme for the sixth symposium because of its importance in quantum science and technology. This topic was reexamined from all aspects, not only in terms of quantum optics and mesoscopic physics, but also in terms of the physics of precise measurement, macroscopic quantum phenomena, complex systems, and other fundamental problems in quantum physics. Two new important fields were added in the 1998 symposium: the field of quantum computing, including quantum teleportation, quantum information, and cryptography, and the field of laser cooling, including Bose-Einstein condensation and atom interferometry. We were delighted that many active and well-known researchers in these fields accepted our invitation.

We are now very happy to offer the fruits of the symposium in the form of these proceedings to a wider audience. As shown in the table of contents, the proceedings include 73 refereed papers in twelve sections: quantum computing, quantum teleportation, and entanglement; quantum information and cryptography; quantum optics; laser cooling, ion trapping, and atom interferometry; Bose-Einstein condensation; mesoscopic magnets; quantum transport; superconductors and single electronics; nanoscale physics and atomics; mesoscopic semiconductors and normal metals; precise measurements; complex systems and fundamental problems in quantum physics. Here we will just mention some of the important key words to give the flavor of the proceedings: quantum computation, Schrödinger's cat, Bose-Einstein condensates, Coulomb blockade, single-electron transistor, mesoscopic ferromagnetic metals, nanostructures, superfluid interferometry, and dynamics of vortices in high-temperature superconductors. We hope that these proceedings will not only serve as a good introductory book on quantum coherence and decoherence for newcomers in this field, but also as a reference book for experts, just as the Proceedings of ISQM-Tokyo '95 did.

We plan to continue to hold ISQM in the new style set in 1995. The next meeting is scheduled to be held in the summer of 2001 in Hatoyama, with financial and other support from Hitachi.

In conclusion, we thank the participants for their contribution to the symposium's success. Thanks are also due to all the authors who prepared manuscripts, and to the referees who kindly reviewed the papers. We also thank the members of the Advisory Committee and Organizing Committee, without whose kind cooperation the symposium would not have been a success. Finally, we would like to express our deepest gratitude to the Advanced Research Laboratory, Hitachi, Ltd. and its General Manager, Dr. Katsuki Miyauchi, for providing us with financial support and an environment ideal for lively discussion, and to his staff members, in particular Naoyuki Chino, Mariko Hotei, and Akemi Tsuchida, for their efforts in making the symposium enjoyable as well as fruitful.

September 1999

Y.A. Ono
K. Fujikawa

The Sixth International Symposium on Foundations of Quantum Mechanics (ISQM-Tokyo '98)

COMMITTEES

Chair:

H. Fukuyama, University of Tokyo

Advisory Committee:

H. Ezawa, Gakushuin University
R.J. Glauber, Harvard University
Y. Imry, Weizmann Institute
S. Kobayashi, The Institute of Physical and Chemical Research (RIKEN)
A.J. Leggett, University of Illinois at Urbana-Champaign
K. Miyauchi, Advanced Research Laboratory, Hitachi, Ltd.
M. Namiki, Waseda University
H.A. Weidenmüller, Max-Planck-Institut für Kernphysik
Y. Yamamoto, Stanford University and NTT
C.N. Yang, State University of New York at Stony Brook
A. Zee, University of California at Santa Barbara

Organizing Committee:

K. Fujikawa, University of Tokyo
Y. Iye, University of Tokyo
S. Nakajima, Superconductivity Research Laboratory,
            International Superconductivity Technology Center
F. Shimizu, University of Electro-Communications
Y.A. Ono, Advanced Research Laboratory, Hitachi, Ltd.

Sponsors:

The Physical Society of Japan
The Japan Society of Applied Physics
Advanced Research Laboratory, Hitachi, Ltd.

TABLE OF CONTENTS

QUANTUM COMPUTING, QUANTUM TELEPORTATION, AND ENTANGLEMENT

QUANTUM INFORMATION AND CRYPTOGRAPHY

QUANTUM OPTICS

LASER COOLING, ION TRAPPING, AND ATOM INTERFEROMETRY

BOSE-EINSTEIN CONDENSATION

MESOSCOPIC MAGNETS

QUANTUM TRANSPORT

SUPERCONDUCTORS AND SINGLE ELECTRONICS

NANOSCALE PHYSICS AND ATOMICS

MESOSCOPIC SEMICONDUCTORS AND NORMAL METALS

PRECISE MEASUREMENTS

COMPLEX SYSTEMS AND FUNDAMENTAL PROBLEMS
IN QUANTUM PHYSICS

Quantum Coherence and Decoherence - ISQM - Tokyo '98
Y.A. Ono and K. Fujikawa (Editors)
© 1999 Elsevier Science B.V. All rights reserved.

Opening Address

Hidetoshi Fukuyama

Chairman of the Organizing Committee, ISQM-Tokyo '98

Good morning, ladies and gentlemen. On behalf of the Organizing Committee, I would like to welcome you to this International Symposium on the Foundations of Quantum Mechanics in the Light of New Technology (ISQM-Tokyo '98). This is the 6th in a series, which started in 1983. It is actually a great pleasure and honor for me to open this symposium, to which more than 130 scientists have pre-registered. In particular, I would like to thank the thirty-five participants who flew across the ocean from twelve different countries in spite of this being the holiday season.

I think some of you who participated in earlier symposia might feel a bit uneasy that Professor Nakajima is not making the opening address this time. Professor Nakajima chaired all five previous ISQM-Tokyo symposia, but, at the end of 1997 he announced his resignation from the chair post and assigned it to me instead. Of course I urged him to stay in the post, but it was no use. That is why I am standing here today.

Here I think it is appropriate to tell you briefly how this unique forum started fifteen years ago. I think, and I believe you will also agree with me, that this forum is very unique in the sense that scientists of very different disciplines, who otherwise would never meet, have this opportunity to convene and exchange information. Our common interest is to gain a deeper understanding of the implications of quantum mechanics. Actually, all of the properties and functionalities of materials around us are due to the different behaviors of electrons and nuclei, whose motions on a microscopic scale are governed by quantum mechanics. Moreover, our understanding of high energy physics, that is, nuclear and elementary particle physics and cosmology, is also based on quantum mechanics. Namely, the understanding of basically all branches of physics is based on quantum mechanics.

In the early 80s, Professor Nakajima, who was then Director of the Institute for Solid State Physics of the University of Tokyo, and Dr. Yasutsugu Takeda, who was General Manager of the Central Research Laboratory, Hitachi, Ltd. of Hitachi were of the same opinion that the potential importance of the interplay between basic science and technology will be achieved by a deeper understanding of quantum mechanics. To encourage this interplay, they decided to hold an international symposium on the foundations of quantum mechanics in the light of new technology (ISQM). Their action was also triggered by a beautiful experiment to prove the existence on the Aharonov-Bohm effect carried out by Dr. Akira Tonomura at the Central Research Laboratory. In his pursuit of this research Dr. Tonomura received much encouragement from Professor Yang of State University of New York at Stony Brook, who has also been helping with this Symposium from the beginning. This is how the ISQM got started.

On this occasion, it would be interesting to look back on past subjects of this series of symposia.

1st ISQM (held at the Central Research Laboratory in 1983)
> Aharonov-Bohm effect, macroscopic quantum tunneling, quantum electronics, theory of measurement, quantum nondemolition measurement, quantum logic, Einstein-Podolsky-Rosen paradox, philosophy of quantum mechanics, quantum optics, neutron interferometry, and quantum Hall effect

2nd ISQM (held at the Central Research Laboratory in 1986)
> Neutron interferometry, single-atom effect and delayed choice experiment, two-photon correlations, non-equilibrium dynamics, Aharonov-Bohm effect and electron interferometry, quantum Hall effect, microfabrication and quantum mechanics, squeezed state of light and quantum jumps, conceptual problems, macroscopic quantum effects, quantum field theory and related problems, and theory of measurement

3rd ISQM (held at the Central Research Laboratory in 1989)
> Quantum interference, topological aspects of quantum mechanics, theories of measurement, mesoscopic systems, optical phenomena, cosmology and irreversibility, superconductivity, information processing

4th ISQM (held at the Central Research Laboratory in 1992)
> Main theme: *Quantum Tunneling*
> Macroscopic quantum tunneling, quantum diffusion, quantum gravity, tunneling phenomena in nuclear physics and cosmology, superconductivity in mesoscopic systems, mesoscopic semiconductors and magnets, quantum Hall effect, quantum wire and quantum dot, dissipation and tunneling, quantum nucleation, and scanning tunneling microscope and spectroscopy

5th ISQM (held at the Advanced Research Laboratory in 1995)
> Main theme: *Quantum Coherence and Decoherence*
> Quantum optics, precise measurements, macroscopic quantum phenomena, single electronics, mesoscopic semiconductors,

superconductors and SQUID, mesoscopic normal metals,
mesoscopic magnets, complex systems, quantum information
and gravity, and fundamental problems

For the present ISQM, the 6th of this series, we have chosen again the main
theme of *Quantum Coherence and Decoherence* and we will discuss the following
topics:

Quantum computing, quantum teleportation, and entanglement,
quantum information and cryptography, quantum optics, laser
cooling, ion trapping and interferometry, Bose-Einstein condensation,
mesoscopic magnets, quantum transport, superconductors and single
electronics, mesoscopic semiconductors and normal metals, nanoscale
physics and atomics, precise measurements, complex systems and
fundamental problems in quantum physics

From this list we can see that there has been a steady interest in the past fifteen
years in the field of mesoscopics in general, where the basic science and
technology really meet. In recent years there has been growing interest in such
areas as nanostructures of not only conventional semiconductors, but also metals,
especially of superconducting and magnetic materials. In addition to these topics,
this symposium introduces new subjects: Bose-Einstein condensation and
quantum computing, where the basic and applied sciences again merge.

Here I should mention that the present symposium would not be possible without
a generous support from the Advanced Research Laboratory, Hitachi, Ltd. headed
by   Dr. Miyauchi, and special thanks are due to him.

We now all agree with that the decision made by Professor Nakajima and Dr.
Takeda fifteen years ago was indeed correct.

I hope all of you enjoy discussing recent scientific progress and making friends for
promoting further international collaborations.

Thank you for your kind attention.

Quantum Coherence and Decoherence - ISQM - Tokyo '98
Y.A. Ono and K. Fujikawa (Editors)
© 1999 Elsevier Science B.V. All rights reserved.

Welcome Address

Katsuki Miyauchi

Advanced Research Laboratory, Hitachi, Ltd.
Hatoyama, Saitama 350-0395, Japan

Good morning, ladies and gentlemen. On behalf of our laboratory, I would like to extend our warmest welcome to you for coming to Hatoyama to attend ISQM-Tokyo '98, or the Sixth International Symposium on Foundations of Quantum Mechanics in the Light of New Technology.

This is the second time we have held an ISQM meeting here at Hatoyama. The 5th ISQM meeting, which was held here three years ago under the theme of *Quantum Coherence and Decoherence*, was a great success because of the lively discussions on basic problems of common interest in quantum science as well as quantum technology among theorists and experimentalists from different fields of research. And this ISQM meeting will focus on the same theme again, *Quantum Coherence and Decoherence*. In the last three years there has been a great deal of development in many areas of quantum physics, such as Bose-Einstein condensation and quantum computing. I am looking forward to hearing the results made possible by new methods based on new technology. Looking over the names of the speakers and the session titles, I am convinced that this meeting is going to be a very exciting one.

Let me talk a little bit about Advanced Research Laboratory, Hitachi, Ltd. (HARL), which was founded in 1985 for the specific purpose of conducting advanced research to push back the frontiers of science and technology. Currently, about 100 scientists are working on electron and photon physics, nanotechnology, biomolecular science and advanced computing. In particular, nano-fabrication using scanning tunneling microscopes, observation of vortex dynamics using coherent electron waves, and laser-cooling based technologies are closely related to the themes of this year's ISQM meeting. And this why we are hosting and sponsoring the meeting here again.

Responding to the requests from participants at the previous meeting, poster and discussion sessions are scheduled for the evening so that you will have ample time to discuss the physics in detail raised in the morning and afternoon sessions as well as the poster presentations.

It would please me greatly if all of you find it rewarding to be here over the next four days. Before concluding, I would like to thank all the members of the Organizing Committee and Advisory Committee for putting together such an exciting program. I also would like to thank all the invited speakers for coming to share your latest findings with the participants here, who represent 14 nations of the world. I very much look forward to a meeting made successful thanks to your active involvement.

Thank you very much for your attention.

Quantum Coherence and Decoherence - ISQM - Tokyo '98
Y.A. Ono and K. Fujikawa (Editors)
© 1999 Elsevier Science B.V. All rights reserved.

Decoherence and Recoherence in Quantum Computation

David P. DiVincenzo* and Artur Ekert**

*IBM Research Division, T. J. Watson Research Center
P. O. Box 218, Yorktown Heights, NY 10598, USA

**Centre for Quantum Computation, Clarendon Laboratory
Oxford University, Parks Road, Oxford OX1 3PU, UK

Quantum computation, in which boolean data can exist and be manipulated in coherent quantum superpositions, is an exciting prospect for further extending the power of physical systems to perform computation. This power will only be realized if quantum decoherence can be brought under control. Some strategies for recohering quantum systems are introduced, both using quantum error correction and symmetisation. Prospects for the ultimate achievement of quantum computation in the laboratory are discussed.

1. INTRODUCTION

Every 18 months microprocessors double their speed and, it seems, the only way to make them faster is to make them smaller. Today's advanced lithographic techniques can etch logic gates and wires less than a micron across onto the surfaces of silicon chips. Soon they will yield even smaller components, until we reach the point where logic gates are so small that they consist of only a few atoms each. If computers are to continue to become faster (and therefore smaller), new, *quantum* technology must replace or supplement what we have now, but it turns out that such technology can offer much more than smaller and faster microprocessors. It can support entirely new modes of computation, with new quantum algorithms that do not have classical analogues.

Quantum computers can accept input states which represent a coherent superposition of many different possible inputs and subsequently evolve them into a corresponding superposition of outputs. Computation, i.e. a sequence of unitary transformations, affects simultaneously each element of the superposition permitting massively parallel data processing within one piece of quantum hardware. Quantum interference among these parallel computations permits quantum computers to solve efficiently some problems that are believed to be intractable on any classical computer. The most striking example is the factoring problem: to factor a number N of L digits on any classical computer apparently requires an execution time that grows exponentially with L. In contrast, Shor[1] has shown that quantum computers require an execution time that grows only as a polynomial function of L ($\approx L^2$). A noteworthy consequence for cryptology is the possibility of breaking public key cryptosystems such as RSA[2]. In this context a practical implementation of quantum computation is a most important issue.

In principle we know how to build a quantum computer; we start with simple quantum logic gates and connect them up into quantum networks[3, 4]. A quantum logic gate, like a classical gate, is a very simple computing device that performs one elementary quantum operation, usually on two qubits, in a specified time interval. Of course, quantum logic gates differ from their classical counterparts in that they can create quantum superpositions and perform operations on them. However, as the number of quantum gates in a network increases, we quickly run into some serious practical problems. The more interacting qubits are involved, the harder it tends to be to engineer the interaction that would result in the desired quantum interference taking place. Apart from the technical difficulties of working at single-atom and single-photon scales, one of the most important problems is that of preventing the surrounding environment from being affected by the interactions that generate quantum superpositions. The more components there are, the more likely it is that quantum information will spread outside the quantum computer and be lost into the environment, spoiling the computation. This is the unwelcome phenomenon of *decoherence*. Thus our task is to engineer sub-microscopic systems in which qubits affect each other but not the environment.

In this paper we briefly outline some techniques that have been developed by the quantum-computation community. After setting up the basic formalism of decoherence, we indicate how recoherence can be brought about by the use of a certain kind of quantum coding. We will indicate that this coding does not work well in all situations, as when an unknown quantum state evolution is to be protected. But we indicate another kind of quantum coding, based on the use of completely symmetric subspaces, which will produce a recoherence in this scenario.

2. DECOHERENCE

Decoherence is the simplest interesting case of the qubit–environment evolution in which the environment effectively acts as a measuring apparatus:

$$|0\rangle|R\rangle \longrightarrow |0\rangle|R_0(t)\rangle, \tag{1}$$
$$|1\rangle|R\rangle \longrightarrow |1\rangle|R_1(t)\rangle. \tag{2}$$

This kind of evolution is generated by the Hamiltonian of the form $\sigma_z \otimes H$, where H acts only on the states of the environment. It entangles a qubit in a pure state $\alpha|0\rangle + \beta|1\rangle$ with the environment,

$$(\alpha|0\rangle + \beta|1\rangle)|R\rangle \longmapsto \alpha|0\rangle|R_0(t)\rangle + \beta|1\rangle|R_1(t)\rangle. \tag{3}$$

In terms of the qubit density operator,

$$\begin{pmatrix} |\alpha|^2 & \alpha\beta^* \\ \alpha^*\beta & |\beta|^2 \end{pmatrix} \longmapsto \begin{pmatrix} |\alpha|^2 & \alpha\beta^*\langle R_0(t)|R_1(t)\rangle \\ \alpha^*\beta\langle R_1(t)|R_0(t)\rangle & |\beta|^2 \end{pmatrix}, \tag{4}$$

where $\langle R_0(t)|R_1(t)\rangle$ will be assumed to be real and written as $r(t)$. The exact form of $r(t)$ depends on the details of the qubit-environment interaction. For short times $r(t)$ is parabolic in t, which can be seen by writing

$$|R_0(t)\rangle = e^{-iHt/\hbar}|R\rangle = (1 - \frac{i}{\hbar}Ht - \frac{1}{\hbar^2}H^2t^2 + \ldots)|R\rangle \tag{5}$$
$$|R_1(t)\rangle = e^{+iHt/\hbar}|R\rangle = (1 + \frac{i}{\hbar}Ht - \frac{1}{\hbar^2}H^2t^2 + \ldots)|R\rangle \tag{6}$$

which gives

$$r(t) = \langle R_0(t)|R_1(t)\rangle = 1 - 2\frac{t^2}{\hbar^2}\langle R|H^2|R\rangle \ldots \tag{7}$$

By short we mean short compared to the inverse of the frequency bandwidth of the environment $(1/(\langle H^2\rangle - \langle H\rangle^2)$, or simply $1/\langle H^2\rangle$ as we can assume that $\langle R|H|R\rangle = 0)$ which is usually of the order of the typical resonant frequency of the qubit (e.g 10^{-15}sec. for optical systems). Let us mention in passing that from a purely mathematical point of view we have assumed here that the expression $(\langle H^2\rangle - \langle H\rangle^2)$, i.e. the variance of the energy in the initial state $|R\rangle$, is finite. Needless to say, in reality it *is* always finite but there are mathematical models in which, due to various approximations, this may not be the case (e.g. the Lorentzian distribution which has no finite moments). For longer times $r(t)$ is usually approximated by an exponential in t.

3. DECOHERENCE AND RECOHERENCE

3.1. Scenario 1

Let us start with decoherence. Suppose we are given an unknown quantum state of the form

$$|\Psi(0)\rangle = \frac{1}{\sqrt{2}}(|0\rangle + e^{i\varphi}|1\rangle) \quad \text{or} \quad \rho = (0) = \frac{1}{2}(1 + \cos\varphi\sigma_x + \sin\varphi\sigma_y) \tag{8}$$

and expose it to decoherence for some period of time t. The qubit state after decoherence is

$$\rho(t) = \frac{1}{2}(1 + r(t)(\cos\varphi\sigma_x + \sin\varphi\sigma_y)). \tag{9}$$

The fidelity with respect to the initial state is: $\mathrm{Tr}\rho(0)\rho(t) = \frac{1}{2}(1 + r(t))$.

3.2. Scenario 2

We encode $\rho(0)$ into three qubits, expose the three qubits to decoherence for the same period of time t and then decode the state to obtain $\rho'(t)$. Encoding means that we map the state Eq. (8) by a unitary transformation (using quantum computing gates) into a three-qubit state for which $|0\rangle$ is replaced by $|000\rangle + |011\rangle + |101\rangle + |110\rangle$ and $|1\rangle$ is replaced by $|001\rangle + |010\rangle + |100\rangle + |111\rangle$. Decoding simply means an inverse mapping. The decoherence still acts independently on each of the three qubits; it is this locality that permits the coded state to be more immune to decoherence. Details have been given in [5]; the fidelity of this error-corrected state is much better:

$$\mathrm{Tr}\rho(0)\rho'(t) = \frac{1}{2}(1 + r'(t)), \tag{10}$$

$$1 - r'(t) = \frac{3}{2}(1 - r(t))^2 - \frac{1}{2}(1 - r(t))^3. \tag{11}$$

So, $r'(t)$ decreases from 1 initially much slower than $r(t)$ itself.

3.3. Scenario 3

Apart from decoherence our qubits undergo an unknown unitary evolution $U(t)$. For simplicity we take this evolution to be $|0\rangle \longrightarrow |0\rangle, |1\rangle \longrightarrow e^{i\omega t}|1\rangle$. In the absence of decoherence our initial state

$$|\Psi(0)\rangle = \frac{1}{\sqrt{2}}(|0\rangle + e^{i\varphi}|1\rangle),$$

(12)

evolves into

$$|\Psi(t)\rangle = \frac{1}{\sqrt{2}}(|0\rangle + e^{i(\varphi+\omega t)}|1\rangle),$$

(13)

which can also be written as the density operator

$$\sigma(t) = \frac{1}{2}(1 + \cos(\varphi + \omega t) + \sigma_x + \sin(\varphi + \omega t)\sigma_y).$$

(14)

If decoherence is taken into account the final state after time t is

$$\tilde{\sigma}(t) = \frac{1}{2}(1 + r(t)(\cos(\varphi + \omega t) + \sigma_x + \sin(\varphi + \omega t)\sigma_y)).$$

(15)

Now suppose we want to recover $\sigma(t)$ rather than $\rho(0)$ (typical case e.g. frequency standards). We want to maximise

$$\mathrm{Tr}\sigma(t)\tilde{\sigma}(t) = \frac{1}{2}(1 + r(t)).$$

(16)

But if we repeat the three qubit encoding i.e. we encode state $\rho(0)$, let it evolve and decohere for time t and then we decode and correct it to obtain $\tilde{\sigma}'(t)$, then

$$\mathrm{Tr}\sigma(t)\tilde{\sigma}'(t) = \frac{1}{2}(1 + \mathrm{Re}\ r'(t)),$$

(17)

$$r'(t) = \frac{3}{4}(1 + e^{2i\phi})r(t) - \frac{1}{8}(e^{-2i\phi} + 3e^{2i\phi})(r(t))^3.$$

(18)

$r'(t)$ is far from one for most parameter values (it is not even real), indicating that this encoding technique does not help. But there are other ways of achieving recoherence in this case, and we show that recoherence can in fact be performed via symmetrisation.

3.4. Scenario 4

This is the same as Scenario 3, but instead of the three qubit encoding we use symmetrisation. Suppose we start with two qubits initially in state $\rho(0) \otimes \rho(0)$ which evolve and decohere into $\tilde{\sigma}(t) \otimes \tilde{\sigma}(t)$. We project this state into its symmetric component (triplet, symmetric subspace spanned by $|00\rangle, |11\rangle, |01\rangle + |10\rangle$). The projection entangles the two qubits.

$$\tilde{\sigma}(t) \otimes \tilde{\sigma}(t) \longrightarrow S(\tilde{\sigma}(t) \otimes \tilde{\sigma}(t))S^\dagger$$

(19)

where $S = \frac{1}{2}(1 + P_{21})$ and P_{21} is the permutation operator (simple transposition in this case). After symmetrising and taking partial traces we obtain

$$\tilde{\sigma}''(t) = \mathrm{Tr}_2\left[S(\tilde{\sigma}(t) \otimes \tilde{\sigma}(t))S^\dagger\right] = \frac{\tilde{\sigma}^2(t) + \tilde{\sigma}(t)}{\mathrm{Tr}(\tilde{\sigma}^2(t) + \tilde{\sigma}(t))} \tag{20}$$

which can also be written as

$$\tilde{\sigma}''(t) = \frac{1}{2}(1 + r''(t)(\cos(\varphi + \omega t) + \sigma_x + \sin(\varphi + \omega t)\sigma_y)). \tag{21}$$

with $r''(t) = \frac{4r(t)}{3 + r^2(t)}$. This gives the fidelity

$$\mathrm{Tr}\sigma(t)\tilde{\sigma}''(t) = \frac{1}{2}\left(1 + \frac{4r(t)}{3 + r^2(t)}\right) \geq \mathrm{Tr}\sigma(t)\tilde{\sigma}(t) = \frac{1}{2}(1 + r(t)). \tag{22}$$

The snag is that the probability of successful projection on the symmetric subspace is $\frac{3 + r^2(t)}{4}$. Projection on the antisymmetric part (singlet) generates the maximally mixed state of each qubit. Taking this into account the average fidelity is exactly equal to $\mathrm{Tr}\sigma(t)\tilde{\sigma}(t)$. It has been shown, however, that if the projections are done frequently enough, then the cumulative probability that they all succeed can be made as close as desired to unity. This is a consequence of the "quantum watch-dog effect" or the "quantum Zeno effect". Details can be found in [6]. This technique, originally proposed by David Deutsch in his talk at the Rank Prize Funds Mini–Symposium on Quantum Communication and Cryptography, Broadway, England in 1993, has been subsequently applied to improve the accuracy of atomic clocks (building upon ideas presented in [7].

4. CONCLUDING REMARKS

This brief contribution has only scratched the surface of the many activities that are presently being pursued under the heading of quantum computation; but for the most part, we would not be contemplating them were it not for the methods that have been introduced for recohering quantum states. Theoretically, the situation for coherent quantum computation is very good, provided that the decoherence satisfies some assumptions. These assumptions include 1) that decoherence occurs independently on each of the qubits of the system; 2) that the performance of gate operations on some qubits do not cause decoherence in *other* qubits of the system; 3) that reliable quantum measurements can be made so that error *detection* can take place; 4) and that systematic errors in the unitary operations associated with quantum gates be made very small (see below for a number).

If all these assumptions are satisfied, then we have a very strong result, one which says that fault-tolerant quantum computation is possible[8,9]. That is, efficient, reliable quantum-coherent quantum computation of arbitrarily long duration is possible, even with faulty and decohering components. Thus, errors can be corrected faster than they occur, even if the error correction machinery is faulty. There is a threshold associated with this result, fault tolerance is possible if the ratio of the quantum-gate operation time to the decoherence time is smaller than some value (now estimated at around 10^{-4}), and if the size of the systematic error in the gates' unitary transformation is also no greater than 10^{-8} (possibly a pessimistic estimate[9].

As various experimental workers who attended this meeting have been aware, these requirements for the physical implementation are very stringent (see [10, 11] for further discussion of this); nevertheless, there seems to be some hope for their eventual achievement in various branches of experimental physics in which quantum coherence has been of interest, including ion traps (see the paper of D. Wineland), cavity quantum electrodynamics (J. Kimble), and even in some areas of solid state physics (Y. Nakamura and C. Cosmelli in superconducting devices, R. Webb in Coulomb blockade structures). We will see what develops by the time of the ISQM-Tokyo '01!

APPENDIX

The symmetric projections were calculated as follows. Let ρ be $\sum_i p_i |a_i\rangle\langle a_i|$ with $\langle a_i|a_j\rangle = \delta_{ij}$. Then $\rho \otimes \rho = \sum_{ij} p_i p_j |a_i\rangle\langle a_i| \otimes |a_j\rangle\langle a_j|$ and if we apply $S = \frac{1}{2}(1 + P_{21})$ to both sides of $\rho \otimes \rho$ we obtain:

$$\frac{1}{4}\sum_{ij} p_i p_j (|a_i\rangle\langle a_i| \otimes |a_j\rangle\langle a_j| + |a_i\rangle\langle a_j| \otimes |a_j\rangle\langle a_i| + |a_j\rangle\langle a_i| \otimes |a_i\rangle\langle a_j| + |a_j\rangle\langle a_j| \otimes |a_i\rangle\langle a_i|) \quad (23)$$

and after tracing over the second qubit we obtain

$$\frac{1}{2}\sum_i p_i|a_i\rangle\langle a_i| + p_i^2|a_i\rangle\langle a_i| = \frac{1}{2}(\rho + \rho^2). \quad (24)$$

ACKNOWLEDGEMENTS

We are grateful for the Oxford University/IBM Quantum Information Theory NATO Collaborative Grant.

REFERENCES

1. P. W. Shor, *SIAM J. Comput.* **26** (1997) 1484; in *Proc. 35th Annu. Symp. Foundations of Computer Science* (IEEE Computer Soc., Los Alamitos, CA, 1994), p. 124.
2. G. Brassard, "Searching the Quantum Phone Book," *Science* **275** (1997) 627.
3. A. Barenco, C. H. Bennett, R. Cleve, D. P. DiVincenzo, N. Margolus, P. Shor, T. Sleator, J. A. Smolin, and H. Weinfurter, *Phys. Rev.* **A52** (1995) 3457.
4. D. P. DiVincenzo, *Science* **270** (1995) 255.
5. D. P. DiVincenzo, *J. Appl. Phys.* **81** (1997) 4602; cond-mat/9612125.
6. A. Barenco, A. Berthiaume, D. Deutsch, A. Ekert, R. Jozsa, and C. Macchiavello, *SIAM J. Comput.* **26** (1997) 1541.
7. S. F Huelga, C. Macchiavello, T. Pellizzari, A. K. Ekert, M.B. Plenio, and J.I. Cirac, *Phys. Rev. Lett.* **79** (1997) 3865.
8. P. Shor, in *Proc. 37th Annu. Symp. Foundations of Computer Science* (IEEE Computer Soc., Los Alamitos, CA, 1996); see http://xxx.lanl.gov/quant-ph/9605011.
9. J. Preskill, *Proc. R. Soc. Lond.* A **454** (1998) 385.
10. D. Loss and D. P. DiVincenzo, *Phys. Rev.* A **57** (1998) 120.
11. D. P. DiVincenzo and D. Loss, *Superlattices and Microstructures* **23** (1998) 419.

Quantum Coherence and Decoherence - ISQM - Tokyo '98
Y.A. Ono and K. Fujikawa (Editors)
© 1999 Elsevier Science B.V. All rights reserved.

Engineering entanglement between atoms and photons in a cavity: non-locality, Schrödinger cats and decoherence

S. Haroche

Laboratoire Kastler Brossel, Département de physique de l'Ecole Normale Supérieure, 24 rue Lhomond 75231 Paris Cedex 05 France

We perform tests of quantum theory with single atoms crossing a cavity containing a few photons. Various kinds of controlled non-local entanglement are realized. Entangling an atom to a field suspended in a superposition of states with different phases, we have obtained a laboratory version of Schrödinger's cat. We have observed how this superposition evolves, due to decoherence, into a mere statistical mixture of states. These experiments, which explore basic aspects of measurement theory, can also be viewed as demonstrations of elementary steps in quantum information processing using atoms and photons as qubits.

1. INTRODUCTION

Many of the « Gedanken experiments » illustrating the basic features of quantum mechanics can now be realized in the laboratory, owing to the recent development of powerful techniques to manipulate and control simple quantum systems. In the experiments performed at Ecole Normale Supérieure, excited Rydberg atoms are sent one by one across a very high Q cavity containing a few microwave photons [1]. The coupling of the atoms to the field results in an entanglement between matter and radiation which survives after the atom leaves the cavity. The entanglement can be controled by adjusting the parameters of the atom- field interaction. This « engineered entanglement » is different from the situations realized in previous experiments where the correlations between pairs of particles were produced in spontaneous uncontrolled events [2].

With resonant atom field interaction, we can entangle an atom to a field made of zero or one photon [3]. We can also entangle two atoms crossing successively the cavity [4]. We thus realize non-local quantum states of two atoms and illustrate with massive particles the paradoxical situation first discussed by Einstein, Podolsky and Rosen [5] (EPR), which had been studied so far mostly with massless photons [2]. Using non-resonant atom field coupling, we have also prepared so-called « Schrödinger cat » states of radiation and studied their progressive decoherence as the number of photons is increased [6]. In this experiment, the field in the cavity acquires two phases at the same time and the atom-field system evolves into a linear superposition of two states corresponding to different field phases. Each field phase component is entangled to a different energy state of the atom. Here again, we get an EPR like situation but now, one member of the pair, the field, contains many quanta. This experiment constitutes an illustration of the quantum measurement theory [7]. The field in the cavity can be seen as a « mesoscopic » meter which « points » towards the energy of the atom. Decoherence is the process by which, due to the coupling with the environment, the entanglement between the pointer and the system states gets transformed into a statistical mixture. This process, which has been theoretically analyzed by many authors [8], has been observed for the first time in this experiment.

2. EXPERIMENTAL SET-UP:

It is shown in Figure 1. Rydberg atoms are crossing one at a time a microwave cavity C made of two superconducting Niobium mirrors. The atoms, emitted by the oven O, are velocity selected by optical pumping in zone V. They are then prepared in box B by a pulsed combined laser and radiofrequency excitation[9] into a circular Rydberg state of principal quantum number n = 51 or 50 (called e and g respectively in the following). These states[10], in which the valence electron is revolving on a circle centered at the atomic nucleus, are very long lived and have a very strong coupling to millimeter wave radiation. The cavity C (volume of the order of $1cm^3$), sustains a single mode of the field, with a long relaxation time T_{cav} of the order of 100 to 200 µs. The field is excited either by the atoms themselves, or by a classical source of radiation S, coupled into C by a waveguide. This field is either resonant or slightly off-resonant with the transition between the states e and g (51 GHz, 6 mm wavelength). The resonant condition can be modified by applying a pulse of electric field on the mirrors, making use of the Stark effect which tunes the atomic transition in or out of resonance with the cavity mode.

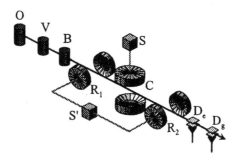

Figure 1. The atom-cavity experimental set-up

At exact resonance, the single atom-cavity coupling is defined by the vacuum Rabi frequency $\Omega/2\pi$ = 50 Khz, which represents the rate at which the atom reversibly emits and reabsorbs a single photon in C. The coherent atom-field coupling largely dominates all causes of field and atom relaxation, whose rates are in the hundred of Hz to kHz range.

The atom-cavity interaction lasts for a time t of the order of 20 µs. The circular atoms then drift out of the cavity and are counted by two field ionization detectors D_e and D_g sensitive to atoms in levels e and g respectively. It is possible to mix coherently levels e and g with the help of classical microwave pulses of radiation before they interact with C (radiation applied in zone R_1) and after this interaction, just before detection (zone R_2). These auxiliary microwave fields are generated by the source S'. The set-up is cooled to 0.6 K to suppress blackbody radiation and optimize the superconducting mirrors reflectivity. The atomic excitation is reduced to a level such that at most one atom is prepared in a single pulse of Rydberg state excitation. Sequences of atoms with variable delay can be sent through the set-up.

3. RESONANT COUPLING: ENGINEERED ENTANGLEMENT

When the cavity mode is tuned in resonance with the e → g atomic transition, photons are emitted or absorbed by each atom in C. The atom-field exchange of energy exhibits an oscillatory behaviour, the Quantum Rabi oscillation [2]. States |e, 0 > and |g, 1>, corresponding to the atom in level e and g with 0 or 1 photon respectively in C, are coherently mixed, providing a simple way of entangling atom and field variables. Starting for example with an atom in level |e > and a cavity in the vacuum state |0 >, the system evolves within the atom-cavity crossing time t into the superposition |ψ >= cos (Ωt/2)|e, 0 >+ isin(Ωt/2)|g, 1>, which corresponds to a non-separable atom-field state, surviving even after the atom and the cavity have separated. This atom-field entanglement can be changed into an atom-atom correlation by sending a second atom which picks the photon left by the first atom.

A first atom, prepared in level e, is sent across the initially empty cavity, and the atom-cavity interaction time is set so that Ωt= π/2. The atom has then equal probabilities to stay excited or to decay to level g by emitting a photon. The system ends up in the state (1/√2) (| e, 0 > + i | g ,1 >). The entanglement is transformed into an atom-atom correlation by sending a second atom initially prepared in g, setting in this case the interaction time so that Ωt = π. The photon left by the first atom is then absorbed by the second one with unit probability, leaving the cavity empty and the atoms in the entangled state |Ψ_{EPR}> = (1/√2) (| e_1, g_2> -| g_1, e_2 >) where the indices 1 and 2 label the first and the second atom respectively. The cavity has served as a « catalyst » to entangle the atoms. The process by which the atom-field entanglement has been transformed into an atom-atom one is a kind of entanglement swapping[11]. The two particles, which have never directly interacted, have got correlated by interacting separately with the cavity field.

This experiment is to our knowledge the first one in which atoms have been entangled at a distance (the maximum separation was 1.5 cm). By combining resonant and non-resonant atom-field interactions, one could generalize this scheme to more than two atoms and prepare, for example, triplets of entangled atoms of the form | e,e,e > - | g,g,g > [12]. The manipulation by cavity QED techniques of many atom entanglement opens the way to new studies of quantum non-locality and tests of generalized Bell's inequalities[13].

4. DISPERSIVE COUPLING: SCHRÖDINGER'S CAT STATES.

In dispersive experiments[1], the atomic transition frequency ω_0 and the field frequency ω differ by a small quantity δ, large compared to Ω and to the cavity line width. A pulsed microwave source S is used to inject a small coherent field in C (average photon number between 0 and 10) and the non-resonant interaction of this field with the atoms is subsequently studied. Photon exchange is then forbidden by energy conservation, but the atom and field subsystems experience dispersive frequency pulling shifts $\Delta\omega = \pm\, \Omega^2 / 4\delta$ [1]. Relative shifts $\Delta\omega/\omega$ of the order of 10^{-7} can be achieved. This corresponds to a huge « single atom index effect » (note that the atomic « density » is of the order of 1atom per cm^3 only). The phase shift $\phi = \Delta\omega t$ accumulated by the field while the atom crosses C is typically of the order of 1 radian. Moreover, this atom index is inherently a quantum effect. The frequency pulling of the cavity takes opposite values for atoms in levels e or g.

The quantum index can be used to generate quantum superposition of field states with different phases [15]. A single atom is prepared in a linear superposition of states e and g by a $\pi/2$ pulse in R_1, while a coherent field corresponding to a complex amplitude a is injected in C. When the atom crosses C, it imparts to the field two opposite phase kicks, $\pm\,\phi$, depending upon whether it is in e or g. The atom-field state becomes $| \psi > = (1/\sqrt{2}\,) (\; | \; e, \alpha e^{\iota\phi} > + | \; g, \alpha e^{-\iota\phi} >)$, an entangled atom-cavity state in which the e and g atomic levels are correlated to coherent states of the field with different phases. The field's phase appears in fact as a "meter" which assumes two different values when the atom is in e or g. One can say that the dispersive interaction realizes in this way an essential step in a "measurement" process [7]. One can also adopt Schrödinger's metaphore [15] and say that the $+\phi$ and $-\phi$ field components are laboratory versions of the "live" and "dead" states of the famous cat trapped in a box with an atom in a linear superposition of its excited and ground states. Since the field in the cavity may contain several photons, these superpositions can be considered as mesoscopic.

If we detect directly the atom after it leaves C and find it in level e or g, the state of the field immediately reduces either to $|\alpha e^{+\iota\phi}>$ or to $|\alpha e^{-\iota\phi} >$, and the quantum ambiguity is lost. In order to preserve this ambiguity, we must mix again, after the atom has crossed C the two levels e and g by applying a $\pi/2$ pulse in R_2. In this way, the detection of the atom does not tell us anymore whether it has crossed C in e or g. Since we do not know the state of the atom in C, the field remains in a superposition of the two « field meter states ». We have performed such an experiment, submitting the atom before and after it crosses C, to two pulses of microwaves in R_1 and R_2. The analyzis of the atomic signal then shows that the field left in the cavity has acquired two distinct phases at once, a trade mark of Schrödinger cat states. The details of this experiment can be found in [6].

5. DECOHERENCE

Theory predicts that coherent mesoscopic field states superpositions of this kind are very fragile and subject to decoherence, when the number of photons, or the angle ϕ between the field components becomes large [8]. In order to check the coherence of the superposition and to study how it gets transformed with time into a mere statistical mixture, we have probed the "cat state" with a second atom, crossing the cavity after a delay [6], according to a scheme first proposed in [16]. The probe atom is identical to the first one and produces the same phase shifts. Since it is also prepared into a superposition of e and g, it again splits into two parts each of the field components produced by the first atom.

The final field state exhibits then four components, two of which coincide in phase. Whether the two atoms have crossed C in the e,g combination, or in the g,e one, the net result is indeed to bring back in both cases the phase of the field to its initial value. After the atomic states have been mixed again in R_2, there is no way to tell in which state the atoms have crossed C (e,g or g,e combination). As a result, two "pathes" associated with the atom pair are undistinguishable. The contributions corresponding to the e,g and g,e pathes lead to the presence of interfering terms in the joint probabilities to detect both atoms in any of the four possible combination of levels. By combining these joint probabilities, one can build a two-atom correlation signal η which is directly linked to the quantum interference resulting from the overlap of the components in the final field in C. If the state superposition survives during the time interval between the atoms, η is different from zero, whereas it vanishes when the state superposition is turned into a statistical mixture. We have measured η versus the time interval T between the two atoms. We have shown that the correlation does indeed vanish at a rate which becomes faster when the separation between the components of the superposition increases. The agreement between experiment and theory is quite good.

In this experiment, the decoherence process is due to the loss of photons escaping from the cavity via scattering on mirror imperfections. Each escaping photon can be described as a small "Schrödinger kitten" copying in the environment the phase information contained in C [18]. The mere fact that this "leaking" information could be read out to determine the phase of the field is enough to wash out the interference effects related to the quantum coherence of the "cat" state. In this respect, we understand that decoherence is a consequence of complementarity [19]. The short decoherence time of our Schrödinger cat, of the order of T_{cav}/n where n is the average photon number in C, is explained by this approach. The larger the photon number, the shorter is the time required to leak a single "photon-copy" in the environment. This experiment verifies the basic features of decoherence and clearly exhibits the fragility of quantum coherences in large systems.

6. CONCLUDING REMARKS

We are presently trying to extend these experiments to more complex situations, involving more atoms, more photons or more cavities. Particularly interesting and intriguing will be the study of entangled states involving mesoscopic fields with many photons entangled between two spatially separated cavities [20]. All these experiments can be related to the active field of quantum information processing. Two-level atoms can indeed be considered as « qubits » carrying quantum information, on which elementary logical operations can be performed. By submitting trains of atoms to well defined resonant or non-resonant interactions, either in the cavity C or in the auxiliary cavities R_1 and R_2, it is possible to map information from one qubit to another (thus realizing a quantum memory[21]) or to produce elementary quantum gates (blue prints for such gates are described in [22]). Note that there are similarities between the microwave cavity QED experiments discussed here and optical cavity QED experiments performed at Caltech [23]. Strong similarities also exist between our experiments and the one being performed with ions oscillating in a trap [24]. In the latter case, the internal states of the ions are entangled to the vibrational degrees of freedom of the ions, which replace the field excitation of the Cavity QED experiments. The extreme efficiency of decoherence is a formidable obstacle for practical use of these systems for actual quantum computation. Even so, experiments performed with isolated atoms in cavities or small number of ions in traps provide fascinating ways to explore the most intriguing aspects of quantum mechanics.

REFERENCES

1. S.Haroche and J.M. Raimond in « Cavity Quantum Electrodynamics », P.Berman, editor, Academic Press, New York (1994).
2. S.J. Freedman and J.F. Clauser, Phys.Rev.Lett. 28, 938 (1972); J.F. Clauser, Phys.Rev.Lett.,36, 1223 (1976); E.S. Fry and R.C. Thompson, Phys.Rev.Lett., 37, 465 (1976); A. Aspect *et al*, Phys.Rev.Lett. 47, 460 (1981); A. Aspect *et al*, Phys.Rev.Lett. 49, 1804 (1982); Z.Y. Ou and L. Mandel, Phys.Rev.Lett. 61,50 (1988); P.G. Kwiat *et al*, Phys.Rev.Lett. 75, 4337 (1995).
3. M.Brune *et al*, Phys.Rev.Lett. 76, 1800 (1996).
4. E. Hagley *et al*, Phys.Rev.Lett. 79, 1 (1997).
5. A. Einstein, B. Podolsky and N. Rosen, Phys.Rev.47, 777 (1935).
6. M. Brune *et al*, Phys.Rev.Lett. 77, 4887 (1996)
7. J.A. Wheeler and W.H. Zurek, « Quantum Theory of measurement », Princeton University Press, Princeton, New Jersey (1983).
8. W.H. Zurek, Physics Today 44, 36 (1991); W.H. Zurek, Phys.Rev.D 24, 1516 (1981); 26, 1862 (1982); A.O. Caldeira and A.J. Legget, Physica A121, 587 (1983); E. Joos and H.D. Zeh, Z.Phys.B 59, 223 (1985); R. Omnès, « The interpretation of Quantum Mechanics », Princeton University Press, Princeton, N.J. (1994).
9. P.Nussenzveig *et al* , Phys.Rev.A 48, 3991 (1993).
10. R.G. Hulet and D. Kleppner, Phys.Rev.Lett 51, 1430 (1983).
11. Jan-Wei Pan et al, Phys.Rev.Lett, 80, 3891 (1998).
12. D.M. Greenberger, M.A. Horne and A. Zeilinger, Am.J.Phys. 58, 1131 (1990).
13. J.S. Bell,Physics (Long Island City, New York) 1, 195 (1964).
14. M. Brune, Phys.Rev.A 45, 5193 (1992).
15. E. Schrödinger, Naturwissenschaften 23, 807, 823,844 (1935); reprinted in english in (7).
16. L. Davidovich *et al*, Phys.Rev A 53, 1295 (1996).
17. J.M. Raimond, M. Brune and S. Haroche, Phys.Rev.Lett. 79, 1964 (1997).
18. W.H. Zurek, Physics World, p.25, Jan 1997.
19. M.O. Scully *et al*, Nature (London), 351, 111 (1991); S. Haroche *et al*, Appl.Phys.B 54, 355 (1992); T. Pfau *et al*, Phys.Rev.Lett. 73, 1223 (1994); M.S. Chapman *et al*, Phys.Rev.Lett. 75, 3783 (1995).
20. L. Davidovich *et al*, Phys.Rev.Lett 71, 2360 (1993).
21. X. Maître *et al* , Phys.Rev.Lett. 79, 769 (1997).
22. A. Barenco *et al*, Phys.Rev.Lett. 74, 4083 (1995); T. Sleator and H. Weinfurter, Phys.Rev.Lett. 74, 4087 (1995); P. Domokos *et al*, Phys.Rev.A 52, 3554 (1995).
23. Q.A. Turchette *et al*, Phys.Rev.Lett. 75, 4710 (1995).
24. C. Monroe *et al*, Phys.Rev.Lett. 75, 4714 (1995), C. Monroe *et al*, Science 272, 1131 (1996); Q.A.Turchette et al, Phys.Rev.Lett. 81, 3631 (1998).

Quantum Coherence and Decoherence - ISQM - Tokyo '98
Y.A. Ono and K. Fujikawa (Editors)
© 1999 Elsevier Science B.V. All rights reserved.

Three- and Four-Photon Correlations and Entanglement: Quantum Teleportation and Beyond

Anton Zeilinger

Institut für Experimentalphysik, University of Vienna
Boltzmanngasse 5, 1090 Wien, Austria

Entanglement is at the root of many of the most fundamental quantum phenomena. Two-photon entangled states, which can readily be made with high quality in the laboratory, are a rich resource for producing quantum phenomena involving more particles. The general concept is to apply quantum erasure techniques jointly to some of the photons thus erasing their source information. This has been used successfully to achieve quantum teleportation, entanglement swapping and, most recently, to observe GHZ-states. While in all such schemes the desired quantum state is obtained only conditioned on its detection, this feature is of no significance in many quantum communication protocols.

1. INTRODUCTION

It has been known for a long time [1] that quantum entanglement is at the heart of fundamental phenomena such as the Einstein-Podolsky-Rosen paradox [2], Bell's theorem [3] or the measurement problem [4]. Most recently it turned out that entanglement also plays a key role in the discussion of quantum information and quantum computation procedures [5]. While sources of entangled pairs of particles are readily available, more advanced schemes involve more than two quanta. This is the case, for example, for quantum teleportation [6], entanglement swapping [7] and multiparticle phenomena involving so-called GHZ states [8]. It has turned out that one possible avenue to realize such phenomena in experiment is to use entangled photon pairs as a resource. The general concept involves sophisticated quantum erasure techniques. These are in general measurements of one or more photons which are performed in such a way that it is not possible to know, not even in principle, to which pair the measured photon(s) belong. The remaining photons then exhibit the desired phenomenon. A general feature of the existing schemes is that, so far, experimental confirmation necessitates detection of all photons involved. We will analyze this latter condition separately for each experiment performed up to now. This, we suggest, neither implies that the respective quantum phenomena have not been observed nor does it impart significant limitations on many possible quantum information protocols.

2. QUANTUM TELEPORTATION AND ENTNAGLEMENT SWAPPING

The aim of our experimental realization [9] of the original proposal [6] for quantum teleportation was to teleport the quantum state of a single photon. In the experiment shown in Fig. 1 we employed the double use of type II parametric down-conversion [10]. We will now analyze in detail the meaning of possible events that are to happen. First, we note that for most UV-pulses no photon pair was created. In fact, in our experiment the probability for an individual UV-pulse to create a photon pair was of the order of 10^{-4}. Secondly, even when a photon pair was created, in most cases it was not accompanied by a second pair. Yet, if we adopt a protocol which only accepts teleportation under the natural condition that a two-photon coincidence in the Bell-state analyzer occurred, these cases can readily be ruled out. We now turn to a careful analysis of those cases where such coincidences (between detectors $f1$ and $f2$) are observed. This necessitates that at least two down-conversion processes have occurred. We note that we can reasonably neglect the case of more than two down-conversions because such events are again sufficiently unlikely (the probability is again by a factor 10 smaller than the probability for a creation of two pairs).

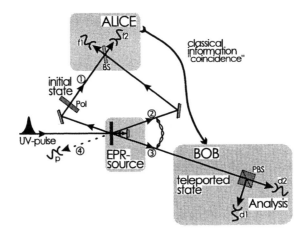

Fig. 1 Experimental Setup of Quantum Teleportation [9]. A UV-pulse creates a pair of entangled photons propagating to the right. One of these photons is directed to Alice while the other will become Bob's teleported photon. The pulse passes once more through the crystal creating another pair propagating to the left. One of those photons can either be prepared in any possible state or left undefined in entanglement swapping [11]. It is finally subject by Alice to a Bell-state measurement jointly with her entangled one. The other photon travelling to the left serves as a trigger.

Two pairs can be created in three different ways, either both during the first passage of the pulse, or both during the second, return passage of the pulse or one upon each passage. We note that the first case is irrelevant because it does not lead to a trigger event. The last case leads to the desired teleportation.

Finally, in the case of both down-conversions occuring on the return passage, one may obtain both a photon registration at the trigger and a coincidence event at both Bell state analyzer detectors.

As a detail we note that in the cases of successful teleportation the antisymmetric state

$$|\psi>_A= \frac{1}{\sqrt{2}}(|H>|V> - \quad |V>|H>)$$ \hfill (1)

was identified at Alice's Bell state analyzer [12]. Thus, even for perfect detections only one out of four possible Bell-state measurement results led to teleportation. This is just a limitation of the efficiency of the chosen procedure and it has no implications for the fact that in the successful cases the teleportation protocol has properly been realized.

Any state, pure or mixed, can be teleported with that procedure. Yet it is necessary to demonstrate that teleportation is successfully achieved. Thus one simply prepares the initial photon travelling to the Bell state analyzer in any arbitrary superposition state

$$|\psi>= \alpha|1>_H +\beta|1>_V$$ \hfill (2)

where $|1>_H$ ($|1>_V$) designs one photon in polarization mode $H(V)$ and $|\alpha|^2 + |\beta|^2 = 1$. Bob then proves that his photon is in that state. In the experiment there was no need for Bob to apply a unitary transformation to his photon because the original entangled state created was also chosen to be $|\psi>_A$. Thus, implicitly, the necessary transformation would just be the identity operation.

We now turn to analyzing the case when both down-conversions happen on the return path of the pump pulse. In this case the trigger detector may certainly fire, this time indicating that two photons are on their way to the Bell-state analyzer. These two photons propagate to the two separate detectors of the Bell-state analyzer with a probability of 50%. The registration of these two detectors may falsely be interpreted by Alice as indicating projection on a Bell state and hence teleportation. Yet Bob does not in fact receive a photon and Braunstein and Kimble [13, 14] suggested the term "a posteriori" teleportation indicating that in the setup used the experimentalists only know that teleportation occurred if Bob's detector eventually registers. I will now argue that this feature is not a limitation of the procedure but actually underlines a significant advantage. Just consider the state after the polarizer

$$|\psi>'= \frac{1}{\sqrt{2}}(\alpha|2>_H +\beta|2>_V)$$ \hfill (3)

which in that case indicates that we now have a two-photon state. Evidently, this state is orthogonal to the state of Eq. (2). Now, since the purpose of the experiment is to teleport any general one-photon state, it is actually an advantage that in the case of the state of Eq. (2) teleportation does not occur. Thus, we arrive at the conclusion that the procedure itself acts as a powerful filter which enables only teleportation of the desired one-photon states as intended. This selectivity of the procedure, I suggest, may be of significant advantage in future quantum computation applications of teleportation.

In entanglement swapping [7],[11], one simply removes the polarizer in the setup of Fig. 1. Then, upon registration of two photons in the Bell-state analyzer, the outgoing two photons, the one propagating to the trigger detector and the one propagating to Bob, are entangled. Here too, it can happen that both photon pairs are created in one single passage. In that case, one registers two photons in the Bell state analyzer without one photon each in the outgoing beam for a certain fraction of all cases. An analogous interpretation problem arose in some early tests of Bell's inequality[15]. Yet, the cases where indeed one photon each is registered in each outgoing beam can be valid tests of quantum nonlocality and Bell's inequality if the probability for a photon to propagate to either detection station is independent of the parameter setting chosen at that detector [16, 17].

3. OBSERVATION OF GHZ STATES

It is well-known [8, 18, 19] that entangled states of three or more quanta can lead to stronger violations of local realism than two-quanta systems. Additionally, in recent years it turned out that such states play crucial roles in many quantum information schemes [20]. While the existence of such states was recently demonstrated experimentally for three nuclei in a single molecule using nuclear magnetic resonance (NMR) techniques [21], no experimental verification for spatially separated quanta existed hitherto. Very recently a first demonstration of GHZ states was obtained [22] for three photons.

The principle of the experiment [23] is to start again, as in the experiments discussed above, from two entangled pairs. One then subjects one photon to a measurement such that all information to which pair it belongs is completely erased. This projects the remaining three photons into the desired state.

The experimental principle is shown in Fig. 2. One again creates two pairs of photons by a sufficiently strong pump pulse.

The state of each pair created by the source is

$$|\psi>_A = \frac{1}{\sqrt{2}}(|H>_1 |V>_2 + e^{i\chi}|V>_1 |H>_2) \tag{4}$$

where the internal phase χ depends on the specific details of the experiment.

Restricting ourselves to four-fold coincidences at all four detectors T, D_1, D_2 and D_3, we obtain a GHZ-state through the following reasoning. The photon at T is always horizontally (H) polarized and therefore one of the photons in beam b must be vertically (V) polarized. If this photon ends up in detector D_2 then the one registered in D_1 must also be vertically polarized resulting in the photon at D_3 being horizontally polarized (H). Thus we obtain the first term in our quantum state

$$|H>_T |V>_1 |V>_2 |H>_3 \quad . \tag{5}$$

On the other hand, if the vertical photon of beam b is registered in detector D_3 then the photons registered in D_1 and D_2 must both carry H polarization resulting in the term

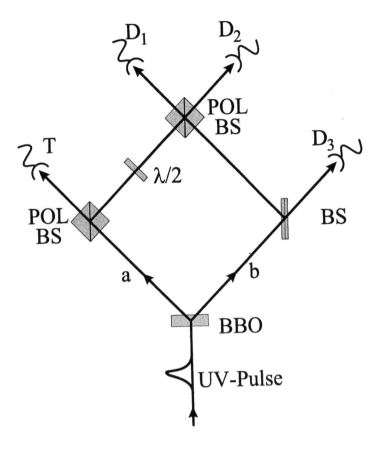

Fig. 2 Principle of the experimental observation of 3-photon GHZ-states [22]. A nonlinear BBO crystal is pumped by a strong UV-pulse. In beam a the photons encounter a two-channel polarizer which reflects vertical (V) polarization and transmits horizontal (H) polarization, the latter being registered by the trigger detector T. The vertical polarization is then rotated by 45° by a $\lambda/2$ plate oriented at 22.5°. In the b beam the photons encounter a standard 50/50 semireflecting beam splitter, with one of the outgoing beams heading towards detector D_3. Finally, detectors D_1 and D_2 observe the outputs of the final polarizer.

$$|H>_T |H>_1 |H>_2 |V>_3 \quad . \tag{6}$$

If the experimental arrangement is such that these two cases are coherent, the state

$$\frac{1}{\sqrt{2}}(|H>_T (|V>_1 |V>_2 |H>_3 + |H>_1 |H>_2 |V>_3) \tag{7}$$

results, exhibiting for photons 1 and 2 the desired GHZ correlations. The specific internal phase in Eq. (7) is a consequence of the two down-conversions being identical.

In the experiment, verification of the state (7) was done by measuring one of the photons at 45° linear polarization. Such a measurement results in an entangled two-photon state of the two remaining photons, as predicted by the concept of entangled entanglement [24]. This is because the GHZ part of state (7) can be rewritten as

$$
\frac{1}{2} \mid +45° >_1 (|H>_2 |V>_3 + |V>_2 |H>_3) +
$$
$$
\frac{1}{2} \mid -45° >_1 (|H>_2 (V>_3 - |V>_2 |H>_3) \tag{8}
$$

Demonstration of the resulting entanglement of photons 2 and 3 is straightforward. It just entails the use of the specific rotational properties of two-particle entangled states which can be observed for example by measuring photons 2 and 3 in a basis that is also rotated by 45°.

We note that here too, it can happen that only one down-conversion occurs. That situation is readily identified by the fact that together with the trigger detector T only one of the other detectors fires. This also holds for the case of two photons heading towards any of the detectors. We note again that, as in the cases of teleportation and of entanglement swapping, this in no way restricts the usefulness of the scheme for tests of quantum nonlocality. The basic reason for that fact is that one can, as is well known, base a nonlocality argument just on an analysis of the events observed. No reference to the underlying quantum state is necessary. Such an analysis was recently performed for our GHZ state.

We also note that all errant cases analyzed here are signified by the fact that not all intended detectors register a photon. From a practical point of view this is the same as non-firing due to the unavoidable detection inefficiency.

4. FINAL REMARKS

We might finally venture into discussing some possible avenues for future research, as far as they can be identified at the moment. Considering the status of technology available at present, a dedicated technological program for the development of novel detectors would be extremely useful. The detectors I envisage should be able to distinguish between one-photon events and two-photon events. These detectors should have little cross-talk, that is, they should confuse the two kinds of events with very small probability even though the detection efficiencies themselves need not be large. I submit that such detectors and their generalizations to $N-$photon detection will be very useful for many future quantum information protocols involving a higher number of photons.

Such a higher number of photons will be employed in many future experiments. A specifically interesting experiment would be to demonstrate coding schemes in multi-photon states, particularly states demonstrating non-trivial error correction schemes.

Another very interesting future avenue will be to realize generalizations[25] of quantum teleportation and of entanglement swapping to more complicated cases like quantum switchboards which again involve multi-photon states.

Presently the most successful line of research towards multiphoton entanglement is to exploit the easily realizable two-photon entanglements as resources for more complicated cases. This will generalize the experimental approaches of the type used in experiments discussed above in the near future.

ACKNOWLEDGMENTS

I would like to thank the theorists M.A. Horne, D.M. Greenberger and M. Żukowski and the experimentalists D. Bouwmeester, J.-W. Pan and H. Weinfurter for useful discussions. I am also grateful to the organizers of ISQM-Tokyo '98 for their kind invitation to present some of our results. This work was supported by the Austrian Science Foundation FWF, project S06502, and by the U.S. National Science Foundation, grant No. PHY 97-22614.

REFERENCES

1. E. Schrödinger, *Naturwissenschaften* **23** (1935) 807, English translation in *Proceedings of the American Philosophical Society,* **124** (1980) 323.
2. A. Einstein, B. Podolsky and N. Rosen, **Phys. Rev. 47** (1935) 777.
3. J.S. Bell, *Physics* 1 (1964) 195, reprinted in J.S. Bell, *Speakable and Unspeakable in Quantum Mechanics* (Cambridge U.P., Cambridge, 1987).
4. E.P. Wigner, *Am. J. Phys.* **31** (1963) 6.
5. *Quantum Information,* special issue, *Physics World* **11** (March 1998) No. 3.
6. C.H. Bennett, G. Brassard, C. Crepeau, R. Josza, A. Peres and W.K. Wootters, *Phys. Rev. Lett* **70** (1993) 1895.
7. M. Żukowski, A. Zeilinger, M.A. Horne and A. Ekert, *Phys.Rev.Lett.* **71** (1993) 4287.
8. D. Greenberger, M.A. Horne, and A. Zeilinger, in *Bell's Theorem, Quantum Theory, and Conceptions of the Universe* , ed. M. Kafatos (Kluwer, Dordrecht, 1989), p. 74.
9. D. Bouwmeester, J.W. Pan, K. Mattle, M. Eibl, H. Weinfurter, A. Zeilinger, *Nature* **390** (1997) 575.
10. P.G. Kwiat, H. Weinfurter, T. Herzog, A. Zeilinger and M. Kasevich, *Phys. Rev. Lett.* **74** (1995) 4763.
11. J.W. Pan, D. Bouwmeester, H. Weinfurter and A. Zeilinger, *Phys. Rev. Lett.* **80**, (1998) 3891.
12. A. Zeilinger, *Physica Scripta* **T76** (1998) 203.
13. S.L. Braunstein and H.J. Kimble, *Nature* **394** (1998) 840.
14. D. Bouwmeester *et al.*, *Nature* **394** (1998) 841.
15. C.O. Alley, Y.H. Shih, in *Proc. Symp. on Foundations of Modern Physics*, eds. P. Lahti, P. Mittelstaedt, (World Scientific, Singapore, 1985), p. 435; Y.H. Shih, C.O. Alley, *Phys. Rev. Lett.* **62** (1988) 2921.
16. S. Popescu, L. Hardy and M. Zukowski, *Phys. Rev.* **A 56** (1997) R4353.
17. M. Żukowski, Violations of Local Realism in the Innsbruck GHZ experiment, preprint, 1998.
18. D.M. Greenberger, M.A. Horne, H. Shimony and A. Zeilinger, *Am. J. Phys.* **58** (1990) 1131.

19. N.D. Mermin, *Rev. Mod. Phys.* **65** (1993) 803; *Am. J. Phys.* **65** (1997) 476.
20. R. Cleve, H. Buhrman, *Phys. Rev.* **A 56** (1997) 1201; D. Bruss, D. DiVincenzo, A. Ekert, C. Fuchs, C. Macchiavello, and J. Smolin, *Phys. Rev.* **A 57** (1998) 2368.
21. S. Lloyd, *Phys. Rev.* **A 57** (1998) R1473; R. Laflamme, E. Knill, W.H. Zurek, P. Catasti, S.V.S. Mariappan, *Phil. Trans. R. Soc. Lond.* **A 356** (1998) 1941.
22. D. Bouwmeester, J.-W. Pan, M. Daniell, H. Weinfurter, A. Zeilinger, submitted.
23. A. Zeilinger, M.A. Horne, H. Weinfurter, M. Żukowski, *Phys. Rev. Lett.* **78** (1997) 3031.
24. G. Krenn, A. Zeilinger, *Phys. Rev.* **A** (1996) 1793.
25. S. Bose, V. Vedral, P. Knight, *Phys. Rev.* **A** (1998) 822.

Quantum Coherence and Decoherence - ISQM - Tokyo '98
Y.A. Ono and K. Fujikawa (Editors)
© 1999 Elsevier Science B.V. All rights reserved.

Quantum manipulations of small Josephson junctions and the measurement process performed with a single-electron transistor

Alexander Shnirman[a], Yuriy Makhlin[b,c] and Gerd Schön[b]

[a]Dept. of Physics, University of Illinois at Urbana-Champaign, Urbana, IL 61801-3080, U.S.A.
[b]Institut für Theoretische Festkörperphysik, Universität Karlsruhe, D-76128 Karlsruhe, Germany
[c]Landau Institute for Theoretical Physics, 117940 Moscow, Russia

Nano-scale electronic devices provide physical realization of the elements required for quantum computation [1–3]. Small Josephson junctions, where Coulomb blockade effects allow the control of individual Cooper-pair charges, constitute quantum bits (qubits), with logical states differing by the charge on one island. Single- and two-bit operations can be performed by applied gate voltages. The phase coherence time is sufficiently long to allow a series of these steps. In addition to the manipulation of qubits the resulting quantum state has to be read out. This can be accomplished by coupling a single-electron transistor to the qubit [4]. We describe this quantum measurement process by considering the time-evolution of the density matrix of the coupled system. The transistor destroys the phase coherence of the qubit only when a transport voltage is turned on. The process is characterized by three time scales: the dephasing time, the 'measurement time' when the signal resolves the different quantum states, and the mixing time after which the measurement process itself destroys the information about the initial state.

1. Introduction

The investigation of nano-scale electronic devices, such as low-capacitance tunnel junctions or quantum dot systems, has always been motivated by the perspective of future applications. By now several have been demonstrated, e.g. the use of SETs as ultra-sensitive electro-meters and single-electron pumps. From the beginning it also appeared attractive to use these systems for digital operations needed in classical computation [5]. Obviously *single-electron* devices would constitute the ultimate electronic memory. Unfortunately, their extreme sensitivity makes them also very susceptible to fluctuations, either due to the external circuit or microscopic sources such as electron hopping in the substrate. Due to these problems – and the continuing progress of conventional techniques – the future of SET devices in *classical* digital applications remains uncertain.

The situation is different when we turn to elements for quantum computers. They could perform certain calculations which no classical computer could do in acceptable times by exploiting the quantum mechanical coherent evolution of superpositions of states [6]. Here conventional systems provide no alternative. In this context, ions in a trap, manipulated

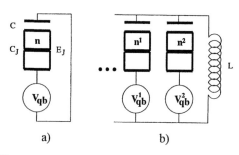

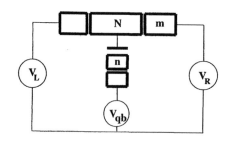

Figure 1. The proposed 1-bit and multi-bit Josephson systems.

Figure 2. The circuit consisting of a qubit plus a SET used as a measuring device.

by laser irradiation are the best studied system. However, alternatives need to be explored, in particular those which are more easily embedded in an electronic circuit. From this point of view nano-electronic devices appear particularly promising.

The simplest choice, normal-metal single-electron devices are ruled out, since – due to the large number of electron states involved – different, sequential tunneling processes are incoherent. Ultra-small quantum dots with discrete levels and spin degrees of freedom are candidates [3], but the strong coupling to the environment renders the phase coherence time short. More attractive are systems of Josephson contacts, where the coherence of the superconducting state can be exploited. Quantum extension of elements based on a single flux logic have been suggested, however, an essential process, the coherent oscillation of the flux between degenerate states, has not yet been observed. We suggest the use of low-capacitance Josephson junctions, where Cooper-pair charges tunnel coherently and can be controlled by applied gate voltages [1].

In addition to controlled manipulations the quantum computation requires a quantum measurement to read out the final state. The requirements for both steps appear to contradict each other. During manipulations the dephasing should be minimized, whereas a measurement should dephase the state of the qubit as fast as possible. The option to couple the measuring device to the qubit only when needed is hard to achieve in nanoscale systems. The alternative, which we suggest [4], is to keep the measuring device permanently coupled to the qubit in a state of equilibrium during the quantum operations. The measurement is performed by driving the measuring device out of equilibrium, which dephases the quantum state of the qubit. Similar nonequilibrium dephasing processes [7] have recently been demonstrated experimentally [8].

2. Josephson junction qubits

First we discuss the properties and quantum manipulations of Josephson qubits shown in Fig. 1. Each qubit consists of superconducting island and lead coupled by a Josephson junction. The coupling to the external circuit is purely capacitive and does not involve dissipative currents. The qubit's state is characterized by n, the number of extra Cooper pairs on the island. The scale of the energy splitting between consecutive charge states is the charging energy E_{qb}; the precise value can be tuned by an applied gate voltage V_{qb}. We concentrate here on the regime where the charging energies of two states, $|\downarrow\rangle \equiv |n\rangle$ and $|\uparrow\rangle \equiv |n+1\rangle$, are close. Choosing the Josephson coupling sufficiently weak, $E_J \ll E_{qb}$, we ensure that the coherent Cooper-pair tunneling only mixes these two states. The effective Hamiltonian thus reduces to $H = \frac{1}{2}E_J(\cot\eta\,\sigma_z - \sigma_x)$. The mixing angle η depends on the ratio of energy scales E_J/E_{qb} and the applied gate voltage V_{qb}. Here it is sufficient to note that the latter allows the tuning of H.

The system is thus equivalent to a spin in a magnetic field with constant x-component, while the z-component can be varied by V_{qb}. Hence, the standard techniques of spin manipulations by time-dependent fields ($\pi/2$-pulses, ...) are available for manipulations of the qubit [1]. By varying the gate voltage we can put the system in a regime where the mixing is strong, which is of advantage when performing quantum manipulations. On the other hand, between manipulations we keep it at the "idle point", $\eta_{idle} \ll 1$, far from degeneracy. The eigenstates of the qubit at this point constitute its logical basis: $|0\rangle = \cos\frac{\eta_{idle}}{2}|\downarrow\rangle + \sin\frac{\eta_{idle}}{2}|\uparrow\rangle$ and $|1\rangle = -\sin\frac{\eta_{idle}}{2}|\downarrow\rangle + \cos\frac{\eta_{idle}}{2}|\uparrow\rangle$. The energy splitting between the logical states is then $\Delta E \equiv E_J/\sin\eta_{idle} \gg E_J$.

Since the splitting ΔE is non-zero, the relative phase of two components of the state evolves between manipulations (the spin precesses). The state is preserved, however, in the interaction representation (in terms of spins, in the rotating frame). In this representation, if the voltage is changed at t_0 for some time τ to the value V defining a new η, the quantum state of the qubit evolves according to the unitary transformation $\mathcal{U}(t_0, \tau, \eta) = \exp(iH(\eta_{idle})(t_0 + \tau)) \cdot \exp(-iH(\eta)\tau) \cdot \exp(-iH(\eta_{idle})t_0)$. With proper choice of t_0, τ, and V any 1-bit operation can be realized. E.g., $\mathcal{U}(t_0, \tau = \pi\hbar\cos\eta_{idle}/E_J, \eta = \eta_{idle} \pm \pi/2)$ is a NOT operation, exchanging $|0\rangle$ and $|1\rangle$, while $\mathcal{U}(t_0, \tau = 2\pi N\hbar\sin\eta/E_J, \eta)$ is a shift of their relative phase by $\phi = 2\pi N\sin\eta/\sin\eta_{idle}$ for any integer N.

To perform 2-bit operations, we couple the qubits, as shown in Fig. 1b, by an oscillatory mode of an inductance L and the total capacitance of the qubits. The oscillator frequency should be much larger than ΔE. Then, the oscillator stays in the ground state, while its zero-point fluctuations induce a coupling between the qubits of the form $E_L\sigma_y^1\sigma_y^2$. The energy scale $E_L \propto E_J^2 L/\Phi_0^2$, with $\Phi_0 \equiv h/2e$ being the flux quantum, should be small compared to ΔE. When, by varying the gate voltages, the states $|\downarrow^1, \uparrow^2\rangle$ and $|\uparrow^1, \downarrow^2\rangle$ are tuned close to degeneracy, this interaction produces coherent "flip-flop" transitions. One can compensate for the effect of unwanted phase shift using single-bit operations. Away from degeneracy, the inductor produces only a weak perturbation for each qubit.

The electromagnetic circuit induces fluctuations of the bias voltage V_{qb}, which couple to the charge of the qubit σ_z. As a result the diagonal elements of the density matrix, in the basis of the qubit's eigenstates, relax to their equilibrium values with a rate Γ_r, while the off-diagonal elements decay with the dephasing rate Γ_ϕ. In the low-frequency regime the system consisting of qubit and environment is described by the spin-boson model at finite bias [9], and we can take over the established results. At the idle point the two rates are $\Gamma_r = \frac{\pi}{\tilde{g}}\frac{\Delta E}{\hbar}\coth\left(\frac{\Delta E}{2k_BT}\right)\sin^2\eta_{idle}$ and $\Gamma_\phi = \frac{1}{2}\Gamma_r + \frac{\pi}{\tilde{g}}\frac{k_BT}{\hbar}\cos^2\eta_{idle}$. The parameter $\tilde{g} \approx (C_J/C)^2 h/(4e^2 R)$ characterizes the coupling to the electromagnetic environment, modeled here by a resistor R. The rate Γ_r corresponds to real transitions between the two eigenstates of the qubit, while the second term in the expression for Γ_ϕ ($\propto T$) originates from the fluctuations of the energy splitting between the eigenstates and describes "pure" dephasing in the absense of transitions ($E_J = 0$). The decay rates are small if the resistance of the circuit is low compared to the quantum resistance, $h/e^2 \approx 26k\Omega$. Furthermore, a low gate capacitance C reduces the coupling of the qubit to the environment. With suitable parameters ($R \leq 100\Omega, C/C_J \leq 0.1$) at low temperatures the number of operations which can be performed before the environment destroys the coherence may be as large as $10^3 - 10^4$.

3. The quantum measurement process

The most elementary system proposed for the quantum measurements is shown in Fig. 2. It is a SET, with a qubit inserted into the control gate circuit. At the stage of the quantum manipulations the SET is kept deeply in the off-state ($N = 0$), no dissipative currents flow in the system, and dephasing effects due to the transistor are minimized. When one drives the SET out of equilibrium, the resulting normal current depends on the state of the qubit, since different charge configuration induce different voltages on the middle island of the SET. The picture gets complicated by various noise factors (shot noise) and by the measurement induced transitions between the states of the qubit. The former set the lower time limit after which we can extract the information from the experimental data, while the latter destroy the quantum state to be measured.

The total system is characterized by three charging energies: the typical energy splitting between consecutive charge states (N) of the SET, E_{set}, the charging energy scale of the qubit, E_{qb}, and the Coulomb repulsion between charges on the qubit and the middle island of the SET, E_{int}. To drive the system out of equilibrium one applies a transport voltage $V_{tr} = V_L - V_R$, so that tunneling to another charge state, say $N = 1$, becomes possible. We choose E_{set} to be the largest energy scale, $E_{set} \gg \Delta E \gg E_J, E_{int}$, and the transport voltage large enough to overcome the Coulomb energy gap between the states $N = 0$ and $N = 1$ by an amount of order E_{set}, but small enough to insure that only these two states are involved in the transport. The actual value of the Coulomb gap differs by E_{int} for different charge states of the qubit. As a result, an electron finds it easier to tunnel to the middle island of the SET when there are less Cooper pairs on the upper island of the qubit. Thus, the transport rates Γ_0 and Γ_1, corresponding to the qubit's states $|0\rangle$ and $|1\rangle$, differ by $2\pi\alpha E_{int}$, where $\alpha \equiv \hbar/(4\pi^2 e^2 R_T)$ characterizes the tunneling resistance of the normal tunnel junctions. Conversely, when the SET is in the state $N = 1$, the energy splitting between the states $|0\rangle$ and $|1\rangle$ is shifted by E_{int} and, therefore, an additional relative phase is acquired. Since the tunneling events in the SET are random, they are a source of dephasing. Moreover, when $N = 1$, the states $|0\rangle$ and $|1\rangle$ are no longer the exact eigenstates. This gives rise to the measurement induced transitions.

To describe the dissipative current in the SET we introduce the variable m which counts the number of electrons which has arrived in the right lead. Thus the total system is described by a reduced density matrix $\hat{\rho}_{i,N,m;i',N',m'}(t)$, where i and i' stand for the quantum states of the qubit, $|0\rangle$ or $|1\rangle$, the variables N and m have been introduced above, and all other degrees of freedom are traced out. The off-diagonal elements in N and m may be eliminated from a closed set of equations [10]. Therefore, we need to consider only the elements $\hat{\rho}_{i,i'}^{N,m} \equiv \hat{\rho}_{i,N,m;i',N,m}$. We assume in the following that at time $t = 0$, as a result of previous quantum manipulations, the qubit is prepared in the superposition state $a|0\rangle + b|1\rangle$, and we switch on a transport voltage to the SET.

To proceed we further reduce the density matrix in two ways to obtain dual descriptions of the measurement. The first procedure is to trace over N and m, which yields a reduced density matrix of the qubit $\hat{\rho}_{i,j} \equiv \sum_{N,m} \hat{\rho}_{i,j}^{N,m}$. Starting in a superposition of states ($\hat{\rho}_{0,0} = |a|^2$, $\hat{\rho}_{1,1} = |b|^2$, $\hat{\rho}_{0,1} = ab^*$) the questions are how fast the off-diagonal elements of $\hat{\rho}_{i,j}$ vanish due to dephasing, and how fast the diagonal elements change their original values (for instance due to transitions induced by the measurement). This description is enough when one is interested in the quantum properties of the qubit only and the measuring device is used as a source of dephasing [7,8]. It does not tell us much, however, about the quantity measured in an experiment, namely the current flowing trough the SET.

The second procedure is to evaluate the probability distribution of the number of electrons m which have tunneled trough the SET during time t: $P(m,t) \equiv \sum_{N,i} \hat{\rho}_{i,i}^{N,m}(t)$. At $t = 0$ no electrons have tunneled, so $P(m,0) = \delta_{m,0}$. Then this peak starts to shift towards positive m, and, at the same time, it widens due to shot noise. Since two states of the qubit correspond to different conductivities, one may hope that after some time the peak splits into two. If after sufficient separation of the two peaks their weights are still close to $|a|^2$ and $|b|^2$, a good quantum measurement has been performed. Unfortunately, there exist processes which destroy this idealized picture. After a long time the two peaks transform into a broad plateau, since transitions between the qubit's states are induced by the measurement. One should find an optimum time for the measurement, so that, on one hand, the two peaks are separate and, on the other hand, the induced transitions have not yet happened. To this end we derive a master equation for $\hat{\rho}_{i,j}^{N,m}$, using the real time diagrammatic technique [10,4], and we analyze it both analytically and numerically.

First, we assume that the Josephson coupling is switched off during the measurement. Then the master equation splits into three independent groups of equations: two for the diagonal matrix elements $\hat{\rho}_{0,0}^{N,m}$ and $\hat{\rho}_{1,1}^{N,m}$ and the third for the off-diagonal elements. The first two groups describe propagation of two independent peaks with velocities Γ_0 and Γ_1. The weights of the peaks are $|a|^2$ and $|b|^2$ and their widths at time t are given by $\sqrt{\Gamma_0 t}$ and $\sqrt{\Gamma_1 t}$ (shot noise). Thus one arrives at the criterion for the peaks' separation: $|\Gamma_0 - \Gamma_1| t \geq \sqrt{\Gamma_0 t} + \sqrt{\Gamma_1 t}$, which yields the time of measurement: $t_{ms} = |\sqrt{\Gamma_0} - \sqrt{\Gamma_1}|^{-2}$.

The third group of equations describes the decay of the off-diagonal matrix elements $\hat{\rho}_{0,1}^{N,m}$. We find the dephasing time τ_ϕ to be parametrically different from the measurement time t_{ms}. In the range of validity of our approach t_{ms} exceeds τ_ϕ, consistent with the fact that a quantum measurement implies a complete dephasing of a quantum state. For instance, in the limit $\Gamma_{0,1} \gg E_{int}$ we have $\tau_\phi \approx (2\pi\alpha)^2 t_{ms}$. The faster dephasing indicates that during the measurement some additional uncontrolled parts of the environment "observe" the quantum state of the qubit. For example, information may be lost into the unspecified microscopic states of the SET.

The residual Josephson coupling ignored so far generates mixing transitions between the states of the qubit. We estimate the mixing time for the concrete physical situation discussed above $t_{mix}^{-1} \approx 2\pi\alpha E_{int}^2 E_J^2 E_{set}/(\Delta E)^4$, while the measurement time is given approximately by $t_{ms}^{-1} \approx 2\pi\alpha E_{int}^2/E_{set}$. The results of numerical simulations of the master equation in the regime $t_{ms} \ll t_{mix}$, showing the peak separation, are presented in Fig. 3.

Here we have described that the current through a SET measures the quantum state of the qubit, in the sense that for a superposition of two eigenstates it yields one or the other result with the appropriate probabilities. This should be distinguished from another

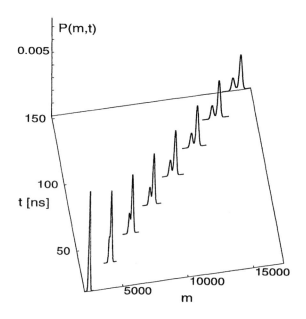

Figure 3. $P(m, t)$, the probability that m electrons have tunneled during time t (measured in nano-seconds). The initial amplitudes of the qubit's states are $a = \sqrt{0.75}$, $b = \sqrt{0.25}$.

question, namely whether it is possible to demonstrate that an eigenstate of a qubit can actually be a superposition of two different charge states. The latter has been addressed successfully in the experiments of Refs. [11] where a single-Cooper-pair box was coupled to a SET or was itself a part of a superconducting SET.

REFERENCES

1. A. Shnirman, G. Schön, and Z. Hermon, Phys. Rev. Lett. **79**, 2371 (1997).
2. J.E. Mooij, private comm.; D.V. Averin, Solid State Commun. **105**, 659 (1998).
3. D. Loss and D.P. DiVincenzo, Phys. Rev. A **57**, 120 (1998); G. Burkard et al., cond-mat/9808026.
4. A. Shnirman and G. Schön, Phys. Rev. B **57**, 15400 (1998).
5. A.N. Korotkov, R.H. Chen, and K.K. Likharev, J. Appl. Phys. **78**, 2520 (1995).
6. A. Barenco, Contemp. Phys. **37**, 375 (1996); D.P. DiVincenzo, in *Mesoscopic Electron Transport*, NATO ASI Series E **345**, eds. L.L. Sohn et al., Kluwer, 1997.
7. Y. Levinson, Europhys. Lett. **39**, 299 (1997); I.L. Aleiner et al., Phys. Rev. Lett. **79**, 3740 (1997); S.A. Gurvitz, Phys. Rev. B **56**, 15215 (1997).
8. E. Buks et al., Nature, Vol. **391**, 871 (1998).
9. A.J. Leggett et al., Rev. Mod. Phys. **59**, 1 (1987); U. Weiss, *Quantum dissipative systems*, World Scientific, Singapore, 1993.
10. H. Schoeller and G. Schön, Phys. Rev. B **50**, 18436 (1994).
11. V. Bouchiat, D. Vion, P. Joyez, D. Esteve, and M.H. Devoret, to appear in Physica Scripta **T176** (1998); Y. Nakamura, C.D. Chen, and J.S. Tsai, Phys. Rev. Lett. **79**, 2328 (1997).

Acknowledgments. We thank E. Ben-Jacob, C. Bruder, E. Buks, L. Dreher, Z. Hermon, J. König, Y. Levinson, J.E. Mooij, T. Pohjola, and H. Schoeller for stimulating discussions.

Quantum Coherence and Decoherence - ISQM - Tokyo '98
Y.A. Ono and K. Fujikawa (Editors)
© 1999 Elsevier Science B.V. All rights reserved.

Extraction of NMR Quantum Computation Signal with Parity Gate

M. Kitagawa [a*]

[a]Graduate School of Engineering Science, Osaka University
Toyonaka, Osaka 560–8531 Japan

Extraction of NMR signal from a group of molecules which belong to·an effective pure state labeled by ancilla is studied and a new quantum logical gate named parity gate is proposed for the purpose. A parity gate reverses a signal qubit if and only if the parity of control qubits is odd. It is another extension of controlled-NOT gate to the multi control bits but scales much better than (controlled)[n]-NOT for increasing number of control bits.

1. INTRODUCTION

Recently quantum computation using nuclear magnetic resonance (NMR) has been proposed [1,2] and 2-qubit experiments have been done [3–6]. The essential discovery in the proposal [1] is that the effective pure state labeled by ancilla can be prepared from the thermal equilibrium mixed state by unitary transformation. However it has not been elaborated on how to extract the signal from an effective pure state [1].

In this *Paper*, we discuss how to extract the desired signal in logically labeled NMR bulk quantum computation and propose a practical strategy based on parity gate.

2. EFFECTIVE PURE STATE BY LOGICAL LABELING [1,7]

The state after the initialization steps called *logical labeling* can be expressed as,

$$\widehat{\rho} = \sum_{j=0}^{M-1} |j\rangle\langle j| \otimes \widehat{\rho}_j, \tag{1}$$

where j denotes a state of m-bit ancilla , $M = 2^m$ the number of blocks, $\widehat{\rho}_j$ the state of j-th block. At least one block (say $\widehat{\rho}_i$) can·be made effectively pure in the sense that any NMR signal from the block coincides with that of a pure state.

The effective pure block consists of equally populated states except for the one with excess (or deficient) population which solely accounts for the signal from the block. The maximum number of pure qubits, k, distilled from n-qubit thermal state is determined from $2^k \leq {}_nC_{n/2}+1<2^{k+1}$ and is shown in Table 2 and for larger n [1],

$$k = n - m, \quad m \approx (\log n + \log \pi - 1)/2. \tag{2}$$

---

*E-mail: kitagawa@ee.es.osaka-u.ac.jp

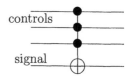

Figure 1. Marking by (controlled)^m-NOT

Table 1. Effective pure state from n-qubit thermal state

total	n	3	4	5	6	7	8	9	10	11	12	13
pure	k	2	2	3	4	5	6	6	7	8	9	10
ancilla	m	1	2	2	2	2	2	3	3	3	3	3

3. CONDITIONAL MEASUREMENT

The state could be reduced to an effective pure state by measuring ancilla bits if it were a single quantum system. However a huge number (10^{23}) of uncorrelated molecules (=quantum computer) in liquid phase NMR reduce independently and uniformly populate all blocks labeled by ancilla even if efficiently strong measurement is imposed on ancilla bits. Therefore "reduction by measurement" cannot be used for state extraction.

The effective pure state in Eq. (1) cannot be extracted but only is labeled. The quantum computation can be done from the initial state Eq. (1) without mixing up the blocks if the evolution of the ancilla is frozen [1]. Then, the state after the computation is also given by Eq. (1) with each $\widehat{\rho}_j$ evolved independently. The readout measurement on the state gives the total signal, $T = \sum_{j=0}^{M-1} S_j$, where $S_j = \text{Tr}(\widehat{\rho}_j \widehat{\sigma}_z)$ denotes the signal from j-th block. The desired signal S_i from the effective pure block is buried in the sum of the signals from all M blocks.

It is essential that the operation conditional to ancilla is made locally within each molecule before readout measurement. For example, we can reverse a signal from the i-th block with (controlled)^m-NOT gate feeding each of ancilla (or its NOT) as control bits and readout bit as a signal bit as shown in Figure 1 and perform a measurement to get, $T_i = \sum_{j=0}^{M-1} S_j(-1)^{\delta_{ij}}$, where δ_{ij} is the Kronecker's delta.

We run two experiments with and without the conditional operation and subtract the former signal from the latter to extract the signal $T-T_i=2S_i$ [7].

Unfortunately, marking a single block by (controlled)^m-NOT is not as simple as it conceptually is and requires lots of operations in NMR [8]. The number of 2-bit controlled operations required for (controlled)^m-NOT gate is $2^{m+1}-3$ for small m up to several qubits where savings due to working bit may not be expected. For larger m, the number may asymptotically become linear in m but still the coefficient is very large, say 80. This may be deadly as these operations must complete before dephasing, although 2^m is still linear in n from Eq. (2) and does not ruin the efficiency of computation.

4. MARKING BY PARITY

The above concept can be generalized to marking multiple blocks in orthogonal ways in more than two experiments. We propose a marking by parities of ancilla. By taking bitwise AND of the ancilla j and m-bit parity mask i, the parity is given by, $\chi_{ij} = \bigoplus_{b=0}^{m-1} i[b]j[b]$. We can specify a single block by specifying parities for all $M=2^m$ masks.

The signal we observe when we use the mask i is, $T_i = \sum_{j=0}^{M-1} S_j(-1)^{\chi_{ij}}$. We can extract a signal from every single block by running M experiments with different parity masks and linearly combining the results as $S_j = \frac{1}{M}\sum_{j=0}^{M-1} T_i(-1)^{\chi_{ij}}$. In terms of the Hadamard matrix $\mathbf{P}=[(-1)^{\chi_{ij}}]$, the relation between the observed $\mathbf{T}=[T_i]$ and the emitted $\mathbf{S}=[S_j]$ signal vectors is $\mathbf{T}=\mathbf{PS}$ and is solved as $\mathbf{S}=\mathbf{PT}/M$. $\mathbf{P}$ for m=1, 2 and 3 are as follows,

$$\mathbf{P}_1 = \begin{bmatrix} + & + \\ + & - \end{bmatrix}, \quad \mathbf{P}_2 = \begin{bmatrix} + & + & + & + \\ + & - & + & - \\ + & + & - & - \\ + & - & - & + \end{bmatrix}, \quad \mathbf{P}_3 = \begin{bmatrix} + & + & + & + & + & + & + & + \\ + & - & + & - & + & - & + & - \\ + & + & - & - & + & + & - & - \\ + & - & - & + & + & - & - & + \\ + & + & + & + & - & - & - & - \\ + & - & + & - & - & + & - & + \\ + & + & - & - & - & - & + & + \\ + & - & - & + & - & + & + & - \end{bmatrix}. \tag{3}$$

5. PARITY GATE

A parity gate which reverses a signal qubit if and only if the parity of control qubits is odd is proposed. It can be implemented using rotations and conditional rotations available in NMR experiments. Figure 2 shows a quantum circuit for m=3. The signal qubit is first rotated $\pi/2$ about y-axis. Then it experiences conditional rotation $(-1)^{\epsilon_p}\pi/2$ about z-axis depending on the state of the p-th control bit $|\epsilon_p\rangle$ (ϵ_p=0, 1). By repeating it for every control bit, the total rotation angle of the signal qubit becomes,

$$\sum_{p=0}^{m-1}(-1)^{\epsilon_p}\pi/2 = \begin{cases} (m \mod 4)\pi/2 & \text{even,} \\ (m \mod 4)\pi/2 + \pi & \text{odd.} \end{cases} \tag{4}$$

The key is that $\pm\pi/2$ rotations always result in π difference categorizing even or odd. $\pm\pi$ rotations due to 00 and 11 have the same effect while both 01 and 10 give no net rotations. By rotating back $(m \mod 4)\pi/2$ about z-axis, and then back $\pi/2$ about y-axis, the signal state is restored if the parity is even while it is reversed if the parity is odd. The last two rotations may be merged or even omitted if the gate is used for conditional readout. The parity gate with m control bits only requires m conditional rotations and two rotations and has very simple circuit. It scales linearly and straightforwardly with m.

It is another extension of controlled-NOT gate to the multi control bits but is much simpler and easier than (controlled)^m-NOT to implement in NMR.

6. DISCUSSIONS AND CONCLUSIONS

There are two kinds of procedures involved in NMR bulk quantum computation; quantum computation experiments and classical post data processing. The former requires quantum coherence and must complete within dephasing time. The latter is not restricted in time. As long as the total number of steps remains polynomial, it is practical to save the steps in the quantum part.

The conditional readout using parity gate requires $M=2^m$ experiments. However this does not ruin the efficiency of computation since it only increases the number of steps by a factor of n. Total number of steps are not appreciably saved from (controlled)^m-NOT scheme. However, the number of conditional steps in each quantum experiment is drastically saved from $2^{m+1}-3$ down to m. We have successfully traded quantum

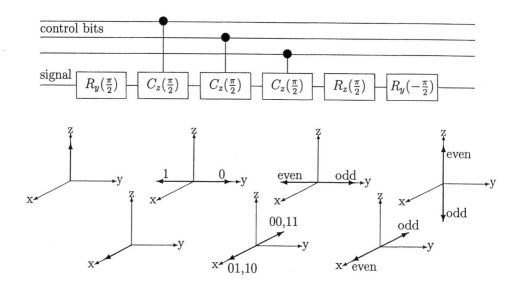

Figure 2. Parity gate. Example of 3 control bits. Generalization is straightforward. R_y is rotation about y-axis. C_z is conditional rotation about z-axis.

complexity with classical one by loosening the marking condition of logically labeled states from the sharpest δ_{ij} to the loosest parities. It makes much sense in decoherence limited experiments.

In summary, we have studied readout schemes for logically labeled bulk NMR quantum computation and proposed promising scheme based on the parity gate which reverses a signal qubit depending on the parity of the control bits. The scheme greatly saves quantum conditional steps while preserving total computational efficiency.

REFERENCES

1. N.A. Gershenfeld and I.L. Chuang, Science **275**, pp. 350–356 (1997).
2. D.G. Cory, A.F. Fahmy and T.F. Havel, Proc. Natl. Acad. Sci. USA **94**, pp. 1634–1639 (1997).
3. I.L. Chuang, N. Gershenfeld and M. Kubinec, Phys. Rev. Lett. **80**, pp. 3408–3411 (1998).
4. J.A. Jones and M. Mosca, LANL preprint quant-ph/9801027.
5. I.L. Chuang, L.M.K. Vandersypen, X. Zhou, D.W. Leung and S. Lloyd, Nature **393**, pp. 143–146 (1998).
6. J.A. Jones, M. Mosca and R.H. Hansen, *ibid.*, pp. 344–346 (1998).
7. I.L. Chuang, N. Gershenfeld, M. Kubinec and D.W. Leung, Proc. R. Soc. Lond. A**454**, pp. 447–467 (1998).
8. A. Barenco, C. H. Bennett, R. Cleve, D. P. DiVincenzo, N. Margolus, P. Shor, T. Sleator, J. A. Smolin, and H. Weinfurter, Phys. Rev. A**52**, pp. 3457–3467 (1995).

Quantum Logical Diagram of Quantum Computing Evolution

Hideaki Matsueda[a]* and David W. Cohen[b]

[a]Department of Information Science, Kochi University,
2-5-1 Akebono-cho, Kochi 780-8520, Japan

[b]Department of Mathematics, Smith College,
Northampton, Massachusetts 01063, USA

A theoretical framework is developed to evaluate the degree of convergence of quantum entangled pure states towards a dispersion-free state of no intrinsic uncertainty. The temporal evolution of states in quantum computing is analyzed diagramatically, providing a visual tool for the refining of quantum algorithms to help achieve minimal uncertainty and maximal efficiency, as well as for better understanding of the quantum entanglements crucial to quantum computing.

1. Introduction

This paper provides a diagramatic method to visualize and assess limitations of quantum algorithms due to intrinsic uncertainty inherent in quantum states. This intrinsic uncertainty is not the same as operational uncertainty, which is the noise accumulated by inefficiencies of physical devices. The purpose is to help build quantum circuit algorithms that converge intrinsic uncertainty to a level low enough so that reduction in operational uncertainty becomes effective.

2. The Quantum Computing Process

The computing process begins with an input row of qubits (quantum bits), consisting of an n-qubit preparation register in a superposition state of maximal uncertainty, together with an input register. We can visualize this qubit row as a row of boxes, with $n = a + f$ boxes representing two preparation registers on the left, and some boxes on the right to represent the input register, as shown in the upper portion of Fig.1.

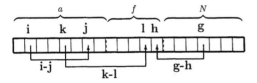

Fig.1 A computational row of quantum bits, and a parallel collection of C-T operations. Each pair connected by an arrow represents a control-bit/target-bit pair.

*Helpful discussions with Huzihiro Araki at Tokyo Science University is acknowledged. This work is partially supported by the Proposal-Based New Industry Creative Type Technology R&D Promotion Program of the New Energy and Industrial Technology Development Organization (NEDO) of Japan.

The computation proceeds via a bank of *row operations*, that are facilitated either by a spatial array of parallel set of quantum gates, or by a temporal series of parallel operations. A row operation consists of a parallel collection of control-bit/target-bit (C-T) operations, such as controlled not (CN), controlled rotation (CR), and controlled phase shift (CPS), as represented in Fig.1. Each C-T operation produces the unitary (or quasiunitary) evolution of the target-bits, initially kept in ground state, driven by the propagation of excited control-bit states [1][2] or by series of external optical excitations.

The overall process of quantum computation may be understood as consisting of four main steps. The first step is to prepare the preparation registers, which is a superposition (minimum entanglement) of the ground state $|0\rangle$ and the excited state $|1\rangle$ in each of a qubits, and a ground state in each of f qubits. The latter preparation register is used to accomodate intermediate results in some algorithms such as for factoring. We'll call the state of each qubit a bit-state.

The state of each row can be represented by an n-dimensional vector $(b_1, b_2, \ldots, b_n)$, where each b_i is a number between 0 and 1, representing the probability that qubit i is in the excited bit-state $|1\rangle$. The number b_i is obtained by $b_i = |e_i|^2$, where each qubit is in a superposition $g_i|0\rangle + e_i|1\rangle$. The initial preparation state is the state of maximum intrinsic uncertainty in the first preparation register, $(\overbrace{\frac{1}{2}, \frac{1}{2}, \ldots, \frac{1}{2}}^{a}; \overbrace{0, 0, \ldots, 0}^{f})$. We'll also call this a state of maximal dispersion, as opposed to a dispersion-free state, where each b_i is either 0 or 1 [3]. It's not hard to see that there are 2^n vectors representing dispersion-free states in the n-dimensional state space.

In the second step, computation proceeds as the sequential (downward) execution of a bank of row operations. Each row operation results in an evolution of the preparation state, generating the quantum mechanical entangled states. A row operation may increase or decrease the intrinsic uncertainty (dispersion) of the preceding state. We'll address that issue in Section 3 of this paper.

The third step in the overall computation is another unitary evolution, converging into the final less dispersive state, or even into a dispersion-free state [4], i.e. a minimal uncertainty state in the first a qubits. This is achieved as the cumulative effects of rotary operations causing interference among the quantum bits. The quantum Fourier transform for the case of the factoring algorithm is an example of such a third step. The iterative combination of two Walsh-Hadamard transforms and a CPS (or conditional phase shift) in a quantum search algorithm, and the building up of the correlation function in a quantum simulator may also be considered as examples of the third step.

The fourth step is the reading or measuring of the final solution state. The less the uncertainty in this final state, the more accurately one can read it.

3. Diagram for the Evolution of Logical States

The third step in our quantum computing schematic is crucial for making quantum computing useful, because without this all we can achieve is an utterly unreadable super-position of all the provisional solutions. We've created a diagram (Fig.2) to visualize the evolution of uncertainty in the quantum logical process of a pair of qubits working as an elemental gate, such as a CN, CR, or CPS gate [2]. Each such gate can be represented as an operation on bit-states by an SU(2) matrix.

Figure 2(a) represents the state space for a single pair of bits, consisting of a control-bit c_i and target-bit t_j. The horizontal axis represents the probability C that the control-bit is in state $|1\rangle$, and the vertical axis represents the same probability T for the target-bit. In the terminology of quantum logic, we are plotting the temporal change of the truth value of the proposition *the target-bit is in state* $|1\rangle$ together with the truth value of the proposition *the control-bit is in state* $|1\rangle$.

The temporal evolution of the state of a provisional solution (an entire row in the quantum computer) may be tracked in the n-dimensional state space by considering the 2-dimensional subspaces generated by control-target pairs for each row. With that in mind we define the *dispersion of the state represented by row-vector* $b = (b_1, b_2, \ldots, b_n)$ as

$$D = \min_d \{\| b - d \| \mid d \text{ is a dispersion–free vector}\}. \tag{1}$$

This is a measure of how close b is to one of the 2^n dispersion-free vectors in the state space. We now consider the effect that row operations have on D.

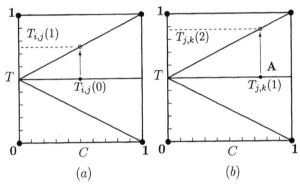

Fig.2 State space diagrams for control-target bit pairs. The temporal evolution of the truth probability T of the statement *the target-bit b_j is in state $|1\rangle$ at time step t* is plotted with an open circle labeled $T_{i,j}(t)$. [The subscript i means that bit b_i is the control-bit for this target.]

(a)   (b)

For a control-target bit (C-T) pair i-j, suppose both the control and target are in an initial superposition of maximal uncertainty: $\frac{1}{\sqrt{2}}(|0\rangle + |1\rangle)$. Then we represent the pair by the closed circle at the center point (0.5, 0.5) in the left diagram, Fig.2 (a). The theoretical dispersion for a pair of qubits is maximal in this initial superposition, because the shortest distance to any one of all the dispersion-free states (0, 0), (0, 1), (1, 0), or (1, 1) is maximal at this middle point. Since we're interested in the evolution of the truth values of target bits, we'll label this first point $T_{i,j}(0)$. We'll follow the evolution of the state as it is affected by controlled rotations.

A controlled rotation by angle β around the y axis $\hat{R}^\beta_{yi,j}$ moves the truth value of the target bit away from the initial superposition point, into some point in the triangular region defined by the oblique $\pm \tan^{-1} \frac{1}{2}$ lines connecting the points $(0, 0.5)$ and $(1, 1)$ or $(0, 0.5)$ and $(1, 0)$, and the vertical line at $C = 1$, as shown in Fig.2 (a) or (b). If the angle of rotation is $\pm \frac{\pi}{2}$, the truth value of the target bit increases or decreases so that the C-T pair falls on one of the $\pm \tan^{-1} \frac{1}{2}$ lines. This is proven elsewhere [2].

Suppose the target is subjected to a y-rotation through angle $\frac{\pi}{2}$. Then the pair state is moved from the center of the left diagram (with $T_{i,j}(0)$) to a state with truth value $T_{i,j}(1)$.

Now assume that target becomes a control-bit b_j with truth value $C_{j,k}(1) = T_{i,j}(1)$ for a new target-bit b_k in a state of maximal dispersion (Point A with $T_{j,k}(1)$ in the right diagram). Another y rotation through angle $\frac{\pi}{2}$ moves the new target to a bit state with probability $T_{j,k}(2)$.

Consecutive similar rotations for C-T pairs, where the control is the target from the preceding rotation, and the target is in a maximally uncertain state, produce targets with truth values converging to 1.

4. Concluding Remarks

We have developed a way to estimate intrinsic uncertainties in provisional results of quantum computations. With this method, we can calculate the reduction of uncertainty resulting from particular controlled rotations applied to control-target bit pairs, and we also can show that other C-T operations do not reduce target bit uncertainty for the maximally uncertain initial superposition.

A lower bound in the number of computational steps to make the provisional solutions converge into a dispersion-free state having minimal uncertainty, could be estimated by this method, within a finite error which decreases with the angle of controlled rotation [2]. In this way the intrinsic or theoretical efficiency could be maximized for any quantum computing process, so that efforts to reduce operational uncertainty should become effective. This method is also helpful for understanding quantum entanglements.

REFERENCES

1. H. Matsueda, and S. Takeno, IEICE Trans. Fundamentals Electron., Commun. and Computer Sci., **E80-A (9)** (1997) 1610-1615; H. Matsueda, in *Proc. European Conference on Circuit Theory and Design (ECCTD'97)* (Budapest, 1997) invited paper, pp.265-270; in *Unconventional Models of Computation*, eds. C. S. Calude, J. Casti, and M. J. Dinneen (DMTCS Series, Springer-Verlag, Singapor, 1998) pp.286-292; in *Proc. 1998 SPIE Quantum Computation Conference, at AeroSense'98* (Orlando, 1998) SPIE Proc. vol.**3385**, pp.84-94; in *Quantum Computing and Quantum Communications*, ed. C. P. Williams, Springer-Verlag Lecture Notes in Computer Science (NASA-QCQC'98), vol.**1509** (1999).
2. H. Matsueda, and D. W. Cohen, Int. J. Theor. Phys., **38**, 701-711 (1999).
3. D. W. Cohen, *An Introduction to Hilbert Space and Quantum Logic* (Springer, New York, 1989) Chaps. 3 and 6; the intrinsic uncertainty is also called ontological uncertainty.
4. This convergence and collapses of wave functions into a limited number of states lead to solutions, from which the period of $f(a)$ $(= x^a \; mod \; N$, for a chosen $x)$ will be derived in an algorithm to factor a number N. The state after the partial collapse into one of the groups of $f(a)$ may be called disentanglement between registers for a and for $f(a)$.

Quantum Coherence and Decoherence - ISQM - Tokyo '98
Y.A. Ono and K. Fujikawa (Editors)
© 1999 Elsevier Science B.V. All rights reserved.

A Generalized EPR Problem and Quantum Teleportation

A.Motoyoshi, T.Yoneda, T.Ogura and M.Matsuoka

Department of Physics, Kumamoto University, Kumamoto 860-8555, Japan

Based on brief discussions of what is the correct understanding of the EPR problem, the method of quantum teleportation is investigated. A few methods for the realization of quantum communication and teleportation by three photons are discussed.

1. INTRODUCTION

New applications of quantum mechanics are expected in an area known as quantum communication, cryptography and computation. In the quantum information processing of these applications, the EPR[1] correlation acts an essential and indispensable role. Quantum communication and teleportation of an unknown state present a generalized EPR problem for three particles. Unfortunately, there are still two deeply rooted and completely different standpoints as concerns the EPR problem.

The experimental confirmation of teleportation will give new tests of *the non-existence of local element of reality* and give an important step toward common understanding of the problems of measurement. As an application, this idea of teleportation will give a method for quantum communication free from eavesdropping, since the disembodiment of the exact quantum state of a particle into classical data and EPR correlations evades eavesdropping. In addition, the verification of the teleportation using photons, especially, may present one way to break through a main difficulty involved in the realization of a quantum computer.

2. A GENERALIZED EPR PROBLEM

The original scheme of teleportation was proposed by C.H.Bennett et al.[2]. From the measurement theoretical point of view, however, their intact scheme of teleportation would be difficult for its realization, since this original scheme of teleportation of an unknown state is based on the existence of non-local long-range correlations between the EPR pair of particles *relevant to the notion of wave function collapse*, which is nothing but the almost direct paraphrase of the von Neumann projection postulate[3]. This projection postulate necessarily leads to the EPR paradox. In the Bennett scheme, the state of the entire three-particle system is represented by

$$|\Psi\rangle_{123} = |\phi\rangle_1 |\Psi^{(-)}\rangle_{23} \longrightarrow \frac{1}{2}\Big\{-|\Psi^{(-)}\rangle_{12}|\phi\rangle_3 - |\Psi^{(+)}\rangle_{12}\big(a|\uparrow\rangle_3 - b|\downarrow\rangle_3\big)$$
$$+ |\Phi^{(-)}\rangle_{12}\big(a|\downarrow\rangle_3 + b|\uparrow\rangle_3\big) + |\Phi^{(+)}\rangle_{12}\big(a|\downarrow\rangle_3 - b|\uparrow\rangle_3\big)\Big\}, \quad (1)$$

where $|\phi\rangle = a|\uparrow\rangle + b|\downarrow\rangle$ with $|a|^2 + |b|^2 = 1$, $|\Psi^{(\pm)}\rangle$ and $|\Phi^{(\pm)}\rangle$ stand for the Bell states.

In order to realize the quantum teleportation and to understand the quantum communication using the idea of teleportation, the correct understanding of the EPR problem and of the measuring process of physical quantities are necessary.

As a solution of the theory of measurement, S.Machida and M.Namiki[4] have constructed an elaborate theory for *the wave function collapse*. In their theory of measurement, which does not rely upon the projection postulate, making use of the statistical operator as the description of the quantum state and introducing the state of an apparatus system, they showed that the resolution of the difficulty concerning *the wave function collapse* does not consist of a demonstration of the disappearance of other branch waves, but follows from *the concept of decoherence*. In the EPR problem, two subsystems and the measuring apparatus form a composite system. In a quantum teleportation scheme of the type proposed by Bennett et al., an unknown state, the EPR pair and the measuring apparatus are the constituents of a composite system. The quantum mechanical state of a composite system and its subsystems cannot be specified without the statistical operator. If one uses the statistical operators to describe the quantum states, and the important role of the conservation law for the identification of the partners of measurement outcomes in one subsystem is confirmed, the formal inconsistency of the EPR problem can be removed within the framework of present-day quantum mechanics[5]. In the Bennett scheme, the transmitted states, which are the partners of measurement outcomes in the subsystem of Alice, are determined by the projection postulate. Therefore, there is no discussion with regard to the law of conservation of the physical quantities subjected to the measurements. Their physical quantities subjected to Alice's measurement also contain no well-known components of spin.

3. QUANTUM TELEPORTATION AND COMMUNICATION

Based on the discussions of what is the correct understanding of the EPR problem[6,7] and making use of a mixture of the triplet state of two spin 1/2 particles as an ancilla, we have proposed a new scheme of teleportation[8] which does not resort to the projection postulate. The statistical operator of our entire system is represented by

$$\hat{\rho}_{1,23} = \frac{1}{4}\{|\psi\rangle_1\langle\psi|\otimes|1,1\rangle_{23}\langle 1,1| + |\psi\rangle_1\langle\psi|\otimes|1,-1\rangle_{23}\langle 1,-1| + 2|\phi\rangle_1\langle\phi|\otimes|\Psi^{(+)}\rangle_{23}\langle\Psi^{(+)}|\}, (2)$$

where $\langle\phi|\psi\rangle = 0$. In order to accomplish this teleportation, Alice may prepare the states $|\Psi^{(+)}\rangle_{23}$, $|1,1\rangle_{23}$ and $|1,-1\rangle_{23}$, and send the particle 3 to Bob. Alice makes a joint measurement on particle 1 and 2. Actually, by partial tracing, we have

$$\begin{cases} \hat{\rho}_1 = \text{Tr}^{(23)}(\hat{\rho}_{1,23}) = \frac{1}{2}\{|\psi\rangle_1\langle\psi| + |\phi\rangle_1\langle\phi|\}, \\ \hat{\rho}_{23} = \text{Tr}^{(1)}(\hat{\rho}_{1,23}) = \frac{1}{4}\{|1,1\rangle_{23}\langle 1,1| + |1,-1\rangle_{23}\langle 1,-1| + 2|\Psi^{(+)}\rangle_{23}\langle\Psi^{(+)}|\} . \end{cases} \quad (3)$$

Recomposing the subsystems 1,23 with 12,3, we obtain $\hat{\rho}_{12,3}$, from which we have

$$\begin{cases} \hat{\rho}_3 & = \mathrm{Tr}^{(12)}(\hat{\rho}_{12,3}) = \frac{1}{2}\big\{|\psi\rangle_3\langle\psi| + |\phi\rangle_3\langle\phi|\big\}, \\ \hat{\rho}_{12} & = \mathrm{Tr}^{(3)}(\hat{\rho}_{12,3}) = \frac{1}{4}\big\{|1,1\rangle_{12}\langle1,1| + |1,-1\rangle_{12}\langle1,-1| \\ & \quad + |\Psi^{(+)}\rangle_{12}\langle\Psi^{(+)}| + |\Psi^{(-)}\rangle_{12}\langle\Psi^{(-)}|\big\}. \end{cases} \tag{4}$$

It can be shown that the quantum mechanical state of the subsystem 3 does not change before and after the measurement in the subsystem 12[8]. Then, we can realize the quantum teleportation of Alice's $|\phi\rangle_1$ to Bob's subsystem 3. Here, use has been made the law of conservation of spin angular momentum to identify the partners of measurement outcomes in the subsystem 12 without resort to the projection postulate. In this scheme, we need not distinguish the difference between $|\Psi^{(+)}\rangle$ and $|\Psi^{(-)}\rangle$ by a measurement, well-known components of spin are measured, and the law of conservation is used to realize the quantum teleportation.

Recently, it is confirmed that the experimental realization of the teleportation is difficult in Bennett's intact scheme[9,10]. Zeilinger[11] says that identification of all four Bell states of the two-photon system "would require a quantum gate that does not exist yet". In the Rome experiment[10], the preparer could not give Alice a photon in an arbitrary state to be teleported. While, the experimental method to realize the above new scheme of teleportation by three photons is under consideration.

Up to now, however, no concrete method to realize the communication using the idea of teleportation has been proposed. Theoretically, any spin $1/2$ particle, such as an electron or neutron etc., can be used for the quantum communication of a qubit. Actually, however, its realization by means of these particles would be difficult and the polarization of photon is the only practical candidate.

According to the above new scheme of teleportation, we have proposed a new method for the realization of quantum communication by three photons in our recent paper[12]. The qubit to be sent is $|\phi\rangle_1 = a|x\rangle_1 + b|y\rangle_1$ with $|a|^2 + |b|^2 = 1$. The states $|x\rangle \equiv |\leftrightarrow\rangle$ and $|y\rangle \equiv |\updownarrow\rangle$ are the horizontal and vertical linearly polarized states of a photon, respectively. The orthogonal state of this qubit is $|\psi\rangle_1 = -b^*|x\rangle_1 + a^*|y\rangle_1$. As a replacement of the above spin state, the triplet state of the polarization of two photons is given by

$$\begin{cases} |1,1\rangle_{23} & \longrightarrow \; |x\rangle_2|x\rangle_3, \\ |1,0\rangle_{23} & \longrightarrow \; \frac{1}{\sqrt{2}}\big(|x\rangle_2|y\rangle_3 + |y\rangle_2|x\rangle_3\big), \\ |1,-1\rangle_{23} & \longrightarrow \; |y\rangle_2|y\rangle_3. \end{cases} \tag{5}$$

The entangled state in the triplet state (5) corresponds to the EPR state. Experimentally, however, if one uses the polarizations of photon as a substitute of the components of spin, one cannot directly construct the mixture of the triplet state of the polarization of two photons such as $\hat{\rho}_{23} = \mathrm{Tr}^{(1)}\{\hat{\rho}_{1,23}\}$ and $\hat{\rho}_{12} = \mathrm{Tr}^{(3)}\{\hat{\rho}_{12,3}\}$ from any light source. Moreover, one can only measure the polarization of each photon. According to these practical constraints, special device is necessary to realize the quantum communication by three photons. With this special device, we can also perform the communication even for the type of Bennett $|\Psi\rangle_{123} = |\phi\rangle_1|\Psi^{(+)}\rangle_{23}$. These experimental implementations for the quantum communication are presented in Ref.[12] and other pages of this issue[13]. In addition, according to these lines of thought, the communication corresponding to the reverse teleportation in the Bennett intact scheme also can be performed, since the structure of the r.h.s. of eq.(1) has resemblance to the structure of the r.h.s. of eq.(2). We have

$$-\frac{1}{2}\left\{|\phi\rangle_1|\Psi^{(-)}\rangle_{23} + \sigma_z|\phi\rangle_1|\Psi^{(+)}\rangle_{23} - \sigma_x|\phi\rangle_1|\Phi^{(-)}\rangle_{23} + i\sigma_y|\phi\rangle_1|\Phi^{(+)}\rangle_{23}\right\} \longrightarrow |\Psi^{(-)}\rangle_{12}|\phi\rangle_3, \quad (6)$$

where σ_x, σ_y and σ_z are the usual Pauli spin matrices. In this communication, well known polarizations of photon are measured and the law of conservation is also guaranteed owing to that three kind of unitary operations σ_z, $-\sigma_x$ and $i\sigma_y$ are performed beforehand. Moreover, in this method, we can sidestep the severe difficulty to perform the complete Bell measurement on a pair of photons. A possible implementation of this communication by three photons is easily devised like in Ref.[12].

4. SYMMETRIC TELEPORTATION IN TIME REVERSAL

In this reverse scheme, one can also perform the teleportation itself. One can easily produce the product states between all four Bell states and their partners in the l.h.s. of eq.(6) by three photons. Preliminary experimental implementation for the realization of this symmetric teleportation in time reversal is presented in other pages of this issue[13]. In this configuration, one can select the reverse teleportation represented by eq.(6) in the Bennett intact scheme by only the measurement on the state $|\Psi^{(-)}\rangle_{12}$, whose experimental method is given in Innsbruck[9]. Detailed discussions of this teleportation will be appeared elsewhere.

REFERENCES

1. A.Einstein, B.Podolsky and N.Rosen, Phys.Rev.,**47**(1935)777.
2. C.H.Bennett, G.Brassard, C.Crépeau, R.Jozsa, A.Peres and W.K.Wootters, Phys.Rev. Lett.,**70**(1993)1895.
3. J. von Neumann,*Mathematical Foundation of Quantum Mechanics* (Princeton University Press, Princeton, N.J.,1955).
4. S.Machida and M.Namiki, Prog.Theor.Phys.,**63**(1980)1457,1833.
5. S.Machida, *Fundamental Quantum Mechanics* (Maruzen Co.,Ltd.,Tokyo,1990), in Japanese.
6. A.Motoyoshi, T.Ogura, K.Yamaguchi and T.Yoneda, Hadronic J.,**20**(1997)117.
7. S.Machida and A.Motoyoshi, Found.Phys.,**28**(1998)45.
8. A.Motoyoshi, K.Yamaguchi, T.Ogura and T.Yoneda, Prog.Theor.Phys.,**97**(1997)819.
9. D.Bouwmeester, J.-W.Pan, K.Mattle, M.Eibl, H.Weinfurter and A.Zeilinger, Nature, **390**(1997)575.
10. D.Boschi, S.Branca, F.De Martini, L.Hardy and S.Popescu, Phys.Rev.Lett.,**80**(1998) 1121.
11. G.P.Collins, Physics Today, February(1998)18.
12. A.Motoyoshi and M.Matsuoka, Prog.Theor.Phys.,**100**(1998) No.2 in print.
13. M.Matsuoka, T.Yamamoto and A.Motoyoshi, ISQM-Tokyo '98(1998).

Purification of Multi-Particle Entanglement*

M. Murao[a], M.B. Plenio[a], S. Popescu[b,c],V. Vedral[a,d] and P.L. Knight[a]

[a]Optics Section, Blackett Laboratory, Imperial College, London SW7 2BZ, U.K.
[b]Isaac Newton Institute for Mathematical Sciences, Cambridge, CB3 0EH, U.K.
[c]BRIMS, Hewlett-Packard Laboratories, Stoke Gifford, Bristol BS12 6QZ, U.K.
[d]Centre for Quantum computing, University of Oxford, Oxford OX1, U.K.

We propose direct and indirect purification protocols for a wide range of mixed entangled states of many particles. These are useful for understanding entanglement, and could be of practical significance in multi-user cryptographic schemes or distributed quantum computation and communication. We investigate purification efficiency and fidelity limits for our multi-particle purification procedures.

1. INTRODUCTION

Entanglement is of central importance for quantum computation [1] and quantum teleportation [2], and certain types of quantum cryptography [3]. For two spin 1/2 particles (qubits), the Bell states are maximally entangled states. For many qubits, the corresponding maximally entangled states are $|\phi^{\pm}\rangle = (|00\cdots0\rangle \pm |11\cdots1\rangle)/\sqrt{2}$, as well as those which are locally unitarily equivalent. The state for each particle is written in the $\{|0\rangle,|1\rangle\}$ basis; for three particles, these are called GHZ states [4]. Entangled states turn into mixed states due to the dissipative effects of the environment, and this is one of the main obstacles for the practical realization of quantum computation and communication. However, the environment does not always completely destroy entanglement. Mixed states resulting from interaction with the environment may still contain some residual entanglement [5]. The task is then to "purify" this residual entanglement to obtain maximally entangled states. These purification procedures use only local operations and classical communication. [6–8].

Disentangled states for two qubits are of the form $\sum p_i\rho_i^1 \otimes \rho_i^2$ where ρ^1 and ρ^2 are the local density matrices [9], and cannot be purified. For many particles the generalisation of disentangled states is not unique. One can define disentangled states as those states from which one cannot purify using local operations a maximally entangled state [10]. This definition gives the investigation of multi-particle purification procedures a fundamental importance for the understanding of entanglement.

Several purification protocols have been proposed [6–8] for the purification of two-particle entangled states. However, for three (many) particles, there is no maximally

*This work was supported in part by the Japan Society for the Promotion of Science, the UK Engineering and Physical Sciences Research Council, the Knight Trust, the European Community and the Alexander von Humboldt Foundation.

entangled state which is invariant under trilateral (multi-lateral) rotations. This is one of the resons why we cannot treat multi-particle entanglement purification protocols as straightforward extensions of the two-particle case. We propose *direct* and *indirect* purification protocols for a wide range of mixed diagonal states having N-particle entanglement. Our aim is to investigate efficiency of these purification procedures and the fidelity limits to make a first step towards a protocol that purifies general mixed states.

2. PURIFICATION PROTOCOLS

Mixed entangled states are likely to appear when one has an ensemble of initially maximally entangled states (for example, $|\phi^+\rangle$) of N particles, and then transmits the N particles to N different parties via *noisy channels*. Consider the effect of a noisy channel, whose action on each particle can be expressed by random rotations about random directions. Each noisy channel causes random rotations (around a random direction and by a random angle) with probability $1-x$, but leaves the particle unaffected with probability x. The state after transmission through such a channel becomes the "Werner-type" state [11,12] given by $\rho_W = x|\phi^+\rangle\langle\phi^+| + (1-x)1/2^N$. The aim of purification is the distillation of a sub-ensemble in the state $|\phi^+\rangle$. The fidelity, $f = \langle\phi^+|\rho_W|\phi^+\rangle$ of the Werner-type state is $f = x + (1-x)/2^N$.

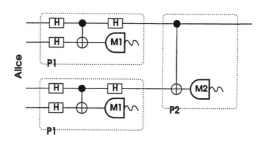

Figure 1. Purification protocol P1+P2. H is a Hadamard transformation, M1 and M2 are local measurement and classical communication. This diagram shows four particles belonging to Alice. Bob and others apply exactly the same procedure.

Now we present a protocol (P1+P2 in Fig 1), which can directly purify a Werner-type state, provided the fidelity of the initial mixed state is higher than a certain critical value. The advantage of this protocol is that Werner-type states for *any number of particles* can be *directly* purified. In the protocol P1+P2, each party (Alice, Bob et al) perform iterations of the operations P1 followed by P2 on the particles belonging to them.

The operation P1 consists of a local Hadamard transformation which maps $|0\rangle \rightarrow (|0\rangle + |1\rangle)/\sqrt{2}$, $|1\rangle \rightarrow (|0\rangle - |1\rangle)/\sqrt{2}$, a local CNOT (Controled NOT) operation and a measurement M1, and another local Hadamard transformation. In M1, we keep the control qubits if an even number of target qubits are measured to be in the state $|1\rangle$, otherwise the control qubits are discarded. For example when purifying for three particles, we only keep $|000\rangle$, $|011\rangle$, $|101\rangle$, $|110\rangle$. The operation P2 consists of a local CNOT operation and a measurement M2 in which we keep the control qubits if all target bits are measured to be in the same state, otherwise the control qubits are discarded. For example, when purifying three particles, we only keep $|000\rangle$ and $|111\rangle$. In this operation, the diagonal and off-diagonal elements of the density matrix are independent of each other, so that the off-diagonal elements do not affect the purification. In addition, there are several type of states which can be purified by the protocol P1 or P2 alone. For example, if the initial mixed state does not have any weight of the pairing state (we call the state $|\phi^-\rangle$ the "pairing state" of $|\phi^+\rangle$) and weights of other states are equal (or even when some weights are zero), iterations of the operation P2 only is sufficient to purify the initial ensemble to the $|\phi^+\rangle$ state. (See [11] for more detail.)

Next we present an indirect purification scheme which purifies many-particle entanglement via two-particle purification. This scheme converts three particle states (Alice Bob, and Claire) into two particle states, then purifies these two particle states, and finally re-converts them to three particle entangled states. The algorithm for this protocol appears more complicated when described in words, so we provide a figure (Fig.2) to help the reader visualise the entire scheme.

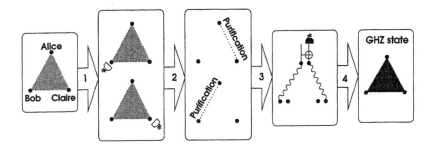

Figure 2. Purification scheme via two-particle purification. Dotted lines represent partial entanglement and wavy lines represent maximum entanglement. The first measurements (represented by white detector symbols) are in the state $|\chi^\pm\rangle = (|0\rangle \pm |1\rangle)/\sqrt{2}$ and the second measurement (represented by a black detector symbol) is in the state $|0\rangle$ or $|0\rangle$. (1) Divide the entire ensemble of the state for three particles into two equal sub-ensembles. (2)Bob and Claire do measurement of their particles of their sub-ensemble. If they obtain $|\chi^-\rangle$, they instruct Alice to perform the σ_z operation on her particles. Otherwise do nothing. (3) Alice and Bob, and separately Alice and Claire perform the two-particle purification protocol in [6,7] to each of the entangled sub-ensembles of two particles. (4) Alice chooses one entangled pair from each sub-ensemble and then performs a CNOT operation on her two particles. Then she measures the target particle. If Alice obtains $|1\rangle$, she instructs Claire to perform the σ_x operation on her particle, and otherwise, do nothing. Then we obtain a sub-ensemble containing the maximally entangled GHZ state. [13,14]

3. EFFICIENCY AND FIDELITY LIMIT

We compare the direct and indirect purification schemes. Any efficient direct three particle purification scheme should perform better than an indirect method via two particles. We note that we only obtain *one* maximally entangled state of three particles from *two* maximally entangled states of two particles by this scheme (in detail see [11]). For purification of N-particle entangled states, we get one maximally entangled state from $N - 1$ maximally entangled states of two particles. In addition, the number of two-qubit CNOT operations, each of which is difficult in practice to carry out to high accuracy, is greater than in our direct scheme. These "inefficiencies" are the main practical disadvantage of the two-particle scheme.

For two-particle entanglement, an initial fidelity $f > 1/2$ is sufficient for successful purification [7] if we have no knowledge of the initial state. The situation is different if we possess additional information about the state, in which case any entangled state can be purified [9]. However, the sufficient condition is not as simple for more than three particles. We have found several different criteria, depending on the type of mixed states [11]. The theoretical fidelity limit for the Werner-type states ρ_W of the purification scheme via two-particle purification is determined by the condition that the fidelity f_r of the reduced two-particle states should satisfy $f_r > 1/2$. The protocol P1+P2 is not optimal for two particles. However, for more than three particles, our observed fidelity limit is lower than that obtained by the purification scheme via two-particle purification.

For states having no weight of $|\phi^-\rangle\langle\phi^-|$ and equal weight of all other states except $|\phi^+\rangle\langle\phi^+|$, the fidelity limit of purification by the protocol P2 is $f > 2^{-(N-1)}$. The fidelity limit obtained by the purification scheme via two-particle purification is worse than that in our protocols [11]. We also see that the fidelity limit of purifiable initial states depends on the distribution of the weight of other diagonal states. This is a condition of a different character from the case of two particles [7]. For two particles, the distribution of weights of other diagonal elements was basically irrelevant for purification, since any distribution of weights of the other diagonal can be transformed into an even distribution by local random rotations of both particles, without changing the amount of entanglement. This suggests that there may be additional structure for many-particle entangled mixed states, which does not exist for two-particle mixed states.

REFERENCES

1. A. Steane, Rep. Prog. Phys. **61** (1998) 117 .
2. C.H. Bennett et al, Phys. Rev. Lett. **70** (1993) 1895.
3. A.K. Ekert, Phys. Rev. Lett. **68** (1991) 661.
4. D.M. Greenberger et al, Am. J. Phys. **58** (1990) 1131.
5. C.H. Bennett et al, Phys. Rev. A **54** (1996) 3824.
6. C.H. Bennett et al, Phys. Rev. Lett **76** (1996) 722.
7. D. Deutsch et al, Phys. Rev. Lett. **77** (1996) 2818.
8. N. Gisin, Phys. Lett. **210** (1996) 151.
9. M. Horodecki et al, Phys. Rev. Lett. **78** (1997) 574.
10. V. Vedral et al, Phys. Rev. Lett. **78** 2275 (1997); V. Vedral et al, Phys. Rev. A **56** (1997) 4452; V. Vedral and M.B. Plenio, Phys. Rev. A **57** (1998) 1619.
11. M. Murao et al, Phys. Rev. A **57** R4075 (1998).
12. R.F. Werner, Phys. Rev. A **40** (1989) 4277.
13. A. Zeilinger et al, Phys. Rev. Lett. **78** (1997) 303.
14. S. Bose et al, Phys. Rev. A **57** (1998) 822.

Quantum Coherence and Decoherence - ISQM - Tokyo '98
Y.A. Ono and K. Fujikawa (Editors)
© 1999 Elsevier Science B.V. All rights reserved.

Quantum Cryptography and Magic Protocols

N. Imoto[1], M. Koashi, and K. Shimizu

NTT Basic Research Laboratories
3-1 Morinosato-Wakamiya, Atsugi-shi, Kanagawa 243-0198, Japan

Quantum information processing is a new research area born by the symbiosis of quantum mechanics and information science. It includes quantum cryptography and quantum computation, and is growing rapidly. Here, it might worth noting that there are two aspects in quantum information processing, namely, *quantum "information processing"* and *"quantum information" processing*. The former is to utilize the bizarre properties of quantum mechanics to realize novel information processing schemes, and the latter is to process the "quantum information", which is a quantum system itself or its equivalent containing enough information to reconstruct the original quantum state. Of course, the use of the latter is essential to the realization of the former, and also, the only information we can directly recognize is classical information. Thus, it is right to say that the latter provides elements for the former. Table I is an attempt to classify several keywords in this context.

Table I. Two aspects of quantum information processing

Quantum "Information Processing" (aimed at practical uses)	"Quantum Information" Processing (element processes)
Quantum cryptography Magic protocols Quantum computing	Cloning Teleportation Entanglement swapping State sharing Quantum error correction Conditional operation

Magic protocols include such protocols as quantum bit commitment, discrete comparison, and zero knoweldge proof, etc. There is a proof that an unconditionally secure quantum bit commitment is impossible [1]. It is not known, however, whether quantum magic protocols are impossible or not in general.

Quantum cryptography schemes basically are key distribution schemes where a random number table is generated at two spacially separated points without communicating the random number table itself. Depending on the modulation scheme and the number

---

[1]Present address: The Graduate University for Advanced Studies, Shonan-Village, Hayama, Kanagawa 240-0193, Japan

of photonic states used, there are several categories of quantum cryptography: four-polarization schemes [2], two-coherent-state schemes [3], cryptography using two-photon interference [4], four-coherent-state schemes [5], and so on.

In all of the above cryptography schemes [2]-[5], a substantial portion of the photons are inevitably discarded since nonorthogonal states are used. An interferometric scheme with delay lines has been proposed [6] in which two orthogonal states are used. In this scheme, however, the timing of the optical pulses must be random, which means that three states are actually used, counting the vacuum between the pulses. Only two of the three states are used for coding.

Fig. 1 shows the scheme proposed in [6], which includes a Mach-Zehnder interferometer with 50/50 beam splitters. Alice, the sender of a crypto-key, sends binary random numbers one by one to Bob, the receiver, with a photon from left for bit "0" or from the bottom for bit "1". With a proper phase shift at Bob's site ($=\pi$ in Fig.1), a "0" photon always appears in the vertical output, and a "1" photon always appears in the horizontal output, which makes Bob possible to determine the bit value with certainty. If this is all of the scheme, Eve, the eavesdropper, can also perform the same thing as Bob and resend fake photons pretending Alice. There are, however, delay lines both at Alice's site and Bob's site as is shown in Fig.1, so that a split wave of a single photon in the lower branch is still at Alice's site when the other wave reaches to Bob's site. This makes Eve impossible to access both waves at the same time, which prohibits Eve from cloning the single photon state carried by the two split waves.

If the timing of photon pulses are open to public, however, there is a successful eavesdropping strategy as is shown in Fig.2 (Strategy 1, hereafter). Calculating the timing of a photon arrival time, Eve sends a fake photon before she receives a signal from Alice as shown in the figure. The split wave in the upper branch leaves Eve immediately, but the other wave is still in the Eve's hands. After measuring Alice's signal by interference, Eve can control the phase of the delayed wave at her hands, which makes Eve possible to flip the state of the single photon before reaching Bob. The authors of Ref. [6] were aware of this eavesdropping strategy, and assumed that Alice sends photons in a random way so that Eve cannot predict the arrival time of the signal. Although this scheme should work well, this introduces the third orthogonal state other than "0" and "1", namely, vacuum states, which are inserted between "0"s and "1"s.

The 4-state cryptography uses two non-orthogonal pairs of states. The 2-state cryptography uses two non-orthogonal states. The scheme in Ref. [6] uses three orthogonal states. A question then arizes here: Is quantum cryptography possible using only two orthogonal states? The firsl topic in this article is to show that it is possible [7].

The point is to use asymmetric beam splitters such as 60/40 or 70/30, etc (of course the same asymmetry should be used at Alice and Bob's sites). In this case, Eve cannot reconstruct the fake signal even with modulating θ due to the asymmetric probability in splitting the photon. It is apparent that the eavesrdropping strategy shown in Fig.2 does not work. Thus, there is no necessity of inserting vacuum states between photons any more. As an extreme case of asymmetry, we can imagine a trivial case with 100/0 or 0/100 beam splitter (which actually is not a beam splitter any more but a mirror). In this case, Eve can adopt a second strategy shown in Fig.3 (Strategy 2, hereafter), where Eve performs a direct photon counting for the first wave only. Since all the information on Alice's signal is obtained at this stage, Eve can send a fake photon without making a mistake. This strategy, again, does not work for a nonzero reflectivity (transmissivity) beam splitter. Not only for strategies 1 and 2, it is possible to prove that any general eavesdropping strategy will fail [7].

In this scheme a complete entanglement is formed by the reflection of a single photon by a 50/50 beam splitter. The use of 100/0 or 0/100 beam splitter leads to no entanglement. The use of intermediate asymmetric beam splitter provides an incomplete entanglement. Both complete entanglement and no entanglement allow Eve to clone the state, while partial entanglement prohibits Eve from cloning the state.

This scheme is also an illustration of the *impossibility of unconditional bit commitment* [1]. A bit commitment is a protocol in which Alice commits a bit to Bob without revealing it (commitment phase). After the commitment phase, Alice cannot change the bit. Bob can read Alice's commitment bit only after receiving a key to open it. A simple example of quantum bit commitment has been analyzed [1], and it became clear that when the marginal density operator in the committed quantum state does not depend on Alice's choice of a committed bit, she can always flip her bit with 100 % certainty by manipulating an ancila system which are entangled to the committed system. The use of entanglement enables a cheating strategy in this bit commitment scheme.

This does not necessarily mean, however, that all magic protocols similar to this bit commitment procedure are proven to be impossible. For example, if Bob (the receiver) first prepares a quantum state unknown to Alice, she cannot prepare an ancila (the auxiliary system) that is completely entangled to the given state, and this may prohibit Alice from cheating. This situation is not ruled out by the assumptions of the proof of the impossibility in [1]. Apparently we need more detailed proof of impossibility (or possibility) for this Bob→Alice→Bob scheme.

A Bob→Alice→Bob protocol goes as follows.

Step 1: Bob determines his "quiz" bit B, and sends Alice a sufficiently long train of $2N$ photons randomly chosen from N right-polarization photons and N vertical-polarization photons when $B = 0$ or N left-polarization photons and N horizontal-polarization photons when $B = 1$.

Step 2: Alice determines her commitment bit A, and measure all the $2N$ photons with either circular polarization base when $A = 0$ or linear polarization base when $A = 1$. Alice gets the value of B from the statistics of the measurement results.

Step 3: Alice reports her answer on B. (Commitment)

Unless Alice determine her bit A in Step 2, she cannot tell the right answer for sure[2].

Step 4: Alice opens A. (Opening phase)

Step 5: Alice reports all of her measurement results if Bob wants. (Check)

---

[2]Alice can of course guess B with half probability without measuring any photon. One way to improve this problem is to use a sequence of "quiz" bits instead of a single bit B.

Alice's cheating stratagy:

Step 2' : Alice does not measure all the N photons but only m $(0 \leq m < 2N)$ photons
with A to determine B.

Step 5' : When she wants to change her bit A to $\overline{A}$, she reports fictitious measurement
results for the m photons. She measure the rest of $2N - m$ photons with $\overline{A}$.

For example, assume $B = 0$ and $A = 0$. If Alice tells $A = 1$ in Step 4, Alice can
always report "vertical" for the m photons.

If $2N$ is sufficiently large, this strategy always works well on the assumption that
Alice can guess the right B by the measurement of m photons. This is fulfilled by taking
sufficiently large m (of course within $m \ll 2N$). Therefore, $2N$ should be finite. In this
case, however, there also should be (hopefully small) probability that an honest Alice gets
the wrong answer for B. Thus,

- Assume p_N to be the probability of getting right B by the measurement of $2N$
  photons. — the probality that an honest Alice passes through the test of Step 3.

- Assume $q_{N,m}$ to be the probability of getting right B by the measurement of m
  photons. — the probability that a cheating Alice passes through the test of Step 3.

- In Step 5', reporting the fictitious results changes the natural statistics (in the
  above example, too many "vertical" photons). Bob should therefore set a tolerable
  deviation of statistics. For the above example, Alice should obtain "vertical" exactly
  for the N "vertical" photons but only about half for the rest of the N "right"
  photons. So, Bob assumes that Alice is honest when the number of "vertical"
  photons in her report is $\leq N + \frac{N+n}{2}$, where the parity of n is chosen to be the same
  as N. Let we assume $p_{N,n}$ to be the probability that an honest Alice passes through
  this test and $q_{N,n,m}$ to be the probability that a cheeting Alice passes through this
  test.

- Assume $P_{N,n}$ to be the probability that an honest Alice passes through both Steps
  3 and 5, and $Q_{N,n,m}$ to be the probability that a cheating Alice passes through both
  Steps 3 and 5. These can be calculated from p_N , $p_{N,n}$, $q_{N,n}$, and $q_{N,n,m}$.

N and n are determined by agreement between Alice and Bob, while m is optimized
by a cheating Alice to make $Q_{N,n,m}$ maximum ($\equiv Q_{N,n}$). Bob's problem is to find N and
n to make $P_{N,n}$ large and $Q_{N,n}$ small.

Fig.4 shows the calculated P and Fig.5 shows the calculated Q as functions of N and
n. No big difference can be recognized at first sight. By subtracting the two probabilities,
however, one can see the large difference between the probability to detect a cheating
Alice and the probability to reject an honest Alice, as is shown in Fig.6. Also, taking
the ratio between $1 - Q$ and $1 - P$, one can see that it is possible to make the ratio of
probability to detect a cheating Alice to that to reject an honest Alice as large as possible
(Fig.7). This means that in the case that this game is repeated many times, one can
always detect a cheating Alice without rejecting an honest Alice.

The above protocol apparently is not an unconditionally secure protocol. Nevertheless, there might be something to be discussed more in detail in the area of quantum magic protocols.

References

[1] Hoi-Kwong Lo and H. F. Chau, Phys. Rev. Lett. **78** (1997) 3410/D. Mayers, *ibid.* 3414.
[2] C. H. Bennett and G. Brassard, *Proc. IEEE Int. Conf. on Computers, Systems, and Signal Processing*, Bangalore, India, IRRR New York (1984) 175.
[3] C. H. Bennett, Phys.Rev. Lett. **68** (1992) 3121.
[4] A. K. Ekert, J. G. Rarity, P. R. Tapster, and G. M. Palma, Phys. Rev. Lett. **69** (1992) 1293.
[5] B. Huttner, N.Imoto, N. Gisin, and T. Mor, Phys. Rev. **A51** (1995) 1863.
[6] L. Goldenberg and L. Vaidman, Phys. Rev. Lett. **75** (1995) 1239.
[7] M. Koashi and N. Imoto, Phys. Rev. Lett. **79** (1997) 2383.

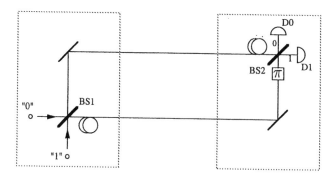

Fig.1. Quantum cryptography using orthogonal states. 50/50 beam splitters are used in [6], and asymmetric beam splitters are used in [7].

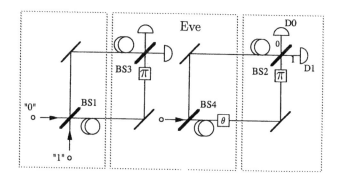

Fig.2. Eavesdropping strategy 1. The strategy is successful for 50/50 beam splitters but not for the asymmetric beam splitters.

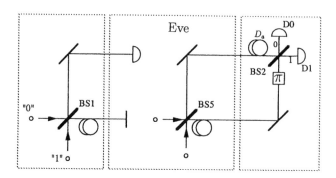

Fig.3. Eavesdropping strategy 2. The strategy is successful for the 0/100 or 100/0 beam splitters but not for the asymmetric beam splitters.

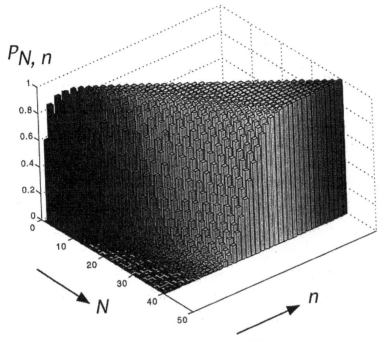

Fig.4. Probability $P_{N,n}$ of accepting an honest Alice.

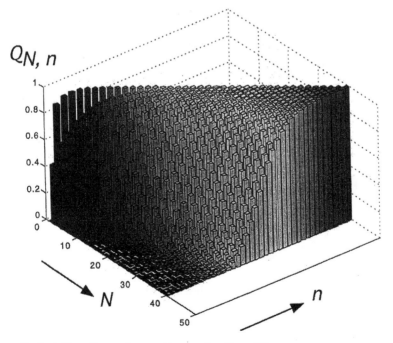

Fig.5. Probability $Q_{N,n}$ of accepting a cheating Alice.

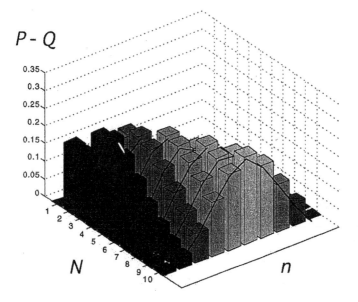

Fig.6. Difference of the accepting probability for an honest Alice to that for a cheating Alice: $P_{N,n} - Q_{N,n}$, which can be made as big as $\simeq 0.25$.

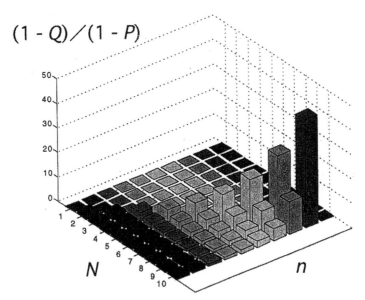

Fig.7. Ratio of the rejecting probability for a cheating Alice to that for an honest Alice: $(1 - Q_{N,n})/(1 - P_{N,n})$, which can be made arbitrary big.

Quantum Coherence and Decoherence - ISQM - Tokyo '98
Y.A. Ono and K. Fujikawa (Editors)
© 1999 Elsevier Science B.V. All rights reserved.

Quantum Entanglement for Secret Sharing and Telesplitting

A. Karlsson[a] * , M. Koashi[a], N. Imoto[a] and M. Bourennane[b]

[a]NTT Basic Research Laboratories, 3-1 Morinosato, Atsugi-shi, Kanagawa, 243-0198, Japan

[b]Department of Electronics, Royal Institute of Technology, 164 40 Kista, Sweden

We show how quantum entanglement can be used to construct protocols for the secret sharing of classical information, and for the splitting of quantum information to several parties by teleportation, where the partial or full reconstruction of the information can be made at one of the receiving sites.

1. Introduction

Entanglement, namely that certain quantum states of two or more particles cannot be written as the product state of the constituent parts, are at the heart of what distinguishes quantum mechanics from classical physics, e.g. the non-locality EPR "paradox" [1]. Not surprisingly, entanglement is also the essential feature of most quantum algorithms [2]. In this contribution, we will present some novel uses of entanglement in quantum cryptography beyond quantum key distribution [3].

2. Two-particle entangled states

An example of an entangled state is the optical quantum state generated from type II parametric down-conversion non-linear optical crystal, which can be written as [4]

$$|\psi\rangle_{A,B} = \frac{1}{\sqrt{2}}(|z+\rangle_A|z-\rangle_B + e^{i\alpha}|z-\rangle_A|z+\rangle_B), \tag{1}$$

where α is a birefringent phase shift of the crystal, and $|z+\rangle$ and $|z-\rangle$ denotes the spin eigenstates, or equivalently the horizontal and vertical polarisation eigenstates, and subscripts A and B denotes the two particles. Here we will be interested in the properties of the following four states, which may be generated from the above state:

$$|\phi^-\rangle_{A,B} = \frac{1}{\sqrt{2}}(|z+\rangle_A|z+\rangle_B - |z-\rangle_A|z-\rangle_B), \quad |\psi^+\rangle_{A,B} = \frac{1}{\sqrt{2}}(|z+\rangle_A|z-\rangle_B + |z-\rangle_A|z+\rangle_B), \tag{2}$$

and

$$|\Psi^+\rangle_{A,B} \equiv \frac{1}{\sqrt{2}}(|\phi^-\rangle_{A,B} + |\psi^+\rangle_{A,B}), \quad |\Phi^-\rangle_{A,B} \equiv \frac{1}{\sqrt{2}}(|\phi^-\rangle_{A,B} - |\psi^+\rangle_{A,B}). \tag{3}$$

---

*Permanent address:Department of Electronics, Royal Institute of Technology, 164 40 Kista, Sweden, email:andkar@ele.kth.se

Now, the set of states $\{\psi^+, \phi^-, \Psi^+, \Phi^-\}$, has the desired feature that $\langle\psi^+|\phi^-\rangle = \langle\Psi^+|\Phi^-\rangle = 0$, which as we will see below allows for a simple encoding of information. Furthermore, all states are not orthogonal, as $|\langle\psi^+|\Psi^+\rangle|^2 = |\langle\psi^+|\Phi^-\rangle|^2 = 1/2$ and $|\langle\phi^-|\Psi^+\rangle|^2 = |\langle\phi^+|\Phi^-\rangle|^2 = 1/2$. This is the crucial feauture that enables the detection of an eavesdropper in the protocols presented below.

3. Quantum entanglement in the service of privacy

3.1. Quantum secret sharing using two and three-particle entanglement

In secret sharing, secret information is sent from a sender Trent, to multiple particants, in our case two; which we by convention call Alice and Bob. The protocol is designed so that neither Alice nor Bob alone can obtain the information sent by Trent. Only if they collaborate can they retrieve the information sent by Trent. In the end of the secret sharing protocol Alice and Bob jointly shares a secret (random) data string with Trent. This string can then be used by Trent to encrypt deterministic data in a one-time-pad cryptography protocol. In classical cryptography applications, secret sharing is an important tool for key management.

3.2. Two-particle quantum entanglement secret sharing protocol

Let us present a protocol for the secret sharing to two persons, in which Alice switches randomly between two sets of states, similar to four-state quantum cryptography [5]. The protocol goes as follows:

1) Trent sends information by using the quantum states $\{\psi^+, \phi^-, \Psi^+, \Phi^-\}$ above, sending one particle to Alice one to Bob. He encodes his information as $\{|0\rangle, |1\rangle\} \Leftrightarrow \{|\psi^+\rangle, |\phi^-\rangle\}$ or $\{|0'\rangle, |1'\rangle\} \Leftrightarrow \{|\Psi^+\rangle, |\Phi^-\rangle\}$, randomly swithing between the $\{|\psi^+\rangle, |\phi^-\rangle\}$ basis and the $\{|\Psi^+\rangle, |\Phi^-\rangle\}$ basis. Alice and Bob detects the states, either in the z or x-basis (the x-basis eigenstates being $|x+\rangle = (|z+\rangle + |z-\rangle)/\sqrt{2}$ and $|x-\rangle = (|z+\rangle - |z-\rangle)/\sqrt{2}$). Note, these measurements are *local*, each made on one particle only.

2) In order to test for the presence of an eavesdropper, or a dishonest party, that is if one of Alice or Bob tries to cheat the protocol and find out the secret alone, Alice and Bob alternates in declaring a set of outcomes for some of the bits sent. After this has been done, they also declare the measurement basis that was used for all bits.

3) After some bits (basis) have been released, Trent reveals to them in which of the two basises the state was sent, but not which state. Also, Trent tells which was the state sent for the released test bits. In half of all cases, Alice and Bob must discard their results, but in the other half they have useful results as is described below in Table 1.

Alice/Bob	z+	z-	x+	x-
z+	ϕ^-	ψ^+	Ψ^+	Φ^-
z-	ψ^+	ϕ^-	Φ^-	Ψ^+
x+	Ψ^+	Φ^-	ψ^+	ϕ^-
x-	Φ^-	Ψ^+	ϕ^-	ψ^+

Table 1: *Correlation between outcomes for Alice and Bob, allowing them to jointly decide which state was sent by Trent, given that they know the choice of basis made by Trent.*

From the table we see that Alice and Bob may together identify which of the states that was sent by Trent. However, neither of them alone may determine if a $|\psi^+\rangle$ or a $|\phi^-\rangle$ was sent in the first basis set, or if a $|\Psi^+\rangle$ or a $|\Phi^-\rangle$ was sent in a second basis set. Hence, in order to know which bit value was sent by Trent, they must collaborate. From the test bits, Alice and Bob can independently make a test for the presence of an eavesdropper, or if one of them is being dishonest. As may be shown, an eavesdropper, or a dishonest party will introduce a 25 % error rate in the outcomes of the test bits.

Quantum Secret sharing may also be realised using three-particle entangled state [6]. Suppose, Trent, Alice and Bob share one particle each from a three-particle entangled Greenberger-Horne-Zeilinger (GHZ)-state [7]

$$|\psi_{GHZ}\rangle = \frac{1}{\sqrt{2}}(|z+\rangle_T|z+\rangle_A|z+\rangle_B + |z-\rangle_T|z-\rangle_A|z-\rangle_B), \tag{4}$$

where the first particle is that of Trent, the second that of Alice and the third that of Bob. Now, they make random measurement, either in the x-direction or the y-direction, where the x eigenstates were defined above, and the y eigenstates are defined $|y+\rangle = (|z+\rangle + i|z-\rangle)/\sqrt{2}$ and $|y-\rangle = (|z+\rangle - i|z-\rangle)/\sqrt{2}$. By re-expressing the GHZ-state in the x and y eigenstates, as was shown in [6], Alice and Bob can construct a table similar to that of table 1 above, allowing Alice or Bob jointly to determine which measurement outcome was observed by Trent if they know in which basis (direction) he measured.

3.3. Quantum telesplitting of information

The secret sharing spreads the information in a classical bits ($\{0, 1\}$) to multiple participants. Quantum key sharing can also extended to spread *quantum information* (quantum superpositions of classical bits). Let us describe a protocol for quantum information splitting by teleportation = "telesplitting" [6,8]. The basic idea is as follows: Trent has a qubit $|Q\rangle = a|z+\rangle + b|z-\rangle$, which he wants to send to either Alice or Bob (both cannot generally have it as that would violate "no-cloning"-theorems). This may be done using a quantum teleportation procedure [9,10] whereby Trent, Alice and Bob initially share a GHZ state. Next, Trent makes a joint Bell state measurement [11] on $|Q\rangle$ and his particle of the GHZ-state. By communicating the outcome (2 bits) to Alice and Bob, their joint state can be rotated to the splitted state $|QS\rangle = a|z+, z+\rangle + b|z-, z-\rangle$, here the notation is that of Alice having the first particle and Bob the second. From this state, Alice may for instance retrieve $|Q\rangle$ if Bob does a measurement in the x basis, and communicates (1 bit) which outcome ($x+$ or $x-$) he obtained. If Bob measures in a different basis than the x basis he may also get partial information on the state, of course at the expence of Alice not being able to fully reconstruct the initial state [8]. Note, that by using quantum controlled-NOT-gates [12], the quantum information in a qubit can also be split to several parties.

The quantum splitting scheme can also be extended to a initial version of a *m-out-of-n*-protocol, where a secret, here a qubit, is divided into n pieces or shares, in such a way that m shares (with $m \leq n$) can be used to reconstruct the secret. Suppose a qubit $|Q\rangle = a|z+\rangle + b|z-\rangle$ is splitted among m parties so they each possess one particle from the entangled state $|QSS\rangle = a|z_1+, .., z_m+\rangle + b|z_1-, .., z_m-\rangle$, the subscript denoting the participants. Now, all m parties are needed if the original qubit is to be reconstructed at (any) one of the locations. However, using entanglement swapping [13,14], we may swap the entanglement from, say participant m, to an "outside" participant k. If the new participant is in possesion of a two-particle maximally entangled state $|N\rangle = (|z+, z+\rangle_k + |z-, z-\rangle_k)/\sqrt{2}$, he may leave one particle at a "swapping center", and keep the second particle for himself. Now, to make the swapping, one performs a joint Bell measurement on particle m and the one particle left by the "outsider" at the "swapping center", giving $|QSS\rangle = (a|z_1+, .., z_{m-1}+\rangle|z+\rangle_k + b|z_1-, .., z_{m-1}-\rangle|z-\rangle_k)$. Effectively, this "teleports" the entanglement to the new participant, without him/her having to be locally present with the particle to be entangled at the swapping instance.

4. Outlook

We have given some examples of cryptographic protocols based on quantum entanglement. In some cases, notably secret sharing, simpler non-entanglement based crypto-protocols are also possible. However, we think the experimental studies of entanglement based primitives, as those discussed above, give interesting insights also for more general quantum computation objectives.

REFERENCES

1. A. Einstein, B. Podolsky, and N. Rosen, Phys. Rev. **47** (1935) 777.
2. A. Ekert and R. Jozsa, Los Alamos Preprint archive,quant-ph/9803072.
3. A. Ekert, Phys. Rev. Lett., **67**, (1991) 661.
4. P.G. Kwiat, K. Mattle, H. Weinfurther and A. Zeilinger, Phys. Rev. Lett., **75**,(1995) 4337.
5. C.H. Bennet, F. Bessete, G. Brassard, L. Salvail, J. Smolin, Journal of Cryptology, **5** (1992) 3.
6. M. Hillery, V. Buzek and A. Bertaiume, Los Alamos Preprint archive,quant-ph/9806063.
7. D. M. Greenberger, M.A. Horne, A. Shimony and A. Zeilinger, Am. J. Phys., **58**, (1990) 1131.
8. A. Karlsson and M. Bourennane, Accepted for publication in Phys. Rev. A, June 26, 1998.
9. C.H. Bennet, G. Brassard, C. Crepeau, R. Jozsa, A. Peres and W.K. Wooters, Phys. Rev. Lett., **70** (1993) 1895.
10. D. Bouwmeister, J.W. Pan, K. Mattle, M. Eible, H. Weinfurther and A. Zeilinger, Nature, **390**, (1997) 575.
11. M. Michler, K. Mattle, M. Eible, H. Weinfurther and A. Zeilinger, Phys. Rev. A **53** (1996) R1209.
12. A. Barenco, D. Deutsch, A. K. Ekert and R. Jozsa, Phys. Rev. Lett., **74** (1995) 4083.
13. M. Zukowski, A. Zeilinger, M. A. Horne and A. K. Ekert,Phys. Rev. Lett., **71**, (1993) 3031.
14. J-W. Pan, D. Bouwmeester, H. Weinfurther and A. Zeilinger, Phys. Rev. Lett., **80**, (1998) 3891.

Quantum Coherence and Decoherence - ISQM - Tokyo '98
Y.A. Ono and K. Fujikawa (Editors)
© 1999 Elsevier Science B.V. All rights reserved.

Communication channels secured from hidden eavesdropping via transmission of entangled photon pairs

K. Shimizu and N. Imoto
NTT Basic Research Laboratories, 3-1 Morinosato Wakamiya,
Atsugi, Kanagawa 243-0198, Japan

We propose a quantum communication scheme for transmitting a definite binary sequence in a secure manner. This scheme is very suitable for sending a ciphertext in a secret-key cryptosystem so that we can detect an eavesdropper who attempts to decipher the key. The security required for a key that is used repeatedly can therefore be enhanced to the level of one-time-pad cryptography. An information sender employs a pair of entangled photon twins as a bit carrier and an information receiver performs Bell state analysis. With these devices, the scheme can be implemented.

1. INTRODUCTION

The security of a secret-key is guaranteed for one-time-pad use in a secret-key cryptosystem. However, repeated use of the same key necessarily degrades the security. This is because Eve who attempts to decipher the key can eavesdrop different texts ciphered in the same key, without being detected by the legitimate users. She may be able to decipher the key by comparing them and finally read all plaintexts. Security degradation is a serious problem and is attributed to the use of classical communication channels for sending ciphertexts. This suggests the possibility of avoiding the problem by using a quantum channel instead of a classical one. Quantum communication schemes proposed for key distribution [1,2], however, cannot be employed for sending a ciphertext or any definite binary sequence which contains intelligible information. This paper, however, shows that the use of polarization-entangled photon twins makes it possible for us to transmit a definite binary sequence in a secure manner.

2. MANIPULATION AND MEASUREMENT FOR PHOTON TWINS

In our proposed scheme, information transmission is separated into two steps. First, Alice encodes a bit of information on a pair of entangled photon twins and sends it to Bob. He measures it with an appropriate basis and obtains partial information. Later Alice notifies Bob of the additional information needed to decode the encoded information completely.

Here we employ the set of four Bell states $\{|P\rangle, |Q\rangle, |R\rangle, |S\rangle\}$ as a normalized orthonormal basis for expressing purely entangled quantum states of photon twins. These are $|P\rangle \equiv |\Psi^+\rangle$, $|Q\rangle \equiv i|\Psi^-\rangle$, $|R\rangle \equiv i|\Phi^+\rangle$ and $|S\rangle \equiv |\Phi^-\rangle$, respectively. Additional $\pi/2$ phase shifts for $|\Psi^-\rangle$ and $|\Phi^+\rangle$ are very convenient. When Alice sends a photon pair to Bob, she prepares a two-term superposition of four Bell states. These are given by

$$|A\rangle \equiv (|P\rangle + |Q\rangle)/\sqrt{2}, \quad |B\rangle \equiv (|R\rangle + |S\rangle)/\sqrt{2},$$
$$|C\rangle \equiv (|P\rangle + |R\rangle)/\sqrt{2}, \text{ and } |D\rangle \equiv (|Q\rangle + |S\rangle)/\sqrt{2}. \quad (1)$$

Here we define the two sets of $\{|A\rangle, |B\rangle\}$ and $\{|C\rangle, |D\rangle\}$ as AB-coding basis and CD-coding basis, respectively. Alice selects either coding basis when she encodes a bit of information on photon twins; $|A\rangle \equiv$ " 0 " and $|B\rangle \equiv$ " 1 " for AB-coding basis, and $|C\rangle \equiv$ " 0 " and $|D\rangle \equiv$ " 1 " for CD-coding basis. She can prepare each state by operating an appropriate one-photon spinol rotation. The use of the two different bases makes it possible for Alice to prepare quantum states which cannot be measured correctly without knowing the correct basis.

Bob, the information receiver, performs a joint measurement of the photon twins with an appropriate measurement basis. We show that Bell state analysis makes it possible for Bob to read out partial information about the quantum states even though he knows nothing about the coding basis. Bell state analysis is a measurement which employs $\{|P\rangle, |Q\rangle, |R\rangle, |S\rangle\}$-basis. If he obtains $|P\rangle$, he has a partial information that either the $|A\rangle$ or $|C\rangle$ state was prepared. No further information is available at this stage. However, he will be able to know the prepared state when Alice eventually notifies him of the selected coding basis.

Thus a bit of information is transmitted by the photon twins as a first step and then another bit of information is transmitted later so that Bob can determine the quantum state sent by Alice. Eve can also obtain the first bit of information by intercepting the photon twins and measuring them with a Bell state analyzer. However, she cannot resend the quantum state correctly because she cannot obtain the second bit of information. So she has to guess but this results in a bit error with a finite probability. This is the basic device for detecting eavesdropping in our proposed communication scheme. Eve can, however, resend her resultant Bell state directly to Bob and in this case Eve can completely escape detection. Our communication scheme, therefore, must be improved to cope with this strategy of Eve.

In the improved scheme, Bob operates a unitary transformation U in a random manner for each photon pair, preceding the Bell state analysis. The U-transformation changes each Bell state as follows; $U|P\rangle = (|P\rangle + |Q\rangle - |R\rangle + |S\rangle)/2$, $U|Q\rangle = (|P\rangle + |Q\rangle + |R\rangle - |S\rangle)/2$, $U|R\rangle = (|P\rangle - |Q\rangle + |R\rangle + |S\rangle)/2$ and $U|S\rangle = (-|P\rangle + |Q\rangle + |R\rangle + |S\rangle)/2$. It does not change the quantum states $|A\rangle$ and $|B\rangle$ as can be easily proven. The states $|C\rangle$ and $|D\rangle$ are transformed as follows;

$$U|C\rangle \equiv (|P\rangle + |S\rangle)/\sqrt{2}, \quad \text{and} \quad U|D\rangle \equiv (|Q\rangle + |R\rangle)/\sqrt{2}, \quad (2)$$

respectively. Although the positions of $|R\rangle$ and $|S\rangle$ are reversed compared with Eq.(1), Bob can decode the quantum information with additional information notifying him of the coding basis. Thus the transmission of the quantum information suffers no disturbance from the random operation of the U-transformation.

By contrast, Eve must abandon her strategy of resending a resultant Bell state because it will lead to a bit error if Bob operates the U-transformation. For example, the above procedure can be explained as follows; (i) Alice sends $|C\rangle$, (ii) Eve intercepts the photon twins and obtains the Bell state $|P\rangle$, (iii) Eve resends $|P\rangle$ to Bob, (iv) Bob operates the U-transformation and obtains the Bell state $|Q\rangle$ or $|R\rangle$ with a probability of 1/2, (v) Alice notifies Bob that she has sent $|C\rangle$, and (vi) Bob can detect eavesdropping because the transformed state $U|C\rangle$ does not contain the Bell states $|Q\rangle$ and $|R\rangle$.

3. COMMUNICATION PROTOCOL FOR CIPHERTEXT TRANSMISSION

Here we describe how to use our proposed scheme for sending ciphertexts in a secret-key cryptosystem. The protocol is shown schematically in Fig.1. A ciphertext is represented by a lengthy definite binary sequence and we define each bit as an intelligible bit. By contrast, we introduce a test bit as a random number for detecting eavesdroppers. Alice randomly inserts some test bits into the intelligible bit sequence. Before the transmission, Alice and Bob decide that AB- and CD- bases will be employed for encoding an intelligible bit and a test bit, respectively. The number of test bits depends on the degree of secrecy to be achieved but it need not be very large compared with the length of the intelligible bit sequence.

Bob performs a Bell state measurement after operating the U-transformation at random. After receiving all the photon pairs, he asks Alice to open all the test bit positions and she does so. He distinguishes intelligible bits from test bits and then decodes all the bit values. They compare the encoded and decoded bit values for all the test bits, so that they can confirm security. We assume that Eve employs an intercept/resend strategy for eavesdropping. As a bit error is regarded as evidence of eavesdropping, she makes an effort to minimize the total frequency of errors in the bit sequence. In particular, errors must be avoided in intelligible bits because it gives rise to an unavoidable disturbance in a deciphered plaintext and Bob necessarily regards the disturbance as explicit evidence of eavesdropping.

Here Eve also should be assumed to know that (i) AB- and CD-coding bases are utilized for intelligible and test bits, respectively and (ii) the number of intelligible bits is far larger than the number of test bits. With these assumptions, she can successfully behave so as not to leave any evidence in an intelligible bit provided she employs the AB-coding basis when she resends the photon twins after interception. Thus she can greatly reduce the total number of bit errors. Nevertheless, she cannot escape detection because she necessarily resends $|A\rangle$ or $|B\rangle$ even though $|C\rangle$ or $|D\rangle$ is prepared by Alice as a test bit. Therefore there is a finite probability that Eve will leave evidence in a test bit.

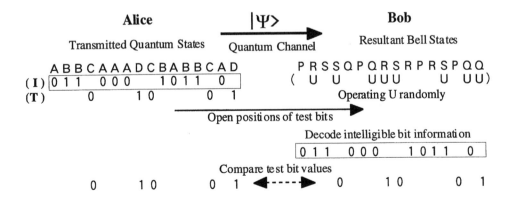

Figure 1.   Basic protocol for secure transmission of ciphertext
(I) Intelligible bit , (T) Test bit.

Alice and Bob can discover Eve as follows ; (i) Alice is supposed to send |C⟩, test bit "0", (ii) Eve intercepts the photon twins with a Bell state analyzer and obtains the Bell state |P⟩, (iii) Eve resends |A⟩ to Bob in accordance with her resend strategy, (iv) Bob obtains the Bell state |Q⟩ with a probability of 1/2, (v) Alice notifies Bob that she has sent a test bit, (vi) Bob considers |Q⟩ to result from test bit "1", and (vii) they finally detect the bit error by comparing their bit values. The error probability per test bit is 1/2. If Eve obtains the Bell state |R⟩ and then resends |B⟩, Bob regards the resultant Bell state |R⟩ or |S⟩ as a bit error, depending on whether he operates the U-transformation or not.

Alice and Bob can expect the error free transmission of a ciphertext if Eve adopts the intercept/resend strategy. In this sense, our communication channel is equivalent to a classical communication channel. However, the use of quantum states unknown to anyone else makes it impossible for Eve to distinguish test bits from large numbers of intelligible bits and she necessarily fails to resend test bits. As the error probability per test bit is 1/2, they can detect Eve with a whole probability of $1 - (1/2)^M$ by inserting M test bits. For example, even with 30 test bits, they can reduce the probability of detection failure to 10^{-9} so Eve cannot escape detection.

Once Alice and Bob have shared a secret-key, they must continue to employ our communication scheme whenever they send any ciphertext. As long as they detect no eavesdropping, they can guarantee the security of the key and plaintexts with a high level of confidence and continue to use the key repeatedly. In contrast, if they detect eavesdropping, they discard the key. Although Eve can intercept all the ciphertext, she can no longer expect to obtain any other ciphertexts ciphered in the same key. Therefore she cannot decipher the key at all. This means the security of repeated key use can be enhanced to the level of one-time-pad cryptography, which greatly reduces the size of the key stock.

4. SUMMARY

We propose a quantum communication scheme for sending a ciphertext in a secure manner. A pair of polarization-entangled photon twins is used to carry a bit which is encoded in a two-term superposition of Bell states. Different bases are employed for encoding an intelligible bit and a random test bit. The photon twins are measured with a Bell state analyzer and any bit can be decoded from the resultant Bell state when the receiver is later notified of the coding basis. By opening the positions and the values of test bits, ciphertext can be transmitted and eavesdropping is detected simultaneously. Finally we should mention that this system can be realized by using a Hong-Ou-Mandel two-photon interferometer and this will be reported elsewhere.

The authors thank Dr.M.Koashi for valuable discussions.

REFERENCES
[1] C. H. Bennett and G. Brassard, in *Proceedings of IEEE International Conference on Computers, Systems and Signal Processing, Bangalore, India* (IEEE, New York, 1984) 175.
[2] C. H. Bennett, G. Brassard, and A. K. Ekert, Sci. Am. **267**, 50 (1992)

Quantum Coherence and Decoherence - ISQM - Tokyo '98
Y.A. Ono and K. Fujikawa (Editors)
© 1999 Elsevier Science B.V. All rights reserved.

Quantum Communication of Qubits by the Polarization of Photons

Masahiro Matsuoka, Takashi Yamamoto and Akio Motoyoshi

Department of Physics, Kumamoto University, Kumamoto 860-8555, Japan

Two schemes of quantum communications are proposed. One is a simple use of the Bennett teleportation to communication of complex coefficients a and b of quantum superposition state. The other is a Bennett-type teleportation with predetermined Bell states.

1. INTRODUCTION

The quantum teleportation proposed by Bennett[1] is to send an unknown quantum superposition state, called a qubit, via two channels, a two-photon EPR channel, called ancilla, and an ordinary classical one. Its experiments have been already performed by Innsbruck[2] and Rome[3] groups. There are a certain difficulty in the original scheme of Bennett in that four Bell states are difficult to be distinguished at the sender, Alice, and a prompt switching of detectors among unitary transformations after receiving the information from a classical channel needs delicate instrumentation at the receiver, Bob.

In the present talk we propose two schemes of quantum communications. The first one is a simple use of this system to communication of complex coefficients a and b of a qubit. The second one is a Bennett-type teleportation in which four Bell states engaged in the teleportation are distinguishable in separate detectors.

2. TELEPORTATION PROPOSED BY BENNETT

In short we explain the original Bennett's scheme. First, Alice has a photon 1 in the state, $|\phi\rangle_1 = a|x\rangle_1 + b|y\rangle_1$ $(|a|^2 + |b|^2 = 1)$, where $|x\rangle$ and $|y\rangle$ are single photon states linearly polarized in the directions x and y. This is the qubit to be teleported. An entangled singlet state of x- and y-polarized photons is prepared by photons 2 and 3 as $|\Psi^{(-)}\rangle_{23} = \left(|xy\rangle_{23} - |yx\rangle_{23}\right)/\sqrt{2}$, in which the photon 2 is sent to Alice and the photon 3 to Bob.

Let us consider the combined system of $|\phi\rangle_1$ and $|\Psi^{(-)}\rangle_{23}$, and describe it by a density operator as

$$\hat{\rho}_{1,23} = |\phi\rangle_1\langle\phi| \otimes |\Psi^{(-)}\rangle_{23}\langle\Psi^{(-)}|$$
$$= \frac{1}{2}\left(|a|^2|x\rangle_1\langle x| + ab^*|x\rangle_1\langle y| + ba^*|y\rangle_1\langle x| + |b|^2|y\rangle_1\langle y|\right) \tag{1}$$
$$\otimes \left(|xy\rangle_{23}\langle xy| - |xy\rangle_{23}\langle yx| - |yx\rangle_{23}\langle xy| + |yx\rangle_{23}\langle yx|\right).$$

In the Bennett's teleportaion, coincidence counting of photons 1 and 2 is made in the Bell basis. Therefore, we rewrite $\hat{\rho}_{1,23}$, using the four Bell-state basis

$$\left|\Phi^{(\pm)}\right\rangle_{12} = \frac{1}{\sqrt{2}}\left(|xx\rangle_{12} \pm |yy\rangle_{12}\right), \quad \left|\Psi^{(\pm)}\right\rangle_{12} = \frac{1}{\sqrt{2}}\left(|xy\rangle_{12} \pm |yx\rangle_{12}\right), \tag{2}$$

and eliminating cross terms between the states $|\Phi\rangle$ and $|\Psi\rangle$, as

$$\hat{\rho}_{12,3\infty} = \frac{1}{4}\Big[\left|\Phi^{(+)}\right\rangle_{12}\left\langle\Phi^{(+)}\right|\otimes|\psi'\rangle_3\langle\psi'| + \left|\Phi^{(-)}\right\rangle_{12}\left\langle\Phi^{(-)}\right|\otimes|\psi\rangle_3\langle\psi|$$
$$+ \left|\Psi^{(+)}\right\rangle_{12}\left\langle\Psi^{(+)}\right|\otimes|\phi'\rangle_3\langle\phi'| + \left|\Psi^{(-)}\right\rangle_{12}\left\langle\Psi^{(-)}\right|\otimes|\phi\rangle_3\langle\phi|\Big], \tag{3}$$

where four states of the photon 3 are defined as

$$|\psi'\rangle_3 = -b|x\rangle_3 + a|y\rangle_3, \qquad |\psi\rangle_3 = b|x\rangle_3 + a|y\rangle_3,$$
$$|\phi'\rangle_3 = -a|x\rangle_3 + b|y\rangle_3, \qquad |\phi\rangle_3 = a|x\rangle_3 + b|y\rangle_3. \tag{4}$$

The suffix ∞ indicates the effective density matrix after the measurement. In Eq. (3) we see that the teleportation of $|\phi\rangle$ is attained at the site of Bob by the photon 3. In addition, if one performs proper unitary transformations on $|\psi'\rangle_3$, $|\psi\rangle_3$, and $|\phi'\rangle_3$, knowing the result of the Bell-basis measurement at Alice, they can be converted to $|\phi\rangle$.

3. QUANTUM COMMUNICATION WITH A SINGLET OR A TRIPLET STATE

If a measurement is made on the $|x\rangle$ and $|y\rangle$ basis rather than on the Bell-state basis and $|\phi\rangle$ and $|\psi\rangle$ basis, the effective density operator can be obtained from Eq. (1) as follows:

$$\hat{\rho}_{12,3\infty}^{\text{diag}} = \frac{1}{2}\Big(|xx\rangle_{12}\langle xx|\otimes|a|^2|y\rangle_3\langle y| + |yy\rangle_{12}\langle yy|\otimes|b|^2|x\rangle_3\langle x|$$
$$+ |xy\rangle_{12}\langle xy|\otimes|a|^2|x\rangle_3\langle x| + |yx\rangle_{12}\langle yx|\otimes|b|^2|y\rangle_3\langle y|\Big), \tag{5}$$

where the off-diagonal elements with respect to $|x\rangle$ and $|y\rangle$ basis are eliminated considering the measurement by detectors of x- and y-polarizations. The first and second terms of Eq. (5) show that the probability that the polarizations of photons 1 and 2 are parallel and that of photon 3 is y-polarized is $|a|^2/2$, and x-polarized is $|b|^2/2$. Similarly, the third and fourth terms show that the probability that the polarizations of photons 1 and 2 are perpendicular and that of photon 3 is x-polarized is $|a|^2/2$, and y-polarized is $|b|^2/2$.

Alice sends information to Bob via classical channel that the polarizations of photons 1 and 2 are just parallel or perpendicular. The probabilities of them are equal, since $\left(|a|^2 + |b|^2\right)/2 = 1/2$. Therefore, there is no possibility of eavesdropping from this channel. The state $|x\rangle$ or $|y\rangle$ takes place randomly, so that repeated measurements are necessary to know how large is $|a|^2$ and how large is $|b|^2$. However, if $|a|^2$ and $|b|^2$ takes 0 or 1, a smaller number of the measurements would be enough.

The schematic of the experiment for the measurement of $|a|$ and $|b|$ with the singlet as ancilla is shown in Figure 1. The photon 1 can be generated by any source. The entangled photons 2 and 3 are generated by a type II parametric down-converter (PDC). A polarization

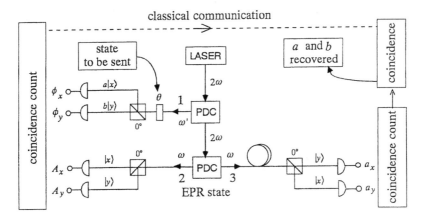

Fig. 1   Quantum communication of coefficients a and b

rotator at an angle θ generates a qubit with $a = \cos\theta$ and $b = \sin\theta$. Coincidence counting is performed separately on x-and y-polarized photons after polarizing beam-splitters for photon 1 and photon 2.

Moreover, the communication of the relative phase of a and b is possible[5]. In this case the basis of the linear polarizations $|x\rangle$ and $|y\rangle$ are converted to those at $\pm\pi/4$, $|x_+\rangle = (|x\rangle + |y\rangle)/\sqrt{2}$ and $|x_-\rangle = -i(|x\rangle - |y\rangle)/\sqrt{2}$, by rotating polarizers and polarizing beam-splitters. Measurement is made on the basis of $|x_+\rangle$ and $|x_-\rangle$. Then, as in Eq. (5), $|a \pm b|$ can be measured, and the relative phase of a and b can be deduced.

The use of a triplet state as an ancilla is proposed[4-6]. This has an advantage that angular momentum measurement alone at Alice determines the states at Bob. In this case communication of the absolute values and the relative phase is possible simultaneously without switching polarizers for two communications.

4. BENNETT-TYPE TELEPORTATION WITH FOUR BELL STATES

One of the four Bell states was used in the above teleportation. If all of the four Bell states and the corresponding four unitary transformed states of $|\phi\rangle_1$ is prepared beforehand, determination of the Bell states and communication of its result via classical channel by Alice will become easier.

We prepare, in this case, four superposition states like Eqs. (4) by a photon 1 at Alice's side. Four Bell states like Eqs. (2) are prepared with photons 2 and 3 by a type II parametric down-converter, and they are sent to Alice and Bob, respectively. The photon 1 of the state $|\phi\rangle_1$ and photon 2 of the state $|\Psi^{(-)}\rangle_{23}$ are lead to the right most beam-splitter (BS) in Fig. 2. They impinge on it from both sides, and detected by pairs of detectors there. In this way, photons 1 and 2 of all four combinations of the Bell states and their partners are lead to separate BS's, and detected there as in Fig. 2. Only the state vectors which include photon 1 and 2 passing the same BS's are detected at the same space points. Therefore, let us expand these combinations of vectors by the orthogonal Bell states of the photons 1 and 2 as follows,

$$|\phi\rangle_1 |\Psi^{(-)}\rangle_{23} = \frac{1}{2}\left(|\Phi^{(+)}\rangle_{12} |\psi'\rangle_3 + |\Phi^{(-)}\rangle_{12} |\psi\rangle_3 + |\Psi^{(+)}\rangle_{12} |\phi'\rangle_3 - |\Psi^{(-)}\rangle_{12} |\phi\rangle_3 \right),$$

$$|\phi'\rangle_1 |\Psi^{(+)}\rangle_{23} = \frac{1}{2}\left(-|\Phi^{(+)}\rangle_{12} |\psi'\rangle_3 - |\Phi^{(-)}\rangle_{12} |\psi\rangle_3 + |\Psi^{(+)}\rangle_{12} |\phi'\rangle_3 - |\Psi^{(-)}\rangle_{12} |\phi\rangle_3 \right),$$

$$|\psi\rangle_1 |\Phi^{(-)}\rangle_{23} = \frac{1}{2}\left(-|\Phi^{(+)}\rangle_{12} |\psi'\rangle_3 + |\Phi^{(-)}\rangle_{12} |\psi\rangle_3 - |\Psi^{(+)}\rangle_{12} |\phi'\rangle_3 - |\Psi^{(-)}\rangle_{12} |\phi\rangle_3 \right),$$

$$|\psi'\rangle_1 |\Phi^{(+)}\rangle_{23} = \frac{1}{2}\left(|\Phi^{(+)}\rangle_{12} |\psi'\rangle_3 - |\Phi^{(-)}\rangle_{12} |\psi\rangle_3 - |\Psi^{(+)}\rangle_{12} |\phi'\rangle_3 - |\Psi^{(-)}\rangle_{12} |\phi\rangle_3 \right).$$

$$(6)$$

It is known that when the photons 1 and 2 in the state $|\Psi^{(-)}\rangle_{12}$ are incident from the opposite side of a BS, the counters at the opposite sides make a coincidence count. In this case the photon 3 is definitely in the state $|\phi\rangle_3$. Instead if the counters were designed to detect $|\Phi^{(+)}\rangle_{12}$, $|\Phi^{(-)}\rangle_{12}$, and $|\Psi^{(+)}\rangle_{12}$, the photon 3 would then be definitely in the states $|\psi'\rangle_3$, $|\psi\rangle_3$, and $|\phi'\rangle_3$, respectively, when they click. We would then be able to arrange the counter for each of the four Bell states as in Fig. 2, and determination of the Bell states and communication of its result via classical channel by Alice would become easier. The classical communication channel serves to send the information which of the counters just clicked.

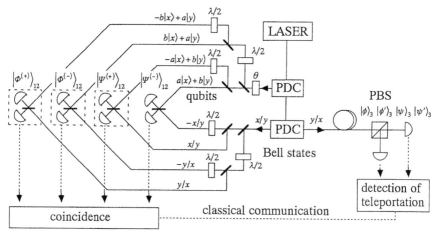

Fig. 2   Teleportation with four Bell states

REFERENCES

1. C. H. Bennett, G. Brassard, C. Crepeau, R. Jozsa, A. Peres and W. K. Wooters, Phys. Rev. Lett. 70 (1993) 1895.
2. D. Bouwmeester, J-W. Pan, K. Mattle, M. Eibl, H. Weinfurter, and A. Zeilinger, Nature, 390 (1997) 575.
3. D. Boschi, S. Branca, F. De Martini, L. Hardy, and S. Popescu, Phys. Rev Lett. 80 (1998) 1121.
4. A. Motoyoshi, K. Yamaguchi, T. Ogura and T. Yoneda, Prog. Theor. Phys, 97 (1997) 819.
5. A. Motoyoshi, M. Matsuoka, Prog. Theor. Phys, in print (August, 1998).
6. A. Motoyoshi, T. Yoneda, T. Ogawa, and M. Matsuoka, ISQM-Tokyo '98.

Quantum Coherence and Decoherence - ISQM - Tokyo '98
Y.A. Ono and K. Fujikawa (Editors)
© 1999 Elsevier Science B.V. All rights reserved.

Optimal Broadcasting of Quantum Information via Teleportation[*]

M. Murao[a], D. Jonathan[a], M.B. Plenio[a] and V. Vedral[b]

[a]Optics Section, Blackett Laboratory, Imperial College, London SW7 2BZ, U.K.
[b]Centre for Quantum Computing, University of Oxford, Oxford OX1, U.K.

We present a new quantum information processing scheme, a quantum telecloning process, which generalizes teleportation by combining it with optimal quantum cloning. This allows the optimal broadcasting of quantum information from one sender (Alice) to M spatially separated recipients. The scheme relies on the establishment of particular multiparticle entangled states, which function as multiuser quantum information channels.

1. INTRODUCTION

Quantum teleportation [1] allows an unknown state $|\phi\rangle_X$ of a quantum system X to be faithfully transmitted between two spatially separated parties (a sender, Alice, and a receiver, Bob) and has become one of the key tools of quantum information theory. The recent successful experiments of Bouwmeester *et al* and Boschi *et al* [2] have demonstrated practical realizations of this idea.

In this paper, we present 'telecloning' [3], a generalization of the teleportation protocol, where the quantum information contained in the input qubit is transferred not to one new qubit, but to a subspace of many. We show that, by an appropriate choice of this subspace, a teleportation-like protocol can be found which reproduces the output of an optimal quantum cloning machine [4,5]. The net result is that optimal copies of an input qubit are simultaneously created at spatially separated locations.

2. OPTIMAL QUANTUM CLONING

Optimal quantum cloning [4,5] seeks to *spread* quantum information among several parties in the most efficient way possible. $N \to M$ Universal Quantum Copying Machines (UQCMs) [5] are unitary transformations that transform N input systems, identically prepared in state $|\phi\rangle_X$, onto M output systems ($M \geq N$), each of which ends up in a mixed state described by the reduced density operator

$$\rho_{out} = \gamma|\phi\rangle_X\langle\phi| + (1-\gamma)|\phi^\perp\rangle_X\langle\phi^\perp| \tag{1}$$

(where $|\phi^\perp\rangle_X$ is a state orthogonal to $|\phi\rangle_X$) [5]. The fidelity factor γ of these imperfect copies has a definite upper limit imposed by quantum mechanics. In the case where each input system consists of one qubit, this optimal value is given by $\gamma = \frac{M(N+1)+N}{M(N+2)}$ [5].

[*]This work was supported in part by EPSRC, JSPS, the Brazilian agency CNPq, European TMR Research Networks, Elsag-Bailey and HP.

Unitary transformations which realize this bound have also been explicitly constructed [5]. In general, they involve the N 'original' qubits, $M - N$ 'blank paper' qubits B (initially prepared in some fixed state $|0 \cdots 0\rangle_B$), and an ancilla system A containing at least $M - N + 1$ levels [5] (also initially in some fixed state $|0 \cdots 0\rangle_A$). In this paper, we shall be mainly interested in the situation where only one original qubit X is available, that is, $N = 1$. In this case, the cloning transformation U_{1M} is defined as follows: for an initial state $|\phi\rangle_X = a|0\rangle_X + b|1\rangle_X$, we have

$$U_{1M}\left(|\phi\rangle_X \otimes |0 \cdots 0\rangle_A |0 \cdots 0\rangle_B\right) = a|\phi_0\rangle_{AC} + b|\phi_1\rangle_{AC}, \tag{2}$$

where

$$|\phi_0\rangle_{AC} = U_{1M}|0\rangle_X|0 \cdots 0\rangle_A|0 \cdots 0\rangle_B = \sum_{j=0}^{M-1} \alpha_j |A_j\rangle_A \otimes |\{0, M - j\}, \{1, j\}\rangle_C, \tag{3}$$

$$|\phi_1\rangle_{AC} = U_{1M}|1\rangle_X|0 \cdots 0\rangle_A|0 \cdots 0\rangle_B = \sum_{j=0}^{M-1} \alpha_j |A_{M-1-j}\rangle_A \otimes |\{0, j\}, \{1, M - j\}\rangle_C, \tag{4}$$

$\alpha_j = \sqrt{\frac{2(M-j)}{M(M+1)}}$ and where C denotes the M qubits holding the copies (originally the X and B qubits). Here, $|A_j\rangle_A$ are M orthogonal normalized states of the ancilla and $|\{0, M - j\}, \{1, j\}\rangle$ denotes the symmetric and normalized state of M qubits where $(M - j)$ of them are in state $|0\rangle$ and j are in the orthogonal state $|1\rangle$.

We note that, even though the minimum number of ancilla qubits required to support the M levels $|A_j\rangle_A$ is of the order of $\log_2 M$, these can be more conveniently represented as the symmetrized states of $(M - 1)$ qubits $|A_j\rangle_A \equiv |\{0, M - 1 - j\}, \{1, j\}\rangle_A$ [5]. In this form, states $|\phi_0\rangle$ and $|\phi_1\rangle$ above become $(2M - 1)$-qubit states, obeying the following simple symmetries

$$\sigma_z \otimes \cdots \otimes \sigma_z |\phi_0\rangle = |\phi_0\rangle, \quad \sigma_z \otimes \cdots \otimes \sigma_z |\phi_1\rangle = -|\phi_1\rangle, \quad \sigma_x \otimes \cdots \otimes \sigma_x |\phi_{0(1)}\rangle = |\phi_{1(0)}\rangle. \tag{5}$$

In other words, the states $|\phi_i\rangle$ transform under simultaneous action of the Pauli operators on all $(2M - 1)$ qubits just as a single qubit transforms under the corresponding single Pauli operator. We also note that these operations are strictly local, that is, factorized into a product of independent rotations on each qubit.

3. TELECLONING

We now show how the existence of these local symmetries allows $1 \rightarrow M$ quantum cloning to be realized remotely, via a 'telecloning' scheme analogous to the teleportation protocol. In this way, quantum information can be optimally broadcast from one sender to many recipients. The scheme works by exploiting the multiparticle entanglement structure of particular joint states of $2M$ particles ('telecloning states'), of the form:

$$|\psi_{TC}\rangle = \left(|0\rangle_P \otimes |\phi_0\rangle_{AC} + |1\rangle_P \otimes |\phi_1\rangle_{AC}\right) / \sqrt{2}, \tag{6}$$

where $|\phi_0\rangle_{AC}$ and $|\phi_1\rangle_{AC}$ are the optimal cloning states given by Eqs. (3) and (4). Here, C denotes the M qubits which shall hold the copies, each of which is held by one of Alice's

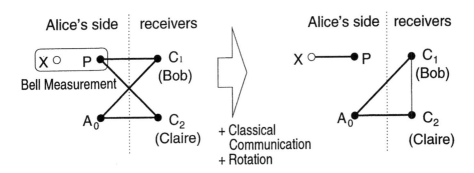

Figure 1. Quantum telecloning $M = 2$ copies of an unknown 1-qubit state denoted X.

associates. For convenience, we shall refer to them collectively as 'the receivers' (though it should be kept in mind that they may all be far away from each other). P represents a single qubit held by Alice, which we shall refer to as the 'port' qubit. Finally, A denotes an $M - 1$ qubit ancilla, which for convenience we will also assume to be on Alice's side (even though, once again, each qubit may in reality be at a different location).

The telecloning of $|\phi\rangle_X$ can now be accomplished by the following generalization of the standard teleportation procedure:

1. Alice performs a Bell measurement of qubits X and P, obtaining one of the four results $|\Psi^\pm\rangle_{XP}$, $|\Phi^\pm\rangle_{XP}$. If the result is $|\Phi^+\rangle_{XP}$, then subsystem AC is projected precisely into the optimal cloning state given in Eq. (2). In this case, our task is accomplished.

2. In case one of the other Bell states is obtained, we can still recover the correct state of AC by exploiting the symmetries of states $|\phi_0\rangle_{AC}$ and $|\phi_1\rangle_{AC}$ under the Pauli matrix operations (Eqs. (5)). Specifically, if $|\Phi^-\rangle_{XP}$ is obtained, we must perform σ_z on each of the $2M - 1$ qubits in AC; similarly, if $|\Psi^+\rangle_{XP}$ or $|\Psi^-\rangle_{XP}$ are obtained, they must all be rotated by σ_x and $\sigma_x\sigma_z = i\sigma_y$, respectively.

We can see that this protocol is formally identical to the standard teleportation. The difference in the present case is that the quantum information contained in the input state $|\phi\rangle_X$ is transferred onto the 'effective qubit' represented by the orthogonal states $|\phi_0\rangle_{AC}$ and $|\phi_1\rangle_{AC}$, and thus becomes spread over several separate 'physical' qubits. This procedure is illustrated in Fig. 5 for the case of $M = 2$ copies. We stress that, apart from Alice's Bell measurement, only local 1-qubit operations are required in the telecloning procedure. In this way, all of the qubits except the input X and the port P can be spatially separated from each other. It is also worthwhile to add that rotating the ancilla qubits in step (2) above is not strictly necessary. The correct copy states of each output (given by Eq. (1)) are obtained at the output regardless of these operations, since local rotations on one qubit cannot affect another qubit's reduced density operator.

Further insight into the telecloning procedure can be gained by examining the detailed entanglement structure of the telecloning state itself. First of all, we note that by rewriting this state so that the qubits on Alice's and the receivers' sides are explicitly separated, we obtain

$$|\psi_{TC}\rangle = \frac{1}{\sqrt{M+1}} \sum_{j=0}^{M} |\{0, M-j\}, \{1, j\}\rangle_{PA} \otimes |\{0, M-j\}, \{1, j\}\rangle_{C}. \qquad (7)$$

This form highlights the high degree of symmetry of the telecloning state: it is completely symmetric under the permutation of any two particles on the same side, and also under the exchange of both sides. This implies that, in fact, any of the $2M$ qubits can be used as the telecloning port, with the clones being created on the opposite side. We can also see that the telecloning state contains only $\log_2(M+1)$ e-bits of entanglement between the two sides, representing a much more efficient use of entanglement than the more straightforward approach where Alice first clones her particle M times and then uses M singlets to transmit these states to the different receivers.

While entanglement between the two sides gives a measure of the resources necessary to accomplish telecloning, the entanglement between each pair of particles helps track how information from Alice's unknown state is conveyed to the clones. Due to the symmetries of the telecloning state, there are only two different classes of pairs: those where both qubits are on opposite sides (Alice's and the receivers') and those where they are on the same side. Using the Peres-Horodecki theorem [6] we find [3] that the first kind of pair are entangled, while the second are disentangled.

These calculations allow us to view the telecloning state as a 'network' of entangled qubits, each of which is only connected to the M qubits on the opposite side (so the total number of 'links' is M^2). Essentially, we may think of these 2-qubit connections as 'communication channels' through which quantum information may travel [1]). In this sense, the multiparticle entanglement structure functions as a *multiuser channel*, allowing quantum information to be broadcast to several receivers.

REFERENCES

1. C.H. Bennett et al, Phys. Rev. Lett. **70**, 1895 (1993).
2. D. Bouwmeester et al, Nature **390** 575 (1997); D. Boschi et al, Phys. Rev. Lett. **80**, 21 (1998).
3. M. Murao et al, e-print quant-ph/9806082.
4. V. Bužek and M. Hillery, Phys. Rev. A **54**, 1844 (1996).
5. N. Gisin and S. Massar, Phys. Rev. Lett. **79**, 2153 (1997); D. Bruß et al, Phys. Rev. A. **57**, 368 (1998); D. Bruß et al, e-print quant-ph/9712019; V. Bužek et al, Fort. Phys. **46**, 521 (1998); V. Bužek and M. Hillery, e-print quant-ph/9801009; R. Werner, e-print quant-ph/9804001; P. Zanardi, e-print quant-ph/9804011.
6. A. Peres, Phys. Rev. Lett. **77**, 1413 (1996); M. Horodecki et al, Phys. Lett. A **223**, 1 (1996).

Quantum Coherence and Decoherence - ISQM - Tokyo '98
Y.A. Ono and K. Fujikawa (Editors)
© 1999 Elsevier Science B.V. All rights reserved.

Semiconductor-based Quantum Information: Noiseless Encoding in Quantum-Dot Arrays

P. Zanardi[a*] and F. Rossi [b]

[a]Istituto Nazionale per la Fisica della Materia (INFM) and Institute for Scientific Interchange (ISI) Foundation, Villa Gualino, IT-10133 Torino, Italy

[b]INFM and Dipartimento di Fisica, Università di Modena, IT-41100 Modena, Italy

A potential implementation of quantum-information schemes in semiconductor nanostructures is discussed. Physical qubits are realized in terms of the two lowest energy levels of a semiconductor quantum dot. Encoding information in suitable singlet states is found to strongly suppress decoherence induced by the coupling to the phononic environment. The proposed strategy allows to realize a quantum-mechanically coherent evolution on a time-scale long compared to the femtosecond scale of modern ultrafast laser technology.

1. Introduction

Ever since the early days of quantum-information theory it has been recognized that the unavoidable open character of any realistic quantum system constitutes the majour problem in realizing any quantum-computing device.[1] This is known as *the decoherence problem*.[2] In order to overcome such limitation, two complementary strategies have been identified: On the one hand, in analogy with its classical counterpart, *error-correcting* schemes have been devised.[3] On the other hand, *error-avoiding* strategies —based on a sort of dynamical stabilization— have been proposed.[4,5]

Due to these severe limitations, the only existing realizations of quantum-computation protocols are based on quantum optics as well as atomic and molecular systems,[1] which are characterized by extremely long decoherence times (compared to those of solid-state physics). It is however generally believed that future applications, if any, of quantum computation may hardly be realized in terms of such systems, mainly due to scalability problems. In contrast, in spite of the "fast" decoherence times, a solid-state implementation of quantum computation (QC) seems to be the only way to benefit from the recent progress in ultrafast optoelectronics [6] as well as in nanostructure fabrication and characterization.[7] To this end, the primary goal is to design quantum structures and encoding strategies characterized by "long" decoherence times, compared to typical time-scale of gating. The first well-defined semiconductor-based proposal of QC [8] relies on spin dynamics in quantum dots, it exploits the low decoherence of spin degrees of freedom in comparison to the one of charge excitations.

---

*Financial support by Elsag-Bailey is gratefully aknowledged

In this paper we study a semiconductor-based implementation of the noiseless quantum-encoding theory proposed in Refs. 4 and 5. The idea is that, in the presence of a sort of "coherent" environmental noise, one can identify states that are hardly corrupted rather than states that can be easily corrected. More specifically, we show that by choosing as quantum register an array of quantum dots [7] and by "preparing" the system in the sub-decoherent multi-dot quantum states suggested in Ref. 4, it is possible to suppress electron-phonon scattering, which is known to be the primary source of decoherence in semiconductors.[9]

2. The Quantum Register

The physical system under investigation consists of an array of N identical quantum dots (QD). We will assume that only the two lowest energy levels in each dot are involved in the QC dynamics; The evolution of this low-energy sector coupled with the phonon modes of the crystal is therefore mapped onto the one of N two-level systems (*qubits*) linearly coupled with the bosonic degrees of freedom of the environment.[4,5] The above theoretical scheme has been applied to state-of-the-art semiconductor QD structures. More specifically, a GaAs/AlGaAs QD similar to that of Ref. 10 has been considered. The three-dimensional confinement potential giving rise to the quasi-0D single-particle states is properly described in terms of a quantum-well (QW) profile along the growth direction times a two-dimensional (2D) parabolic potential in the normal plane. Since the width d of the GaAs QW region is typically of the order of few nanometers, the energy splitting due to the quantization along the growth direction is much larger than the confinement energy E induced by the 2D parabolic potential (typically of a few meV). Thus, the two single particle states —state $|0\rangle$ and $|1\rangle$— realizing the *qubit* considered so far are given by products of the QW ground state times the ground or first excited state of the 2D parabolic potential, and their energy splitting E coincides with the confinement energy $\hbar\omega$ of the 2D harmonic potential.

3. Decoherence Analysis

Let us first discuss the role played by carrier-phonon interaction in a single QD structure with $d = 4\,\text{nm}$. Since in these QD structures the typical value of the energy spacing (of the order of a few meV), is smaller than the optical-phonon energy (36 meV in GaAs), due to energy conservation scattering with LO phonons is not allowed. Therefore, the only phonon mode which contributes to the total carrier-phonon scattering rate is that of acoustic phonons. Again, due to energy conservation, the only phonon wavevectors $\mathbf{q}$ involved must satisfy $|\mathbf{q}| = E/\hbar c_s$, c_s being the GaAs sound velocity. It follows that by increasing the energy spacing E the wavevector q is increased, which reduces the carrier-phonon coupling constant and thus the corresponding scattering rate. This well-established behaviour is known as phonon bottleneck. Indeed, for $E = 5\,\text{meV}$ —a standard value for many state-of-the-art QD structures— the carrier-phonon scattering rate is already suppressed by three orders of magnitude compared to the corresponding bulk values.[9]

We will now show that by applying the subdecoherent encoding strategy proposed in Ref. 4 to a properly designed QD array, one is able to suppress at will phonon-induced

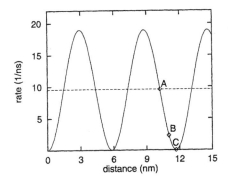

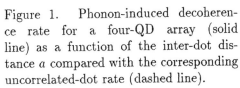

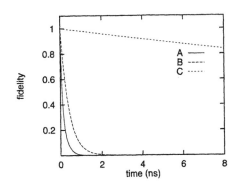

Figure 1. Phonon-induced decoheren-ce rate for a four-QD array (solid line) as a function of the inter-dot distance a compared with the corresponding uncorrelated-dot rate (dashed line).

Figure 2. Fidelity F as a function of time as obtained from a direct numerical solution of the Master equation for the relevant case of a four-QD array (see text).

decoherence processes, thus further improving the single-dot scenario discussed so far. If $|\psi\rangle$ is the initial state preparation, the *fidelity* $F(t) \equiv \langle\psi|\rho(t)|\psi\rangle = 1 - t/\tau_1 + o(t^2)$ describes the corruption of information encoded in $|\psi\rangle$. Here, the density matrix $\rho(t)$ is the time-dependent solution of a corresponding master equation $\dot{\rho} = \mathcal{L}(\rho)$ obtained by tracing out the phonon degrees of freedom (in the Born Markov approximation). The Liouvillian superoperator $\mathcal{L}$ has two components: a unitary one (containing free Hamiltonian and polaronic shifts) and a dissipative part describing the irreversible exchange of energy/information between the register and the phonon bath [see Ref. 5]. The latter has the explicit form

$$\mathcal{L}_d(\rho) = \frac{1}{\hbar} \sum_{ii',\eta=\pm} \Gamma_{ii'}^{(\eta)} \left([\sigma_i^\eta \rho, \sigma_{i'}^{-\eta}] + [\sigma_i^\eta, \rho\,\sigma_{i'}^{-\eta}] \right), \tag{1}$$

where the matrix elements $\Gamma_{ii'}^{(\eta)}$ contain all the information about register-environment couplings, and environment state as well. The decoherence rate for the states $|\psi_\mathcal{D}\rangle \equiv \otimes_{(i,i')\in\mathcal{D}}(|01\rangle - |10\rangle)_{ii'}$ (here, $\mathcal{D}$ is a dimer partition of the qubit array) is given by $\tau_1^{-1} = \sum_{\eta=\pm}(2\tau_\eta)^{-1} f_\mathcal{D}(\Gamma^{(\eta)})$, where $\hbar\,\tau_\eta^{-1} = N\,\Gamma_{11}^{(\eta)}$ is the (maximal) decoherence rate for N uncorrelated qubits and

$$f_\mathcal{D}(\Gamma) = 1 - \frac{2}{N} \Re \sum_{(i,i')\in\mathcal{D}} \Gamma_{ii'}/\Gamma_{11}. \tag{2}$$

The quantity $f_\mathcal{D}$ contains the information about the degree of multi-qubit correlation in the decay process. We have numerically evaluated the decoherence rate τ_1^{-1} for a four QD's array choosing as energy splitting $E = 5\,\mathrm{meV}$ and as initial state the singlet $|\psi_{\mathcal{D}_1}\rangle$, where $\mathcal{D}_1 = \{(1,2), (3,4)\|$. The resulting decoherence rate is shown as solid line in Fig. 1(a) as a function of the inter-dot distance a. The uncorrelated-dot decoherence rate is also reported as dashed line for comparison. In spite of the 3D nature of the carrier-phonon deformation-potential interaction considered, the decoherence rate exhibits a periodic behaviour ($\Gamma_{i,i'}^{(\pm)} \sim \cos[q\,(i - i')\,a]$,) over a range comparable to the typical QD length scale. This effect —which would be the natural one for a 1D phonon system— stems from the exponential suppression, in the carrier-phonon matrix elements, of the contributions of phononic modes with non-vanishing in-plane component. The 1D behavior plays a central role since it allows, by suitable choice of the inter-dot distance a, to realize the symmetric regime in which all the dots experience the *same* phonon field and therefore decohere collectively. Figure 1 shows that for the particular QD structure considered, case C should correspond to a decoherence-free evolution of a singlet state, which is not the case for A and B (see simbols in the figure). In order to extend the above short-time analysis, we have performed a full time-dependent solution of the Master equation for the density matrix ρ based on the theoretical analysis presented in Ref. 5. Starting from the same GaAs QD structure considered so far, we have simulated the above noiseless encoding for a four-QD array.

Figure 2 shows the fidelity as a function of time as obtained from our numerical solution of the Master equation. In particular, we have performed three different simulations corresponding to the different values of a depicted in Fig. 1. Consistently with our short-time analysis, for case C we find a strong suppression of the decoherence rate which extends the sub-nanosecond time-scale of the B case (corresponding to the uncorrelated-dot rate) to the microsecond time-scale. This confirms that by means of the proposed encoding strategy one can realize a decoherence-free evolution over a time-scale comparable with typical recombination times in semiconductor materials.[9] This result may constitute a first step toward a solid-state implementation of quantum computers

REFERENCES

1. For reviews, see D.P. Di Vincenzo, *Science* **270**, 255 (1995); A. Ekert and R. Josza, *Revs. Mod. Phys.* **68**, 733, (1996)
2. W. G. Unruh, *Phys. Rev. A* **51**, 992 (1992) P.W. Shor, W. H. Zurek, I.L. Chuang, and R. Laflamme, *Science* **270**, 1633 (1995)
3. See for example E. Knill and R. Laflamme, Phy. Rev. A **55**, 900 (1997) and references therein
4. P. Zanardi and M. Rasetti, Phys.Rev. Lett. **79**, 3306 (1997); Mod. Phys. Lett. B **25**, 1085 (1997)
5. P. Zanardi, Phys. Rev. A **57**, 3276 (1998)
6. A.P. Heberle, J.J. Baumberg, and K. Kohler, Phys. Rev. Lett. **75**, 2598 (1995).
7. For reviews see e.g. M. A. Kastner, Rev. Mod. Physics **64**, 849 (1992).
8. D. Loss and D. P. Di Vincenzo, Phys. Rev. A **59**, 120 (1998)
9. J. Shah, *Ultrafast Spectroscopy of Semiconductors and Semiconductor Nanostructures* (Springer, Berlin, 1996).
10. S. Tarucha *et al.*, Phys. Rev. Lett. **77**, 3613 (1996).

Quantum Coherence and Decoherence - ISQM - Tokyo '98
Y.A. Ono and K. Fujikawa (Editors)
© 1999 Elsevier Science B.V. All rights reserved.

Quantum Information Enabled by Quantum Optics

H. J. Kimble

Norman Bridge Laboratory of Physics 12-33, California Institute of Technology
Pasadena, CA 91125, USA

Quantum mechanics offers the potential for revolutionary advances in the processing and distribution of information. In the Quantum Optics Group at Caltech, we are attempting to lay the foundations for quantum information science with advances on several fronts in optical physics.

1. QUANTUM INFORMATION SCIENCE

Although the origins of the field date to the early 1980s with seminal work by Benioff, Feynman, and Deutsch,[1] the discovery in 1994 by P. Shor [2] of an efficient algorithm for prime factorization of large numbers has led to an explosive growth in the field of quantum computation. A partial list of ensuing highlights includes the algorithm by Grover for data base searches,[3] the analysis of Lloyd proving Feynman's conjecture about the efficiency of quantum computers for simulation [4] and the discoveries of quantum error correction [5] and of fault tolerant quantum computation.[6]

In addition to these theoretical advances, on the experimental front primitive capabilities for the implementation of quantum logic have been demonstrated in several physical systems, beginning with the first demonstrations [7, 8]. While there are rapid advances in both the diversity and achievements of systems being investigated for quantum computation, there nonetheless remains a tremendous gulf between the capabilities of these systems and those required for large-scale calculations, such as Shor's factoring algorithm. Indeed, the experimental challenges are so daunting that there is no reasonable expectation that actual quantum computers composed of say 10^6 quantum bits (or *qubits*) will be constructed in the foreseeable future.

This prospect leads some to suggest that research investments in quantum computation are premature and best deferred in favor of near-term prospects in other areas. However, such an outlook focuses too strongly on large-scale quantum computation and overlooks the much broader significance of the emerging field of Quantum Information Science, of which quantum computing is but one facet. From a more global perspective, the marriage of quantum mechanics and information science is likely to impact profoundly science and technology on fronts ranging from nano-structure engineering to precision metrology. The time scale for this impact is not the indefinite future, but rather looms on the horizon within the next twenty years. For example, there is a widely held view that the geometrical scaling of capability for silicon technology will end within the next

15 years, with no viable successors having yet clearly emerged. Setting aside the question of quantum computing as a new paradigm for information processing, I believe that there will be a clear and pressing need to understand both the enabling and debilitating nature of quantum coherence in future generations of information processing systems. [9]

Beyond quantum computation *per se*, another very promising area is that of quantum communication. Optical communication networks offer the potential for the development of new paradigms for the reliable transmission and storage of quantum information, which is the subject to which I now turn.

2. QUANTUM NETWORKS

Recent advances in Quantum Information Science presage a revolution in the ability to harness quantum mechanics to realize networks for the distribution and processing of quantum information. For example, quantum cryptography exploits non-orthogonal quantum states for the secure exchange of information between two parties.[10] More generally, with capabilities developed within the setting of cavity quantum electrodynamics (cavity QED), it should be possible to implement an expanded set of fundamental quantum communication protocols and thereby to develop quantum communication networks, as illustrated in Fig. 1. Within this setting, two complimentary paradigms are exploited involving the atom-field interaction. In the first, photons become the carriers of quantum information, with interactions between these "flying" qubits mediated by an atom in a cavity, as in the previously demonstrated quantum-phase gate.[7] In the second approach, the internal states of atoms are employed as the qubits, with interactions proceeding by way of photons in the intracavity field.[11]

These two paradigms are not disjoint, but interactive and can be combined to form quantum networks, as illustrated in Fig. 2.[12, 13] Here, multiple atom-cavity systems

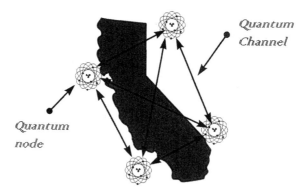

Fig. 1 Schematic of a quantum information network as enabled by capabilities from cavity QED with strong coupling. Internal states of atoms at the quantum nodes are used to generate, process, and store quantum information. Photons propagate along the quantum channels to transport quantum states and distribute quantum entanglement following the protcols of Refs.[12, 13].

located at spatially separated "nodes" are interconnected via optical fibers to create a quantum network (QN) whose unique and powerful properties have been anticipated by recent advances in quantum information theory. Indeed, a complete set of elementary network operations has been proposed and analyzed including local processing of quantum information, transmission of quantum states from one node to another, and the distribution of quantum entanglements. The enabling capability for this work will be the successful trapping and localization of atoms inside high-finesse optical cavities, a front on which rapid progress is being made.[14-16] It should be emphasized that these protocols are fully realistic and within reach of the current technical capabilities.

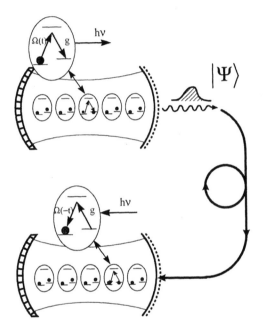

Fig. 2   Illustration of the protocol of Ref.[12, 13] whereby one component (denoted figuratively by $|\Psi\rangle$) of an entangled state for a set of atoms at one site can be transferred to an atom in another set at a remote location. Here $\Omega(t)$ refers to a classical external control field, while g sets the quantum coupling of atom to cavity field.[9] By simple repetition any component of the original state may be so transferred to create nonlocal entanglements.

3. QUANTAM TELEPORTATION

Because quantum information differs so profoundly from classical information (by virtue of the properties, implications, and uses of *quantum entanglement*—the nonlocal correlations among parts of a quantum system), quantum networks can accomplish tasks that are classically impossible. The prime example of the exploitation of entanglement is *quantum teleportation*, the transport of an unknown quantum state to a remote location, aspects of which have first been demonstrated experimentally in the past year,[17, 18] with the first *bona fide* demonstration having been achieved.[19] From a practical point of view, quantum teleportation makes it possible for two individuals to communicate securely without the need for a direct, secure link, and without knowledge of the other's location. Having previously arranged to share entangled pairs of particles (e.g., over a quantum network via the protocol illustrated in Fig. 2), one individual (the sender) need only to broadcast an apparently random data stream over a completely public channel in order to convey his message to the second individual (the receiver). Note that neither sender nor receiver need have knowledge of the location of the other, that the roles of sender and receiver can be reversed for the protocol, and that *only* the sender and receiver share a message (with immunity to interception). Moreover, for quantum networks as illustrated in Fig. 1, quantum teleportation may well be the avenue of choice for the distribution of entanglement for distributed quantum computation and multi-party quantum communication.

In addition to discrete *qubits*, our interest in quantum teleportation has also arisen within the context of quantum information processing with continuous quantum variables. Indeed, following the lead of Professor L. Vaidman,[20] we have developed a theory for quantum teleportation of continuous quantum variables in an infinite dimensional Hilbert space,[21] including a theory for broad bandwidth teleportation.[22] We have also applied this formalism for super-dense quantum coding.[23]

These advances as well as other prospects for quantum information processing with continuous variables [24-27] have motivated an experimental program involving entangled EPR beams generated by parametric down conversion.[19] As illustrated in schematic form in Fig. 3, an unknown quantum state input to "Alice's sending station" is destroyed, and recreated at "Bob's receiving terminal." The quantum nature of the protocol is confirmed by measurements to determine the fidelity F, which quantifies the match between input and output fields. For a pure state $|\psi_{in}\rangle$ as the input, F is given by

$$F = \langle\psi_{in}|\rho_{out}|\psi_{in}\rangle, \tag{1}$$

where ρ_{out} is the density operator for the output field generated at Bob's terminal. In our experiment, fidelity $\bar{F}_{exp} = 0.58 \pm 0.02$ is achieved for the teleportation of coherent states. Note that an average fidelity greater than 0.5 cannot be achieved for the teleportation of coherent states in the absence of shared entanglement.[28]

Full teleportation is thus realized by our experiment, where "full" means that the following three criteria have been met.

1. An unknown quantum state propagates to Alice's station.

2. A "recreation" of this quantum state emerges from Bob's receiving terminal.

3. The fidelity of input and output states is higher than that which could have been achieved if Alice and Bob shared only a classical communication channel.

Note that prior teleportation experiments involving the polarization state of a single photon pulse did not meet criterion (1) [18] or (2) [17]. Furthermore, for polarization states drawn randomly and uniformly from the Poincare sphere, Alice and Bob can in principle achieve an average fidelity $\bar{F} = \frac{2}{3}$ *in the absence of shared quantum entanglement*,

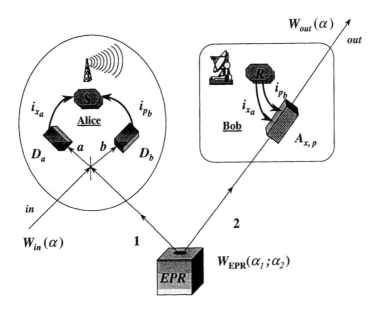

Fig. 3 Quantum teleportation of an unknown state as specified by the Wigner function $W_{in}(\alpha)$ from Alice's sending station to Bob's receiving terminal. The degree of "similarity" between the teleported output $W_{out}(\alpha)$ and the input $W_{in}(\alpha)$ is quantified by the fidelity F. Shared quantum entanglement between Alice and Bob enables quantum teleportation with fidelity greater than that possible with only a classical channel, as was first demonstrated in Ref.[19].

which means that $\bar{F} = \frac{2}{3}$ becomes the line of demarcation between classical and quantum teleportation for these experiments. Because of low efficiencies, neither of the experiments reported in Refs.[17, 18] could approach a nonclassical fidelity for the fields involved.

ACKNOWLEDGMENTS

The experiments described herein have been carried out in the Quantum Optics Laboratory at Caltech, with the personnel responsible for the research including graduate students J. Buck, N. Georgiades, C. Hood, H. Mabuchi (now an assistant professor of physics at Caltech), T. Lynn, J. Sorensen (visitor from Aahrus University), Q. Turchette (now at NIST, Boulder), and D. Vernooy, and undergraduate E. Streed. Senior members of the group include Drs. M. Chapman (now an assistant professor of physics at Georgia Tech), C. Fuchs, A. Furusawa (Nikon Advanced Research Labs), S. van Enk, and J. Ye. We have benefited greatly from ongoing collaborations with the groups of Professors S. L. Braunstein, E. S. Polzik, D. F. Walls, and P. Zoller. This work is supported by DARPA via the QUIC Institute which is administered by ARO, by the National Science Foundation, and by the Office of Naval Research. For more information, please visit our web site at *http://www.cco.caltech.edu/~qoptics.*

REFERENCES

1. P. Benioff, *Phys. Rev. Lett.* **48** (1982) 1581; R. P. Feynman, *Int. J. Theor. Phys.* **21** (1982) 467; D. Deutsch, *Proc. Royal Soc. London* **A400** (1985) 1818.
2. P. Shor, in *Proceedings of the 35th Annual Symposium on Fundamentals of Computer Science,* Los Alamitos, CA, USA, IEEE Press, pp.124-134 (1994).
3. L. K. Grover, in *Proceedings of the 28th ACM Symposium on Theory of Computation,* 212 (1996).
4. S. Lloyd, *Science* **273** (1996) 5278.
5. P. Shor, *Phys. Rev.* **A52** (1995) 2493; A. M. Steane, *Phys. Rev. Lett.* **77** (1996) 793.
6. For a review, see J. Preskill, *Proc. Royal Soc. London* **A454** (1998) 385.
7. Q. A. Turchette, C. J. Hood, W. A. Lange, H. Mabuchi, and H. J. Kimble, *Phys. Rev. Lett.* **75** (1995) 4710.
8. C. Monroe, D. M. Meekhof, B. E. King, W. M. Itano, and D. J. Wineland, *Phys. Rev. Lett.* **75** (1995) 4714.
9. H. J. Kimble, *Physica Scripta* **T76** (1998) 127.
10. C. H. Bennett, G. Brassard, and A. K. Ekert, *Scientific American* **267** (1992) 50.
11. T. Pellizzari, S. Gardiner, J. I. Cirac, and P. Zoller, *Phys. Rev. Lett.* **75** (1995) 3788.
12. J.-I. Cirac, P. Zoller, H. J. Kimble, and H. Mabuchi, *Phys. Rev. Lett.* **78** (1997) 3221.
13. J.-I. Cirac, S. J. Van Enk, P. Zoller, H. J. Kimble, and H. Mabuchi, *Physica Scripta* **T76** (1998) 223.
14. C. J. Hood, M. S. Chapman, T. W. Lynn, and H. J. Kimble, *Phys. Rev. Lett.* **80** (1998) 4157.
15. H. Mabuchi, Q. A. Turchette, M. S. Chapman, and H. J. Kimble, *Opt. Lett.* **21** (1996) 1393.
16. H. Mabuchi, J. Ye, and H. J. Kimble, *Appl. Phys.* **B** (submitted, 1998).
17. D. Bouwmeester, J. W. Pan, K. Mattle, M. Eibl, H. Weinfurter, and A. Zeilinger, *Nature* **390** (1997) 575.
18. D. Boschi, S. Branca, F. De Martini, L. Hardy, and S. Popescu, *Phys. Rev. Lett.* **80** (1998) 1121.
19. A. Furusawa, J. Sorensen, S. L. Braunstein, C. Fuchs, H. J. Kimble, and E. S. Polzik, *Science* **282** (1998) 706.
20. L. Vaidman, *Phys. Rev.* **A49** (1994) 1473.
21. S. L. Braunstein and H. J. Kimble, *Phys. Rev. Lett.* **80** (1998) 869.
22. P. van Loock, S. L. Braunstein, and H. J. Kimble, in preparation.
23. S. L. Braunstein and H. J. Kimble, in preparation.
24. S. Lloyd and S. L. Braunstein, quant-ph/9810082.
25. S. Lloyd and J. J. E. Slotine, *Phys. Rev. Lett.* **80** (1998) 4088.
26. S. L. Braunstein, *Phys. Rev. Lett.* **80** (1998) 4084.
27. S. L. Braunstein, *Nature* **394** (1998) 47.
28. C. Fuchs, S. L. Braunstein, and H. J. Kimble, in preparation.

Quantum Coherence and Decoherence - ISQM - Tokyo '98
Y.A. Ono and K. Fujikawa (Editors)
© 1999 Elsevier Science B.V. All rights reserved.

Manipulation of Quantum Statistics in Mesoscopic Experiments

Yoshihisa Yamamoto

IERATO Quantum Fluctuation Project, Edward L. Ginzton Laboratory
Stanford University, Stanford, CA 94305-4085, USA
and
NTT Basic Research Laboratories
Morinosato, Atsugi-shi, Kanagawa 243-0198, Japan

The indistinguishability of identical quantum particles can lead to quantum interference that profoundly affect their generation and scattering processes. We present the first experimental evidence for a quantum interference effect in the collision of two electrons. Artificial manipulation of quantum statistics for photons and electrons are demonstrated in semiconductor mesoscopic systems.

1. INTRODUCTION

If two particles collide and scatter, the process that results in the detection of the first particle in one direction and the second particle in another direction interferes quantum mechanically with the physically indistinguishable process where the roles of the particles are reversed [1-2]. For bosons such as photons, a constructive interference between probability amplitudes can enhance the probability, relative to classical expectations, that both are detected in the same direction; this effect is known as "bunching" and is the origin for "final-state stimulation." But, for fermions such as electrons, a destructive interference should suppress this probability; this effect is known as "anti-bunching" and is the origin of the "Pauli exclusion principle."

Although two-particle interferences have been shown for colliding photons [3], a similar demonstration for electrons does not exist. Here, we report the realization of this destructive quantum interference in the collision of electrons using a Fermi-degenerate GaAs two-dimensional electron gas (2DEG) [4].

Artificial manipulation of quantum statistics for electrons and photons are at the heart of recent quantum physics experiments. The Pauli exclusion principle, which results in conductance quantization, is a unique feature of fermions. Here, we report the realization of a similar effect for photons; that is, the generation of single photons with a well-defined time interval using the simultaneous Coulomb blockade effect for electrons and holes in a GaAs mesoscopic pn tunnel junction [5]. The final-state stimulation, which results in laser and maser action, is a unique feature of bosons. We also report the final-state stimulation observed for the excitons, i.e., the composite particles consisting of fermionic constituents in a GaAs quantum-well microcavity [6].

2. QUANTUM INTERFERENCE IN THE COLLISION OF TWO IDENTICAL PARTICLES

If N particles are incident one-by-one on a beamsplitter with a transmission coefficient T and scattered into two output ports independently [Fig. 1(a)], the probability for m particles transmitted out of N incident particles is given by a binomial distribution. The normalized variance (Fano factor) for transmitted particles is $F \equiv \langle \Delta m^2 \rangle / \langle m \rangle = 1 - T$. In the special case of a 50-50% beamsplitter, this partition noise is equal to one-half the Poisson limit (half-shot noise). This result holds for any particles, which means that a single-particle partition process does not depend on the quantum statistics of particles.

However, when two identical particles are simultaneously incident on the 50-50% beamsplitter, the partition characteristics critically depend on the quantum statistics of particles. If the two particles are classical particles, they are independently scattered and the

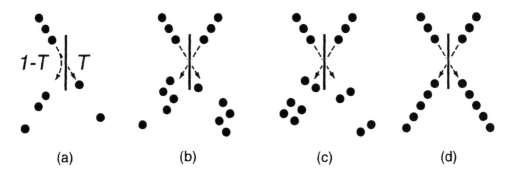

(a)   (b)   (c)   (d)

Figure 1. (a) The single-particle partition process at a beamsplitter with a transmission coefficient $T = 1/2$. (b)-(d) The scattering characteristics for (b) two classical particles, (c) two bosonic particles, and (d) two fermionic particles, incident upon a 50-50% beamsplitter.

output particle flow from the 50-50% beamsplitter carries the half-shot noise [Fig. 1(b)]. If the two particles are identical quantum particles, the scattering properties are governed by quantum interference effects. The incident two-particle wavefunction is

$$|\psi\rangle = \frac{1}{\sqrt{2}} \left(|A : \psi_1, B : \psi_2\rangle \pm |A : \psi_2, B : \psi_1\rangle \right) \tag{1}$$

where the first ket represents particle (A) incident from port 1 and particle (B) from port 2 and the second ket represents the reverse situation. These two states cannot be distinguished. A symmetrized state should be taken for bosons and an anti-symmetrized state for fermions. For the measurement result where two particles are scattered to the same output port, the two indistinguishable probability amplitudes (direct and exchange terms) have the same magnitude and sign, yielding a probability of 1/2 for bosons (constructive interference) and zero for fermions (destructive interference). For the measurement result where one particle is scattered to each output port, the two indistinguishable probability amplitudes (one for both particles is reflected and the other for both particles is transmitted) have the same magnitude and opposite sign, yielding a probability of zero for bosons and one for fermions. This is a simple demonstration of the Pauli exclusion principle for fermion particles and the final-state stimulation for bosonic particles. In this 50-50% beamsplitter, while two bosons always scatter into the same port so that the output flux carries full-shot noise, one fermion always scatters into each port so that the output fermion flux is completely free from partition noise, as shown in Figs. 1(c) and 1(d).

2.1. Collision of two electrons

To observe the quantum interference in the collision of the two fermions shown in Fig. 1(d), the mesoscopic electron collision circuit shown in the Fig. 2(a) was fabricated using a GaAs 2DEG system and operated at about 1.6 K. The two input ports, 1 and 2, are defined by single-mode quantum point contacts in which the input conductance of each port is equal to the quantum unit of conductance $2e^2/h$. The electron wavepacket emitted

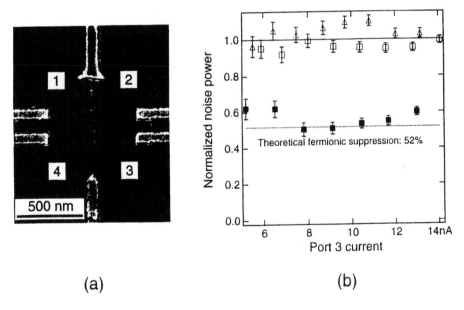

(a)   (b)

Figure 2. (a) Scanning electron microscope (SEM) photo of a 2DEG electron collision circuit. (b) The measured noise power normalized by a half-shot noise value vs. the output current.

from the respective input quantum point contact is split equally into the two output ports, 3 and 4, by the 50-50% electron beamsplitter which is defined by an etched trench and gate electrode. An ac modulation technique is used with a high-input impedance cryogenic cascode preamplifier to improve the signal-to-noise ratio in the noise measurement. The measured noise powers at 15.6 MHz normalized by the half-shot noise value vs. the input current are plotted in Fig. 2(b) for both single and double input cases. When either port 1 (open squares) or port 2 (open triangles) is biased, the single-particle partition noise of the half-shot noise value indicated by the solid line is observed. On the other hand, when both ports 1 and 2 are simultaneously biased (filled squares), the partition noise is suppressed to below the half-shot noise value. Imperfect suppression is due to the existence of non-ideal reflections scattering back to the input ports and the theoretical noise suppression factor, taking into account the finite reflection loss, is shown by the dotted line. This is the first experimental demonstration of the quantum interference effect in the collision of two fermions.

2.2. Scattering of two excitons

The exciton is a bound state of an electron and a hole and is the equivalent of a hydrogen atom in a semiconductor. Having integer spin, it behaves as a boson, but due to its fermionic constitutents, is not ideal. The (weakly) interacting boson is a good approximation when the exciton density is smaller than the Mott density $n_{exc} \simeq a_B^{-d}$, where a_B is the exciton Bohr radius and d is the dimensionality of the system. Here, we consider two-dimensional excitons confined in a GaAs quantum well (QW) embedded in a microcavity which confines the photons (Fig. 3, inset). Due to this confinement, the photon acquires a finite mass, $m_{ph} \simeq 10^{-4} \, m_{exc}$. Typical dispersion relations of excitons and cavity photons are shown in Fig. 3. The exciton and the photon are degenerate at the in-plane momentum $k = 0$ and strong dipole interaction between them results in the formation of hybrid modes, the upper and lower exciton polaritons (UP and LP) [7]. Thus the polaritons are also (weakly) interacting bosons below the Mott density.

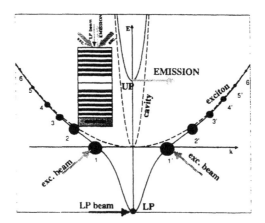

Figure 3. Dispersion of GaAs exciton-polaritons. In the experiment, two exciton beams at large angles and a LP beam in the normal direction are incident upon the top facet. The emission is also collected in the normal direction.

We inject two excitons with opposite in-plane momentum $\pm k$ by an external pump laser. These two excitons are scattered into UP and LP with $k = 0$. This process satisfies the energy and momentum conservation simultaneously. The rate of emission of UP in this scattering process is proportional to $n_{exc}^2(1 + N_{LP})(1 + N_{UP})$, where N_{LP} and N_{UP} are the LP and UP populations at $k = 0$, but are normally much smaller than one due to the weak scattering rates and the short polariton lifetime. However, the final-state stimulation of this scattering process can be induced if the coherent population of LP is created using an external probe laser.

In Fig. 4 (top) we plot the UP emission rate as a function of the pump power I_{exc} for two different probe powers I_{LP}. These curves were fitted with simple parabolas $\frac{d}{dt}N_{UP} = C_1(I_{LP})I_{exc} + C_2(I_{LP})I_{exc}^2$ for five different I_{LP}. The results for C_1 and C_2 are shown in Fig. 4 (bottom). C_1 is independent of I_{LP}, whereas C_2 is linearly dependent on I_{LP}. Since the exciton density n_{exc} and the LP population N_{LP} are proportional to I_{exc} and I_{LP} respectively, this experimental result proves that the exciton-exciton scattering into LP and UP is quadratic in n_{exc}^2 and linear in N_{LP}. This is direct experimental evidence of the final-state stimulation of the exciton-exciton scattering process due to $N_{LP} > 1$. It is concluded that a GaAs exciton polariton behaves as a massive boson in the dilute limit.

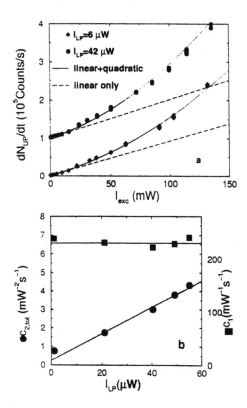

Figure 4. Upper polariton emission rate dependence on I_{exc} and I_{LP}: (a) as a function of I_{exc}, for two different I_{LP}, the highest set shifted by 10^5 for clarity. The dashed lines indicate linear components only. (b) The linear and total quadratic dependence on I_{exc} as a function of I_{LP}. The solid lines indicate the fit with a constant for C_1 and with a line for C_2.

3. SINGLE-PHOTON STATE GENERATION

A Coulomb blockade effect for a single electron and hole in a mesoscopic pn junction leads to the generation of single-photons with a well-regulated time interval. Figure 5(a) illustrates the operational principle of a single-photon turnstile device, in which a single charging energy $\frac{e^2}{C_i}$ ($i = n$ or p) is much greater than the thermal energy $k_B\theta$ and the energy-level broadening $\hbar\Gamma$ of the central QW, where C_i is the tunnel junction capacitance for the n and p sides. At a certain junction voltage, V_0, a single and first electron resonantly tunnels into the conduction sub-band of the central QW from the conduction band of the n-type bulk layer, but subsequent electron tunneling is inhibited due to the Coulomb repulsive force. The junction voltage is then switched to $V_0 + \Delta V$, where a single and first hole resonantly tunnels into the valence sub-band of the central QW from the valence band of the p-type bulk layer with the assistance of the Coulomb attractive force, but subsequent hole tunneling is inhibited due to the absence of the Coulomb attractive force. A single photon is generated per modulation cycle by radiative recombination of a single electron-hole pair. The numerical simulation shown in Fig. 5(b) confirms the operation of this single-photon turnstile device. The sharp spikes in the two stages of modulated junction voltage correspond to a single electron and hole tunneling and the plus signs following the hole tunneling event indicate spontaneous photon generation events.

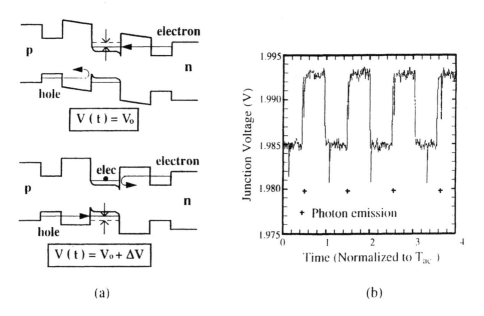

(a) (b)

Figure 5. The structure, operational principle, and numerical simulation of a single-photon turnstile device.

The SEM micrograph of fabricated GaAs/AℓGaAs pn junction posts with a diameter of ~ 200 mm is shown in Fig. 6 (inset). The device was installed in a dilution refrigerator with a base temperature of ~ 50 mK and was biased with a dc and ac voltage source. As shown in Fig. 6(a), the dc current increases linearly with the modulation frequency when the ac voltage is fixed but the dc voltage is changed to three different values. The measured current is in close agreement with the relation $I = ef$, $2ef$, and $3ef$, depending on the dc bias voltage [Fig. 6(b)], which indicates that the charge transfer through the device is locked to the external modulation signal. At the first current plateau $(I = ef)$, a single electron and hole are injected and a single photon is generated per modulation cycle. At the second plateau $(I = 2ef)$, two electrons and holes are injected and two photons are generated per modulation cycle, etc. This is the first demonstration of theheralded single photon flux, which is analogous to the regulated electron flow in conductance quantization.

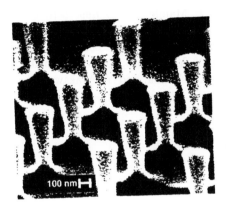

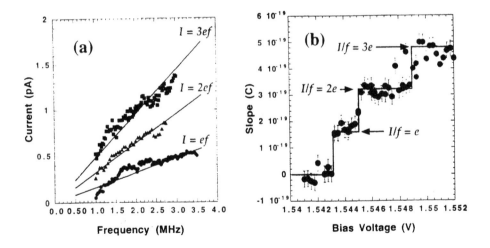

Figure 6. Inset: SEM photo of a fabricated single-photon turnstile device. (a) The tunnel current I vs. modulation frequency f for three different dc bias voltages and a fixed ac modulation voltage. (b) The slope I/f vs. the dc bias voltage.

ACKNOWLEDGEMENTS

The author wishes to thank R. C. Liu, J. Kim, F. Tassone, R. Huang, O. Benson, H. Kan, and A. Imamoglu for their long-time collaborations.

REFERENCES

1. R. P. Feynman *et al.*, *The Feynman Lectures on Physics,* Vol.3: *Quantum Mechanics* (Addison-Wesley, New York, 1965).
2. R. Loudon, in *Coherence and Quantum Optics VI*, eds. J. H. Eberly *et al.* (Plenum Presss, New York, 1989)
3. C. K. Hang, Z. Y. On, and L. Mandel, *Phys. Rev. Lett.* **59** (1987) 2044
4. R. C. Liu, B. Odom, Y. Yamamoto, and S. Tarucha, *Nature* **391** (1998) 263.
5. A. Imamoglu and Y. Yamamoto, *Phys. Rev. Lett.* **72** (1994) 210
6. F. Tassone, R. Huang, and Y. Yamamoto, to be published.
7. C. Wesbuch, M. Nishioka, A. Ishikara, and Y. Arakawa, *Phys. Rev. Lett.* **69** (1992) 3314.

Quantum Coherence and Decoherence - ISQM - Tokyo '98
Y.A. Ono and K. Fujikawa (Editors)
© 1999 Elsevier Science B.V. All rights reserved.

Extension of Nonlocal Response Theory to Raman Process

H. Ajiki and K. Cho

Department of Physical Science, Graduate School of Engineering Science,
Osaka University, Toyonaka, Osaka 560-8531, Japan

A set of self-consistent equations of vector potentials at various space-time positions has been derived by solving the equations of motion in QED for a coupled-matter system. The integral kernels in these equations become the form of transition susceptibility proposed by Born and Huang to treat Raman process if we treat the vector potential as c-number. This scheme provides a smooth extension of semiclassical nonlocal response theory. A discussion about radiative width is given according to this theory.

1. Introduction

Optical properties of a mesoscopic system are characterized by coherent extensions of excited states, which gives rise to spatial nonlocality of susceptibility. A semiclassical framework of optical response theory in such a system, called microscopic nonlocal response theory, has been established.[1–3] It provides a unified treatment of optical responses from microscopic to macroscopic systems. In fact, it reproduces the radiative decay width of a two-level atom obtained from quantum electrodynamics (QED),[1] and the dispersion equations for polariton, X-ray dynamical scattering, and photonic bands in bulk crystals can be derived in the same framework.[4]

So far, the source current density in the nonlocal response theory is defined as the expectation value of current density operator. This scheme works well for the elastic (Rayleigh) scattering processes. For Raman process, however, source current can not be represented by an expectation value, because the initial and final states are different. In this paper we derive a QED based scheme, which corresponds to an extended version of the nonlocal theory including Raman process.

2. Nonlocal Linear Response Theory via QED

We consider a system of an ensemble of charged particles α, having charge e_α and mass m_α, interacting with an EM field. Let $\hat{r}_\alpha$ and $\hat{P}_\alpha$ be the position and momentum operators of the particle α, respectively, and $\hat{A}$ be a vector-potential operator which is transverse in the Coulomb gauge. The Hamiltonian of the system in the Coulomb gauge is given as

$$H = \sum_\lambda \hbar\omega_\lambda |\lambda\rangle\langle\lambda| + \sum_{\mathbf{k}\sigma} \hbar c k \left(a^\dagger_{\mathbf{k}\sigma} a_{\mathbf{k}\sigma} + \frac{1}{2} \right)$$

$$-\sum_{\lambda\lambda'}\sum_{\alpha}\left(\frac{e_\alpha}{m_\alpha c}\langle\lambda|\hat{\boldsymbol{P}}_\alpha\cdot\hat{\boldsymbol{A}}(\hat{\boldsymbol{r}}_\alpha)|\lambda'\rangle-\frac{e_\alpha^2}{2m_\alpha c^2}\langle\lambda|\hat{\boldsymbol{A}}^2(\hat{\boldsymbol{r}}_\alpha)|\lambda'\rangle\right)|\lambda\rangle\langle\lambda'|,\qquad(1)$$

where $|\lambda\rangle$ represents an eigen state in the Heisenberg picture, with energy $\hbar\omega_\lambda$, of charged particles containing Coulomb energy, and $\hat{\boldsymbol{A}}$ is given as

$$\hat{\boldsymbol{A}}(\hat{\boldsymbol{r}}_\alpha)=\sum_{\boldsymbol{k}\sigma}\sqrt{\frac{2\pi c\hbar}{Vk}}\boldsymbol{e}_{\boldsymbol{k}\sigma}e^{i\boldsymbol{k}\cdot\boldsymbol{r}_\alpha}\left(a_{\boldsymbol{k}\sigma}+a^\dagger_{-\boldsymbol{k}\sigma}\right),\qquad(2)$$

with $\boldsymbol{e}_{\boldsymbol{k}\sigma}$ being a unit polarization vector ($\sigma=1,2$), $a^\dagger$ and a being creation and annihilation operators, respectively, and V being the normalization volume.

Solving the Heisenberg equations for $a_{\boldsymbol{k}\sigma}$ and $a^\dagger_{\boldsymbol{k}\sigma}$ without any approximation, we get an operator version of the solution of Maxwell equations as

$$\hat{\boldsymbol{A}}(\boldsymbol{r},\omega)=\hat{\boldsymbol{A}}_0(\boldsymbol{r},\omega)+\frac{1}{c}\sum_{\lambda\lambda'}\int d\boldsymbol{r}'\boldsymbol{G}_{\mathrm{T}}(\boldsymbol{r},\boldsymbol{r}';q)\langle\lambda|\hat{\boldsymbol{j}}(\boldsymbol{r}')|\lambda'\rangle_\omega|\lambda\rangle\langle\lambda'|,\qquad(3)$$

where $\hat{\boldsymbol{A}}_0$ is the vector-potential operator for a free (incident) EM field, $\boldsymbol{G}_{\mathrm{T}}$ is the dyadic Green's function given by

$$\boldsymbol{G}_{\mathrm{T}}(\boldsymbol{r},\boldsymbol{r}';q)=\frac{4\pi}{V}\sum_{\boldsymbol{k}\sigma}\boldsymbol{e}_{\boldsymbol{k}\sigma}\frac{1}{k^2-(q+i\delta)^2}\boldsymbol{e}_{\boldsymbol{k}\sigma}e^{i\boldsymbol{k}\cdot(\boldsymbol{r}-\boldsymbol{r}')},\qquad(4)$$

with $q=\omega/c$ ($\delta=0^+$), and $\langle\lambda|\hat{\boldsymbol{j}}|\lambda'\rangle_\omega$ is a Fourier component of $\langle\lambda|\hat{\boldsymbol{j}}(t)|\lambda'\rangle$ with $\hat{\boldsymbol{j}}$ being a current-density operator given by

$$\hat{\boldsymbol{j}}(\boldsymbol{r})=\sum_\alpha\frac{e_\alpha}{2m_\alpha}[\hat{\boldsymbol{P}}_\alpha\delta(\boldsymbol{r}-\hat{\boldsymbol{r}}_\alpha)+\delta(\boldsymbol{r}-\hat{\boldsymbol{r}}_\alpha)\hat{\boldsymbol{P}}_\alpha]-\sum_\alpha\frac{e_\alpha^2}{m_\alpha c}\hat{\boldsymbol{A}}(\boldsymbol{r})\delta(\boldsymbol{r}-\hat{\boldsymbol{r}}_\alpha).\qquad(5)$$

We calculate the time evolution of the current density up to the first order in $\hat{\boldsymbol{A}}$. The Fourier component of its matrix element is

$$\langle\lambda|\hat{\boldsymbol{j}}(\boldsymbol{r})|\lambda'\rangle_\omega=\langle\lambda|\hat{\boldsymbol{I}}(\boldsymbol{r})|\lambda'\rangle\delta_{\omega\omega_{\lambda'\lambda}}+\frac{1}{c}\sum_{\lambda''}\Big[g_{\lambda''\lambda'}(\omega+\omega_{\lambda\lambda'})\langle\lambda|\hat{\boldsymbol{I}}(\boldsymbol{r})|\lambda''\rangle\hat{F}_{\lambda''\lambda'}(\omega+\omega_{\lambda\lambda'})$$
$$+h_{\lambda''\lambda}(\omega+\omega_{\lambda\lambda'})\langle\lambda''|\hat{\boldsymbol{I}}(\boldsymbol{r})|\lambda'\rangle\hat{F}_{\lambda\lambda''}(\omega+\omega_{\lambda\lambda'}),\Big]\qquad(6)$$

with

$$g_{\lambda''\lambda'}(\omega)=\frac{1}{\hbar\omega_{\lambda''\lambda'}-\hbar\omega-i\delta}-\frac{1}{\hbar\omega_{\lambda''\lambda'}},\qquad h_{\lambda''\lambda}(\omega)=\frac{1}{\hbar\omega_{\lambda''\lambda}+\hbar\omega+i\delta}-\frac{1}{\hbar\omega_{\lambda''\lambda}},\qquad(7)$$

$$\hat{\boldsymbol{I}}(\boldsymbol{r})=\sum_\alpha\frac{e_\alpha}{2m_\alpha}[\hat{\boldsymbol{P}}_\alpha\delta(\boldsymbol{r}-\hat{\boldsymbol{r}}_\alpha)+\delta(\boldsymbol{r}-\hat{\boldsymbol{r}}_\alpha)\hat{\boldsymbol{P}}_\alpha],\qquad(8)$$

$$\hat{F}_{\mu\nu}(\omega)=\int d\boldsymbol{r}\langle\mu|\hat{\boldsymbol{I}}(\boldsymbol{r})\cdot\hat{\boldsymbol{A}}(\boldsymbol{r};\omega)|\nu\rangle,\qquad(9)$$

and $\omega_{\lambda\lambda'} = \omega_\lambda - \omega_{\lambda'}$. The form of the second term of the factors $g_{\lambda''\lambda'}$ and $h_{\lambda''\lambda}$ is obtained from the assumption that the spatial variation of EM field is small.[3]

In the following calculation we will neglect the first term in eq. (6), since the term does not contribute to scattering process. The second term describes induced-current density by field. If we treat the vector potential as a classical field, the induced-current density is given as a vector potential multiplied by the transition susceptibility tensor. The transition susceptibility has the same form as that introduced by Born *et al.*[5] to describe the Raman scattering on the analogy of the Rayleigh scattering. For $\lambda = \lambda'$ the transition susceptibility becomes the usual susceptibility for linear response, which describes the Rayleigh scattering.

From eqs. (3), (6), and (9), we find a set of coupled equations of $\hat{F}_{\mu\nu}(\omega)$ as

$$\hat{F}_{\mu\nu}(\omega) = \hat{F}_{\mu\nu}^{(0)}(\omega) \; + \; \sum_{\lambda\lambda''} g_{\lambda''\nu}(\omega + \omega_{\lambda\nu}) A_{\mu\lambda;\lambda\lambda''}(\omega)\hat{F}_{\lambda''\nu}(\omega + \omega_{\lambda\nu})$$

$$+ \; \sum_{\lambda\lambda''} h_{\lambda''\lambda}(\omega + \omega_{\lambda\nu}) A_{\mu\lambda;\lambda''\nu}(\omega)\hat{F}_{\lambda\lambda''}(\omega + \omega_{\lambda\nu}), \tag{10}$$

with

$$A_{\mu\nu;\mu'\nu'}(\omega) = \frac{1}{c^2} \int d\boldsymbol{r} \int d\boldsymbol{r}' \langle\mu|\hat{\boldsymbol{I}}(\boldsymbol{r})|\nu\rangle \boldsymbol{G}_{\mathrm{T}}(\boldsymbol{r}, \boldsymbol{r}'; q)\langle\mu'|\hat{\boldsymbol{I}}(\boldsymbol{r}')|\nu'\rangle, \tag{11}$$

being a retarded interaction between transition-current densities $\langle\mu|\hat{\boldsymbol{I}}(\boldsymbol{r})|\nu\rangle$ and $\langle\mu'|\hat{\boldsymbol{I}}(\boldsymbol{r}')|\nu'\rangle$ via transverse EM field. The retarded interaction provides a radiative correction.

Note that $\hat{F}_{\mu\nu}(\omega)$ is still an operator in photon space. Eqs. (10) are the linear simultaneous equations of operators $\hat{F}_{\mu\nu}(\omega)$ with the inhomogeneous terms depending on the initial condition. There appear terms with different ω's, in contrast with the corresponding equations in semiclassical nonlocal theory. This is because different states $|\lambda\rangle$ and $|\lambda'\rangle$ are allowed in eq. (6), which leads to Raman scattering process. If we keep only the term with a same ω in eqs. (10), this is the same set of equations in the semiclassical nonlocal theory by just regarding $\boldsymbol{A}(\hat{r},\omega)$ in $\hat{F}_{\mu\nu}(\omega)$'s as classical quantity.[1]

3. Radiative Decay Width

Let us apply the above theory to resonant light scattering by a three-level atom and discuss radiative decay widths by comparing with the results of the QED. The three levels are denoted by e, g_1, and g_0 having energy $\hbar\omega_e > \hbar\omega_{g_1} > \hbar\omega_{g_0}$, and we assume that the transition between g_1 and g_0 is forbidden for simplicity. When an incident field with ω_0 is applied to the atom with the initial state g_0, frequencies of scattered lights are ω_0 with final state g_0 and $\omega_1 \equiv \omega_0 - \omega_{g_1 g_0}$ with the final state g_1. Although there exist scattered fields having various frequencies due to multiple scattering, the frequencies are restricted to ω_0 and ω_1.

Solving the self-consistent equations (10) in this model with the rotating wave approximation, we obtain the scattered field by Rayleigh and Raman processes as follows:

$$\hat{\boldsymbol{A}}(\boldsymbol{r},\omega_0) \; = \; \hat{\boldsymbol{A}}_0(\boldsymbol{r},\omega_0) + \frac{1}{c^2}\int d\boldsymbol{r}' \boldsymbol{G}_{\mathrm{T}}(\boldsymbol{r},\boldsymbol{r}';q_0)\frac{\langle g_0|\hat{\boldsymbol{I}}(\boldsymbol{r}')|e\rangle}{\hbar\omega_{\mathrm{ego}} - \hbar\omega_0 + A_{\mathrm{ego;goe}}(\omega_0)}\hat{F}_{\mathrm{ego}}^{(0)}(\omega_0), \tag{12}$$

$$\hat{\boldsymbol{A}}(\boldsymbol{r},\omega_1) \; = \; \frac{1}{c^2}\int d\boldsymbol{r}' \boldsymbol{G}_{\mathrm{T}}(\boldsymbol{r},\boldsymbol{r}';q_1)\frac{\langle g_1|\hat{\boldsymbol{I}}(\boldsymbol{r}')|e\rangle}{\hbar\omega_{\mathrm{ego}} - \hbar\omega_0 + A_{\mathrm{ego;goe}}(\omega_0)}, \hat{F}_{\mathrm{ego}}^{(0)}(\omega_0) \tag{13}$$

with $q_0 = \omega_0/c$ and $q_1 = \omega_1/c$. This results exhibit that the real and imaginary parts of the retarded interaction $A_{eg_0;g_0e}(\omega_0)$ lead to the line width and resonant-energy shift as a radiative correction, respectively. Note that the radiative widths of the three-level atom have the same form as that of a two-level atom. For a three-level atom, however, each radiative width of Rayleigh and Raman scatterings becomes the summation of the transition rates from e to g_0 and from e to g_1 in the QED.[6] Therefore the radiative width for the three-level atom in the nonlocal theory is smaller than the results of QED by the transition rate from e to g_1. However, the present theory seems to be quite useful because a scattered field is obtained including the effects of multiple (spatial and temporal) scatterings of EM fields, which start to appear significantly in a mesoscopic system.

4. Summary and Conclusion

We have derived a set of self-consistent equations of vector potentials including the effects of multiple (spatial and temporal) scattering by QED. In the formulation Raman process is described as well as Rayleigh process, and the transition susceptibility for Raman scattering has been derived in the semiclassical approximation. Thus this scheme provides a smooth extension of semiclassical nonlocal response theory. The present theory is applied to scattering by a three-level atom. Radiative decay widths obtained in the theory approximately agree with the results by QED, if the transition rates of Raman processes are small.

Acknowledgments

This work was supported in part by the Grant-in-Aid for Scientific Research on Priority Areas "Laser Chemistry of Single Nanometer Organic Particles" and Grant-in-Aid for Scientific Research (A) "Many-Body Effects on Electronic Excited States in Quantum Dots under Multi-photon Excitation".

REFERENCES

1. K. Cho, Prog. Theor. Phys. Suppl. 106 (1991) 225.
2. H. Ishihara and K. Cho, Phys. Rev. B48 (1993) 7960.
3. Y. Ohfuti and K. Cho, Phys. Rev. B52 (1995) 4828; *ibid.* J. Luminesc 66/67 (1996) 94 (Errata).
4. K. Cho, J. Phys. Soc. Jpn. 66 (1997) 2496.
5. M. Born and K. Huang, *Dynamical Theory of Crystal Lattices* (Oxford at the Clarendon Press, 1954).
6. W. Heitler, *The Quantum Theory of Radiation*, 3rd ed (Oxford at the Clarendon Press, 1957).

Quantum Coherence and Decoherence - ISQM - Tokyo '98
Y.A. Ono and K. Fujikawa (Editors)
© 1999 Elsevier Science B.V. All rights reserved.

Quantum Maxwell-Bloch Equations for Spontaneous Emission in Optical Semiconductor Devices

Ortwin Hess[a] and Holger F. Hofmann[a]

[a]Theoretical Quantum Electronics, Institute of Technical Physics, DLR
Pfaffenwaldring 38-40, D–70569 Stuttgart, Germany

We present quantum Maxwell-Bloch equations (QMBE) for spatially inhomogeneous optical semiconductor devices taking into account the quantum noise effects which cause spontaneous emission and amplified emission. Analytical expressions derived from the QMBE are presented for the spontaneous emission factor β and the far field pattern of amplified spontaneous emission in broad area quantum well lasers.

1. INTRODUCTION

In an optical semiconductor device, spontaneously emitted light may have significant influence on the spatiotemporal dynamics of the (coherent) light field. Undoubtedly, the spatial coherence of spontaneous emission and amplified spontaneous emission is most important close to threshold or in devices with a light field output dominated by spontaneous emission such as superluminescent diodes and ultra low threshold semiconductor lasers [1,2]. Even in more common devices such as VCSEL-arrays, gain guided or multistripe and broad area semiconductor lasers, the spontaneous emission factor is modified by the amplification and absorption of spontaneous emission into the non-lasing modes [3] and amplified spontaneous emission may be responsible for multi mode laser operation and for finite spatial coherence.

In the present article, Quantum Maxwell-Bloch Equations (QMBE) are presented, which describe the spatiotemporal dynamics of both, stimulated amplification and (amplified) spontaneous emission on equal footing in terms of expectation values of field-field correlations, dipole-field correlations, carrier densities, fields and dipoles. In particular, the quantum dynamics of the interaction between the light field and the carrier system is formulated in terms of Wigner distributions for the carriers and of spatially continuous amplitudes for the light field. Analytical results for the spontaneous emission factor β and the far field pattern of amplified spontaneous emission in broad area quantum well lasers are discussed.

2. Quantum Maxwell-Bloch Equations

The Quantum Maxwell-Bloch equations for the carrier density $N(\mathbf{r}_{||})$ and the Wigner distributions of the field-field ($I(\mathbf{r}_{||}; \mathbf{r}'_{||})$) and the field-dipole ($C(\mathbf{r}_{||}; \mathbf{r}'_{||}, \mathbf{k}_{||})$) correlation

$$\frac{\partial}{\partial t} N\left(\mathbf{r}_{||}\right) = -\gamma N\left(\mathbf{r}_{||}\right) + D_{amb}\Delta N\left(\mathbf{r}_{||}\right) + j\left(\mathbf{r}_{||}\right) \tag{1}$$

$$+ ig_0\sqrt{\nu_0}\,\frac{1}{4\pi^2}\int d^2\mathbf{k}_{||}\left(C\left(\mathbf{r}; \mathbf{r}_{||}, \mathbf{k}_{||}\right)_{z=0} - C^*\left(\mathbf{r}; \mathbf{r}_{||}, \mathbf{k}_{||}\right)_{z=0}\right)$$

$$\frac{\partial}{\partial t} C\left(\mathbf{r}; \mathbf{r}'_{||}, \mathbf{k}_{||}\right) = -\left(\Gamma + i\Omega(\mathbf{k}_{||}) + i\frac{\omega_0}{2k_0^2}\Delta_\mathbf{r}\right) C\left(\mathbf{r}; \mathbf{r}'_{||}, \mathbf{k}_{||}\right) \tag{2}$$

$$+ ig_0\sqrt{\nu_0}\left(f_{eq}^e\left(k_{||}; N(\mathbf{r}'_{||})\right) + f_{eq}^h\left(k_{||}; N(\mathbf{r}'_{||})\right) - 1\right) I\left(\mathbf{r}; \mathbf{r}'\right)_{z'=0}$$

$$+ ig_0\sqrt{\nu_0}\,\delta(\mathbf{r}_{||} - \mathbf{r}'_{||})\delta(z)\, f_{eq}^e\left(k_{||}; N(\mathbf{r}_{||})\right) f_{eq}^h\left(k_{||}; N(\mathbf{r}_{||})\right)$$

$$\frac{\partial}{\partial t} I\left(\mathbf{r}; \mathbf{r}'\right) = -i\frac{\omega_0}{2k_0^2}\left(\Delta_\mathbf{r} - \Delta_{\mathbf{r}'}\right) I\left(\mathbf{r}; \mathbf{r}'\right) \tag{3}$$

$$- ig_0\sqrt{\nu_0}\,\frac{1}{4\pi^2}\int d^2\mathbf{k}_{||}\left(C\left(\mathbf{r}; \mathbf{r}'_{||}, \mathbf{k}_{||}\right)\delta(z') - C^*\left(\mathbf{r}'; \mathbf{r}_{||}, \mathbf{k}_{||}\right)\delta(z)\right).$$

describe the interaction of a quantum well with a quantized light field, where the two-dimensional coordinates parallel to a quantum well at $z = 0$ are marked by the index $||$. In the QMBE, many particle renormalizations are not mentioned explicitly, but can be added in a straightforward manner [4]. Moreover, the Wigner distributions of electrons (e) and holes (h), have been approximated by corresponding quasi equilibrium Fermi distributions $f_{eq}^{e,h}(\mathbf{k}_{||}, N(\mathbf{r}_{||}))$ which are associated with a local carrier density of $N(\mathbf{r}_{||})$ at a temperature T. In Eq. (1), D_{amb} is the ambipolar diffusion constant, $j\left(\mathbf{r}_{||}\right)$ the injection current density, γ the rate of spontaneous recombinations by non-radiative processes and/or spontaneous emission into modes not considered in $I\left(\mathbf{r}_{||}, \mathbf{r}'_{||}\right)$, and $\Delta_\mathbf{r}$ is the Lapalcian with respect to $\mathbf{r}$. Another major feature of the semiconductor medium is the fact that the difference $\Omega(\mathbf{k}_{||})$ between the dipole oscillation frequency and the band gap frequency ω_0 is a function of the semiconductor band structure. Here we model the band structure by assuming parabolic bands such that $\Omega(\mathbf{k}_{||}) = \left(\hbar/(2m_{eff}^e) + \hbar/(2m_{eff}^h)\right)\mathbf{k}_{||}^2$. Moreover, the carrier dynamics of the interband-dipole $p\left(\mathbf{r}_{||}, \mathbf{k}_{||}\right)$ and the dipole part of the field-dipole correlation $C\left(\mathbf{r}_{||}; \mathbf{r}'_{||}, \mathbf{k}_{||}\right)$ depend on a correlation of the electrons with the holes. The relaxation of this correlation is modeled in terms of a rate of $\Gamma(\mathbf{k})$ which may be interpreted as the total momentum dependent scattering rate in the carrier system.

In the QMBE, the three dimensional intensity function of the light field $I(\mathbf{r}; \mathbf{r}')$ may be interpreted as a single photon density matrix or as a spatial field-field correlation. The light-matter interaction is mediated by the correlation function $C(\mathbf{r}; \mathbf{r}'_{||}, \mathbf{k}_{||})$ which describes the correlation of the complex field amplitude at $\mathbf{r}$ and the complex dipole amplitude of the $\mathbf{k}_{||}$ transition at $\mathbf{r}'_{||}$. The phase of this complex correlation corresponds to the phase difference between the dipole oscillations and the field oscillations. Finally, the source of emission is the imaginary part of the field-dipole correlation $C(\mathbf{r}; \mathbf{r}'_{||}, \mathbf{k}_{||})$. The quantum Maxwell- Bloch equations thus show how this imaginary correlation originates either from stimulation by $I(\mathbf{r}; \mathbf{r}')$ or spontaneously from the product of electron and hole densities.

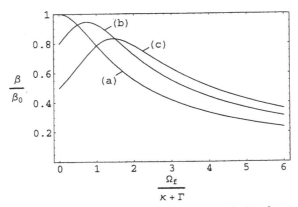

Figure 1. Carrier density dependence of the spontaneous emission factor β for three modes with frequencies above the band gap frequency given by (a) $\omega = 0$, (b) $\omega = 0.5(\Gamma + \kappa)$, and (c) $\omega = \Gamma + \kappa$. $\beta_0 = \beta(\omega = N = 0)$ is determined by the geometry of the laser. The carrier density is given in terms of the transition frequency at the Fermi edge Ω_f.

3. SPONTANEOUS EMISSION FACTOR

The spontaneous emission factor β is generally defined as the fraction of spontaneous emission being emitted into the cavity mode [6]. On the basis of our theory, an analytical expression for β may be obtained for zero temperature where the k-space integrals may be solved analytically on the assumption of Γ being independent of $\mathbf{k}_{\parallel}$. The analytical result reads

$$\beta(\omega, \Omega_f) = \frac{3\sigma}{2\pi \rho_L \, W \, L \, \Omega_f} \left(\arctan\left(\frac{\Omega_f - \omega}{\Gamma + \kappa}\right) + \arctan\left(\frac{\omega}{\Gamma + \kappa}\right) \right), \tag{4}$$

with the density of light field modes (at the band edge frequency ω_0) $\rho_L = \omega_0^2 \pi^{-2} c^{-3} \epsilon_r^{3/2}$. The Fermi frequency $\Omega_f(N) = \pi \hbar \, M \, N$, with the effective carrier mass $M = \left(m_{eff}^e + m_{eff}^h\right) / \left(m_{eff}^e m_{eff}^h\right)$, expresses, in particular, the carrier density dependence of β. For $\Omega_f, \omega \ll \Gamma + \kappa$ we recover the result typically given in the literature (e.g. [6]) being independent of N. Fig. 1 shows the deviation of the spontaneous emission factor from this value as the Fermi frequency Ω_f passes the point of resonance with the cavity mode. Fig. 1 illustrates, in particular, the carrier density dependence of β for three modes with frequencies above the band gap frequency given by (a) $\omega = 0$, (b) $\omega = 0.5(\Gamma + \kappa)$, and (c) $\omega = \Gamma + \kappa$. Most notably, β is always smaller than the usual estimate given by $\beta(\omega = \Omega_f = 0)$, which is based on the assumption of ideal resonance between the transition frequency and the cavity mode.

4. FARFIELD PATTERN OF A BROAD AREA LASER

In large spatially inhomogeneous laser devices, the spatial coherence of amplified spontaneous emission defines an angular distribution of the emitted light in the far field. Spatial coherence increases as the laser threshold is crossed. For $T = 0$ we may obtain an analytical expression for the far-field intensity distribution of a broad area semiconductor laser [5]. Figure 2 shows the far field intensity distribution for different carrier densities below

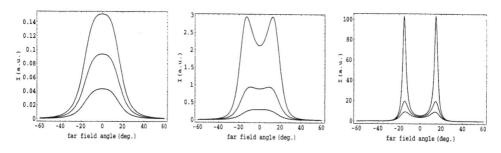

Figure 2. Far field intensity distributions for increasing values of the carrier density $\mathcal{N} = N/N_p$. (a) $\mathcal{N} = 0.05, 0.1, 0.15$, (b) $\mathcal{N} = 0.25, 0.5, 0.75$, and (c) $\mathcal{N} = 0.90, 0.95, 0.99$. The peaks appear at emission angles of $\pm 15°$.

threshold defined by a pinning carrier density N_p. Figure 2 (b) shows the intensity distribution for carrier densities halfway towards threshold. Already, the intensity maxima move to angles of $\pm 15°$, corresponding to the frequency at which the gain spectrum has its maximum. In the case of Fig. 2(c), the threshold region is very close to the pinning density. The peaks in the far field pattern narrow as the laser intensity is increased. Consequently the far field pattern indeed is a measure of the spatial coherence – similar as the linewidth of the laser spectrum is a measure of temporal coherence. It is therefore desirable to consider quantum noise effects in the spatial patterns of optical systems. In the context of squeezing, such patterns have been investigated by Lugiato and coworkers [7]. The laser patterns presented here are based on the same principles. Usually, however, the strong dissipation prevents squeezing in laser systems unless the pump-noise fluctuations are suppressed [8].

5. CONCLUSIONS

Quantum Maxwell-Bloch equations (QMBE) for spatially inhomogeneous optical semiconductor devices have been presented, which take into account the quantum mechanical nature of the light field as well as that of the carrier system and thus describe the effects of coherent spatiotemporal quantum fluctuations. An example of the spatial coherence characteristics described by the QMBE, an analytic expression for the density dependence of the spontaneous emission factor β and spatial profiles of the far field distribution of a broad area edge emitting laser are discussed.

REFERENCES

1. G. Björk, A. Karlsson, and Y. Yamamoto, Phys. Rev. A **50**, 1675 (1994).
2. Y. Yamamoto and R.E. Slusher, Physics Today, June 1993, 66 (1993).
3. H.F. Hofmann and O. Hess, Opt. Lett. **23**, 391 (1998).
4. ·O. Hess and T. Kuhn, Phys. Rev. A **54**, 3347 (1996).
5. H.F. Hofmann and O. Hess, physics/9807011 (submitted to Phys. Rev. A, 3 Jul 1998).
6. K.J. Ebeling, *Integrated Optoelectronics* (Springer, Berlin 1993).
7. L.Q. Lugiato and F. Castelli, Phys. Rev. Lett. **68**, 3284 (1992).
   A. Gatti *et al.*, Phys.Rev. A **56**, 877 (1997).
8. Y. Yamamoto, S. Machida and O. Nilsson, Phys. Rev. A **34**, 4025 (1986).

Quantum Coherence and Decoherence - ISQM - Tokyo '98
Y.A. Ono and K. Fujikawa (Editors)
© 1999 Elsevier Science B.V. All rights reserved.

Quantum control of atomic systems by time-resolved homodyne detection of spontaneous emission

Holger F. Hofmann [a], Ortwin Hess [a], and Günter Mahler [b]

[a]Institut für Technische Physik, DLR,
Pfaffenwaldring 38-40, 70569 Stuttgart, Germany

[b]Institut für Theoretische Physik und Synergetik,
Pfaffenwaldring 57, 70550 Stuttgart, Germany

We describe the light-matter interaction of a single two level atom with the electromagnetic vacuum in terms of field and dipole variables by considering homodyne detection of the emitted fields. Spontaneous emission is then observed as a continuous fluctuating force acting on the atomic dipole. The effect of this force may be compensated and even reversed by feedback.

1. INTRODUCTION

The spontaneous emission of light from a single atom is usually described as the random appearence of a photon in the vacuum light field. However, this description is only valid if photons are actually detected. The field variables continuously evolve from the dipole dynamics of the atom according to Maxwells equations. If field variables are measured, the spontaneous emission of a single two level atom may be interpreted as the interaction of a fluctuating dipole with a noisy light field [1,2]. Quantum jumps are avoided and the continuous evolution of the atomic system may be controlled by weak coherent feedback fields compensating the observed quantum noise. In the following we describe the evolution of the quantum state of a two level atom conditioned by projective homodyne detection and discuss some of the possible feedback scenarios.

2. HOMODYNE DETECTION OF WEAK FIELDS

In a balanced homodyne detection setup the coherent laser field of a local oscillator interferes with the low intensity input field at a beamsplitter as shown in Figure 1. The difference between the photon numbers registered in detector 1 and detector 2 corresponds to the interference term

$$\Delta \hat{n} = \hat{a}^\dagger \hat{b} + \hat{a} \hat{b}^\dagger, \tag{1}$$

where $\hat{a}$ and $\hat{b}$ represent the annihilation operators of the local oscillator mode and the source field, respectively. The field modes are emitted during the measurement time interval τ and represent wave packets of length $c\tau$. The quantum state emitted by the local oscillator may be represented by a coherent state with an average complex amplitude

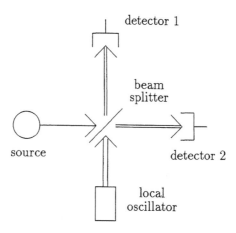

Figure 1. Schematic setup of balanced homodyne detection

of α. The photon number difference Δn then corresponds to $2 \mid \alpha \mid$ times the quadrature component of the source field which is in phase with the local oscillator. For weak fields, the probability distribution of the measurement results Δn is approximately given by the vacuum fluctuations of the observed quadrature component,

$$p(\Delta n) \approx \frac{\exp[-\frac{\Delta n^2}{2|\alpha^2|}]}{\sqrt{2\pi |\alpha^2|}}. \tag{2}$$

Within the measurement time interval τ the dipole fluctuations of a two level atom with a spontaneous emission rate of Γ emit an average light field energy of $\Gamma\tau$ times the quantum fluctuation intensity of $\hbar\omega/2$. If $\Gamma\tau$ is much smaller than one, the dipole radiation emitted by the atom is much weaker than the quantum fluctuations of the electromagnetic vacuum. Therefore, the dipole radiation is obscured by quantum noise and the information about the state of the atomic system obtained in the homodyne detection measurement is extremely small. Nevertheless some information is obtained about the most likely orientation of the atomic dipole and this observation will modify the quantum state of the system as explained below.

3. QUANTUM DIFFUSION OF A TWO LEVEL ATOM

The back action of continuous time-resolved homodyne detection on a quantum system results in a stochastic evolution of the wave function equivalent to a Monte Carlo wavefunction formalism [3,4]. The derivation from a projective measurement base is discussed in [1]. It is convenient to describe the back action in terms of the Bloch vector **s** of the atomic two level system, where s_x is the expectation value of the observed dipole component and s_z is the expectation value of the atomic inversion. The s_y component describes

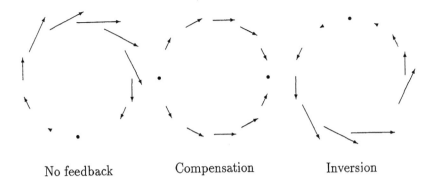

No feedback   Compensation   Inversion

Figure 2. Diffusion of the Bloch vector corresponding to measurement results of $\Delta n > 0$ for different feedback scenarios. The top of the circles represents the excited state $s_z = +1$ and the bottom represents the ground state $s_z = -1$.

the expectation value of the unobserved dipole component. The back action corresponding to a measurement result of Δn for an arbitrary initial Bloch vector $(s_x, 0, s_z)$ in the $s_y = 0$ plane reads

$$\delta s_x = \sqrt{\Gamma \tau} \frac{\Delta n}{|\alpha|} (1 + s_z) s_z$$

$$\delta s_z = -\sqrt{\Gamma \tau} \frac{\Delta n}{|\alpha|} (1 + s_z) s_x. \tag{3}$$

The Bloch vector is thus rotated by an angle of $\sqrt{\Gamma \tau} \frac{\Delta n}{|\alpha|} (1 + s_z)$ around the y-axis in response to the homodyne detection measurement.

4. CONTROLLING THE QUANTUM STATE BY FEEDBACK

Without feedback, the ground state $s_z = -1$ is stationary while diffusion is at a maximum for the excited state $s_z = +1$. It is possible to interpret this back action effect as a sum of Rabi rotations induced by the quantum noise and an epistemological effect of the information obtained about the dipole component s_x of the atom. The ground state is stationary because the dipole emission effects compensate the absorption of vacuum fluctuations. In the excited state the dipole emission effect and the response to the vacuum fluctuations add up and cause twice the diffusion expected from classical field fluctuations.

By applying a negative feedback equal to the observed quadrature component the Rabi rotations induced by the quantum fluctuations of the measured field component may be compensated. The diffusive back action which remains is then associated with the information gained about the atomic system. As mentioned in the previous section, it is possible to identify this back action as a weak measurement effect of the dipole variable s_x. Due to the weak dipole radiation emitted by the atom, $\Delta n > 0$ is more likely for $s_x = +1$

and $\Delta n < 0$ is more likely for $s_x = -1$. Coherent superpositions of dipole eigenstates diffuse because of the modified statistical weight of the dipole eigenstate components. Consequently the dipole eigenstates are stationary while the diffusion is at a maximum for both the ground state $s_z = -1$ and the excited state $s_z = +1$.

By applying twice the negative feedback necessary for compensation it is possible to invert the effects of quantum fluctuations. The excited state $s_z = +1$ becomes stationary and diffusion is at a maximum in the ground state $s_z = -1$. By inverting the sign of the observed quantum fluctuation component the roles of the excited state and the ground state are exchanged. The excited state now absorbs quantum fluctuations, thus compensating the effects of dipole emission, while the ground state amplifies the fluctuations and thus shows twice the average diffusion.

The back action effect of homodyne detection without feedback, with feedback compensating the quantum fluctuations and with feedback inverting the quantum fluctuations is shown in figure 2.

5. CONCLUSIONS

The coherent and excited states of a two level atom may be stabilized by homodyne detection and negative feedback. If the dipole eigenstates are stabilized the back action of homodyne detection corresponds to weak measurements of the dipole component which emits light in phase with the local oscillator. The irreversible nature of spontaneous emission is thus associated with a weak projective measurement of the atomic dipole.

Homodyne detection avoids the discontinuous quantum jumps associated with photon detection. Therefore the stabilization of quantum states does not require short time pulses of high intensity. The feedback amplitudes needed for the stabilization of quantum states by homodyne detection and feedback are of the same order of magnitude as the observed vacuum fluctuations. It may thus be sufficient to couple the local oscillator field to the atomic system as a function of the intensity difference Δn by an optical nonlinearity.

REFERENCES

1. H.F. Hofmann, G. Mahler, and O. Hess, Phys. Rev. A **57**, 4877 (1998).
2. H.F. Hofmann, O.Hess, and G.Mahler, Optics Express **2**, 339 (1998).
3. H.M. Wiseman and G.J. Milburn, Phys. Rev. A **47**, 1652 (1993).
4. H. Carmichael in *Coherence and Quantum Optics VII*, ed. by J.Eberly, L.Mandel, and E.Wolf, Plenum 1996, p.177.

Quantum Coherence and Decoherence - ISQM - Tokyo '98
Y.A. Ono and K. Fujikawa (Editors)
© 1999 Elsevier Science B.V. All rights reserved.

Quantum Computation, Spectroscopy of Trapped Ions, and Schrödinger's Cat *

D.J. Wineland, C. Monroe, W.M. Itano, D. Kielpinski, B.E. King, C.J. Myatt, Q.A. Turchette, and C.S. Wood

National Institute of Standards and Technology (NIST), Boulder, CO, 80303

We summarize efforts at NIST to implement quantum computation using trapped ions, based on a scheme proposed by J.I. Cirac and P. Zoller (Innsbruck University). The use of quantum logic to create entangled states, which can maximize the quantum-limited signal-to-noise ratio in spectroscopy, is discussed.

1. INTRODUCTION

The invention by Peter Shor [1] of a quantum algorithm for factorizing large numbers has stimulated a host of theoretical and experimental investigations in the field of quantum information [2]. In the area of quantum computation, various schemes have been proposed to realize experimentally a model quantum computer [2]. In the ion storage group at NIST, we are trying to realize such a device based on the proposal by Cirac and Zoller [3].

In the Cirac-Zoller scheme, qubits are formed from two internal energy states, labeled $| \downarrow \rangle$ and $| \uparrow \rangle$, of trapped atomic ions. If the ions are laser cooled in the same trap, they form a crystalline array whose vibrations can be described in terms of normal modes. The ground and first excited states of a selected mode can also form a qubit. This qubit can serve as a data bus, since the normal modes are a *shared* property of the ions. An individual ion in the array can be coherently manipulated and coupled to the selected normal mode by using focused laser beams [3]. A universal logic operation, such as a controlled-not (CN) logic gate between ion qubit i and ion qubit j, is accomplished by (1) mapping the internal state of qubit i onto the selected motional qubit, (2) performing a CN between the motional qubit and qubit j, and (3) mapping the motional qubit state back onto qubit i. Each of these steps has been accomplished in the NIST experiments with a single ion [4,5]. We are currently devoting efforts to: (1) scaling quantum logic operations to two or more ions (Sec. 5), (2) applying quantum logic to study fundamental measurement problems on EPR and GHZ-like states, and (3) applying quantum logic to fundamentally improve the signal-to-noise ratio (SNR) in spectroscopy and atomic clocks. In this paper we briefly discuss this last application. We are aware of similar efforts to implement trapped-ion quantum logic at IBM, Almaden; Innsbruck University; Los Alamos National Laboratory; Max Planck Institute, Garching; and Oxford University.

---

*Contribution of NIST; not subject to U.S. copyright

2. ENTANGLED STATES FOR SPECTROSCOPY

A collection of atoms (neutral or charged) whose internal states are entangled in a specific way can improve the quantum-limited SNR in spectroscopy. This application of quantum logic to form entanglement is useful with a relatively small number of atoms and logic operations. For example, for high-accuracy, ion-based frequency standards [6], a relatively small number of trapped ions ($L \leq 100$) appears optimum due to various experimental constraints; with $L = 10 - 100$, a significant improvement in performance in atomic clocks could be expected. In contrast, factoring a number which cannot easily be factored on a classical computer would require considerably more ions and operations.

In spectroscopy experiments on L atoms, in which the observable is atomic population, we can view the problem in the following way using the spin-1/2 analog for two-level atoms. The total angular momentum of the system is given by $\mathbf{J} = \sum_{i=1}^{L} \mathbf{S}_i$, where $\mathbf{S}_i$ is the spin of the ith atom ($S_i = 1/2$). The task is to measure ω_0, the frequency of transitions between the $| \downarrow \rangle$ and $| \uparrow \rangle$ states, relative to the frequency ω_R of a reference oscillator. We first prepare an initial state for the spins. Typically, spectroscopy is performed by applying (classical) fields of frequency ω_R for a time T_R according to the method of separated fields by Ramsey [7]. We assume the same field amplitude is applied to all atoms (the phases might be different) and that the maximum value of T_R is fixed by experimental constraints (Sec. 3). After applying these fields, we measure the final state populations; for example, the number of atoms $L_\downarrow$ in the $| \downarrow \rangle$ state. In trapped-ion experiments, this has been accomplished through laser fluorescence detection with nearly 100% efficiency, which we assume here (see the discussion and references in Ref. [5]). In the spin-1/2 analog, measuring $L_\downarrow$ is equivalent to measuring the operator J_z, since $L_\downarrow = J\mathbb{I} - J_z$ where $\mathbb{I}$ is the identity operator. The SNR (for repeated measurements) is fundamentally limited by the quantum fluctuations in the number of atoms which are observed to be in the $| \downarrow \rangle$ state. These fluctuations can be called quantum projection noise [8]. Spectroscopy is typically performed on L initially unentangled atoms (for example, $\Psi(t = 0) = \prod_{i=1}^{L} | \downarrow \rangle_i$) which remain unentangled after the application of the Ramsey fields. For this case, the imprecision in a determination of the frequency of the transition is limited by projection noise to the "shot noise" limit $(\Delta\omega)_{meas} = 1/\sqrt{LT_R\tau}$ where $\tau \gg T_R$ is the total averaging time [8]. If the atoms can be prepared initially in particular entangled states, it is possible to achieve $(\Delta\omega)_{meas} < 1/\sqrt{LT_R\tau}$.

In optics, squeezed states have been shown to improve the SNR in interferometers beyond the shot noise limit [9,10]. In 1986, Yurke [11] showed how particular entangled states, if they could be created, could be used as inputs to Mach-Zehnder interferometers to approach the Heisenberg limit of SNR. In 1991, Kitegawa and Ueda [12] showed how the Coulomb interaction between electrons in the two arms of an electron interferometer might be used to improve the SNR beyond the shot-noise limit. Because of the formal identity of Mach-Zehnder interferometers and Ramsey spectroscopy [13], similar ideas might be applied to the spectroscopy problem. Reference [13] showed how a Jaynes-Cummings-type coupling between trapped-ion internal states and a normal mode could be used to improve the SNR in spectroscopy beyond the shot-noise limit. The scheme in Ref. [13] has the advantage that the appropriate states can be generated by acting on all the ions at once (thus not requiring focused laser beams), but has the disadvantage that these states are entangled with the motion, thereby requiring small motional decoherence. Reference [14] investigated the use of the generalized GHZ state, sometimes called the maximally entangled state, in spectroscopy. This state has the form

$$\psi_{max} = \frac{1}{\sqrt{2}}\Big(|\downarrow\rangle_1|\downarrow\rangle_2\cdots|\downarrow\rangle_L + e^{i\phi(t)}|\uparrow\rangle_1|\uparrow\rangle_2\cdots|\uparrow\rangle_L\Big), \tag{1}$$

where $\phi(t) = \phi_0 - L\omega_0 t$. After application of the Ramsey radiation, we measure the operator $\hat{O} \equiv \prod_{i=1}^{L} S_{zi}$. The resulting signal gives the exact Heisenberg limit of SNR $((\Delta\omega)_{meas} = 1/L\sqrt{T_R\tau}$ where $\tau \gg T_R)$ in spectroscopy (and interferometry).

The state ψ_{max} can be generated in a straightforward way by the application of L CN gates [3]. An alternative method was suggested in Ref. [14] and in Refs. [5] and [15] methods to generate ψ_{max} with a fixed number of steps (independent of L) are discussed. For all of these methods, the the motion is entangled with internal states during the creation of ψ_{max}, but is not entangled afterwards. Therefore, once ψ_{max} is created, the motion can lose coherence without affecting the entanglement of the internal states.

2.1. Schrödinger's Cat

As L becomes large and more macroscopic, states like ψ_{max} become more like Schrödinger's cat in that they represent coherent superpositions between widely separate regions of a large Hilbert space; for example, $|\uparrow\rangle_1|\uparrow\rangle_2\cdots|\uparrow\rangle_L \Longleftrightarrow$ "live cat;" $|\downarrow\rangle_1|\downarrow\rangle_2\cdots|\downarrow\rangle_L \Longleftrightarrow$ "dead cat". As has been emphasized in many discussions, as L becomes large the coherence between the two components of the cat becomes harder and harder to preserve [16]. This is apparent in Eq. (1) because if, for example, ω_0 fluctuates randomly, the two components of ψ_{max} will decohere relative to each other L times faster than for one ion (ψ_{max} for $L = 1$). Trapped ions are interesting because it may be possible to make L very large without significant decoherence. This is the same property that makes trapped ions interesting as possible frequency standards. For example, in Refs. [17] and [18], coherence times for individual ions ($L = 1$) exceeding 10 minutes were obtained.

3. Applicability

In the above, we have assumed that T_R is fixed, limited by some independent experimental factor. This assumption is warranted in many trapped-ion atomic clock experiments, where, for example, we want to limit the heating that takes place with laser cooling radiation absent. (During application of the Ramsey fields the cooling radiation must be removed to avoid perturbing the clock states.) Additionally, we may want to lock a local oscillator to the atomic reference in a practical time [6,19], thereby limiting T_R.

However, the use of entangled states may not be advantageous, given other conditions. For example, Huelga, *et al.* [20] assume that the ions are subject to a certain dephasing decoherence rate (decoherence time less than the total observation time). In this case, there is no advantage of using maximally entangled states over unentangled states. The reason is that since the maximally entangled state decoheres L times faster than the states of individual atoms, when we use the maximally entangled state, T_R must be reduced by a factor of L for optimum performance. Therefore, the gain from using the maximally entangled state is offset by the required reduced value of T_R.

Reference [5] discusses another case of practical interest. In atomic clocks, the frequency of an imperfect "local" oscillator, whose radiation drives the atomic transition, is controlled by the atom's absorption resonance. Depending on the spectrum of this oscillator's frequency fluctuations (when not controlled) the use of entangled states may or may not be beneficial.

4. Implementations

If we are able to create, with good fidelity, the state ψ_{max} (Eq. (1)), how do we perform spectroscopy? First, we note that ψ_{max} is the state we want *after* the first Ramsey $\pi/2$ pulse. Therefore, if we were to follow as closely as possible the Ramsey technique, we would take ψ_{max} and apply a $\pi/2$ pulse of radiation at frequency ω_0 to make the input state for the Ramsey radiation. However the first Ramsey $\pi/2$ pulse would only reverse this step; therefore, it is advantageous to take the creation of ψ_{max} as the first Ramsey $\pi/2$ pulse. The second Ramsey pulse (after time T_R) can be applied directly with radiation at frequency ω_R. The phase of this pulse (on each ion) must be fixed relative to the phases of the radiation used to create ψ_{max}. In general, the relation between these phases and ϕ_0 (Eq. (1)) will depend on the relative phases of the fields at the positions of each of the ions [5,21]. This will lead to a signal $S = \langle \tilde{O} \rangle \propto cos(L\Delta\omega T_R + \phi_f)$ where $\Delta\omega \equiv \omega_R - \omega_0$ and where ϕ_f depends on all of these phases.

We can extract ω_0 (relative to ω_R) by measuring $\langle \tilde{O} \rangle$ as a function of T_R, with $\Delta\omega$ fixed. This can be further simplified by measuring the signal for two values of T_R, $T_{R2} \gg T_{R1}$, where $\langle \tilde{O} \rangle \simeq 0$. Unfortunately, if the measured signal has a systematic bias as a function of T_R, an error in the determination of $\Delta\omega$ will result. This might happen, for example, if the ions heat up during application of the Ramsey radiation and a loss of signal occurs due to a reduced overlap between the ions and the laser used for fluorescence detection of the states. This problem could be overcome by measuring $\langle \tilde{O} \rangle$ for two values of ω_R, ω_{R1} and ω_{R2} such that $\omega_{R1} - \omega_0 \simeq -(\omega_{R2} - \omega_0)$ (determined by the above method), and two values of T_R, $T_{R1} \ll T_{R2}$. We then iterate the following steps: (1) we make $\langle \tilde{O}((\omega_{R1} - \omega_0)T_{R1}) \rangle \simeq \langle \tilde{O}((\omega_{R2} - \omega_0)T_{R1}) \rangle$ by adjusting the phase of the final $\pi/2$ pulse to make $\phi_f \rightarrow 0$. This will take a negligible amount of time since $T_{R1} \ll T_{R2}$. (2) We make $\langle \tilde{O}((\omega_{R1} - \omega_0)T_{R2}) \rangle \simeq \langle \tilde{O}((\omega_{R2} - \omega_0)T_{R2}) \rangle$ by adjusting ω_{R1} and/or ω_{R2} to force $\omega_{R1} - \omega_0 \rightarrow -(\omega_{R2} - \omega_0)$. This gives ω_0 relative to ω_R even if $\langle \tilde{O} \rangle$ has a systematic bias as a function of T_R.

An alternative solution is suggested by Huelga, *et al.* [20]. After T_R, instead of applying a $\pi/2$ pulse of radiation at frequency ω_R, we apply the time-reversed sequence of operations which created ψ_{max}. This has the advantage of cancelling out all of the CN phases that contribute to ϕ_0 and maps the signal ($\propto cos(L\Delta\omega T_R)$) onto a single ion (whereupon S_z is measured for that ion). This also reduces the problem of detection efficiency to one ion rather than L ions. The disadvantage of this technique is that for large values of T_R, the motional mode used for logic will, most likely, have to be recooled. This would require sympathetic cooling with the use of an ancillary ion which, to avoid the decohering effects of stray light scattering on the logic ions, might have to be another ion species [5].

A more serious limitation to the accurate determination of ω_0 is that, in practice, ψ_{max} will be realized only approximately and the state produced by the logic operations will also be composed of states other than the $|\uparrow\rangle_1|\uparrow\rangle_2\cdots|\uparrow\rangle_L$ and $|\downarrow\rangle_1|\downarrow\rangle_2\cdots|\downarrow\rangle_L$ states; these other states will have a definite phase relation to the $|\uparrow\rangle_1|\uparrow\rangle_2\cdots|\uparrow\rangle_L$ and $|\downarrow\rangle_1|\downarrow\rangle_2\cdots|\downarrow\rangle_L$ states. Consequently, in general, the signal produced with either implemenation will be of the form

$$S = \sum_{p=1}^{L} C_p cos(p\Delta\omega T_R + \xi_p). \tag{2}$$

To accurately determine $\Delta\omega$, it will be necessary to Fourier decompose S. Since this will take more measurements, the advantages of using entangled states will be reduced.

In spite of this, in some applications, it will be useful to determine changes in ω_0 with respect to some external influence. For example, we might want to detect changes in ω_0 caused by changes in an externally applied field. In this case, as long as $|C_p| \ll 1$, for all $p < L$, we derive the benefits of entangled states (assuming the decoherence time is longer than T_R/L) by measuring changes in S for a particular value of T_R.

5. Experiments

As usual, our enthusiasm for implementing these schemes far exceeds what is accomplished in the laboratory; nevertheless, some encouraging signs are apparent from recent experiments. In Ref. [22], all motional modes for two trapped ions have been cooled to the ground state. The non-center-of-mass modes are observed to be much less susceptible to heating, suggesting the use of these modes in quantum computation or quantum state engineering. In Ref. [21], we describe logic operations which enabled ψ_{max} for $L = 2$ to be generated with modest fidelity ($\simeq 0.7$). For small L, it is only necessary to *differentially* address individual ions to create ψ_{max} and for $L = 2$, general logic can be realized even if the laser beams cannot be focused exclusively on the individual ions [21]. For general logic on more than two ions, two avenues are being pursued. For modest numbers of ions in a trap, the Cirac-Zoller scheme of individual addressing with the use of focused laser beams is the most attractive. Current efforts are devoted to obtaining sufficiently strong focusing to achieve individual ion addressing in a relatively strong trap where normal mode frequencies are relatively high ($\simeq 10$ MHz) in order to maximize operation speed. Alternatively, general logic on many ions could be accomplished by incorporating accumulators [5], and using differential addressing on two ions at a time. This idea might be realized by scaling up a version of a linear ion trap made with lithographically deposited electrodes as we have recently demonstrated [16,23]. Concurrently, efforts are being devoted to the investigation (and hopefully, elimination) of mode heating [5] for different electrode surfaces and dimensions.

6. Acknowledgments

We gratefully acknowledge the support of the U.S. National Security Agency, U.S. Army Research Office, and the U.S. Office of Naval Research. We thank J. Bollinger, R. Blatt, D. Sullivan, and M. Young for helpful comments on the manuscript.

REFERENCES

1. P.W. Shor, *Proc. 35th Ann. Symp. Foundations of Computer Science*, S. Goldwasser ed., IEEE Computer Society Press, New York, 1994, p. 124.
2. See, for example Proc. Royal Soc., Math., Phys. and Eng. Sci., **454** (1969) (1998); Fortschritte der Physik **46** (4-5) (1998).
3. J.I. Cirac and P. Zoller, Phys. Rev. Lett. **74**, 4091 (1995).
4. C. Monroe, D.M. Meekhof, B.E. King, W.M. Itano, and D.J. Wineland, Phys. Rev. Lett. **75**, 4714 (1995).
5. D.J. Wineland, C.R. Monroe, W.M. Itano, D. Leibfried, B.E. King, and D.M. Meekhof, NIST J. Research **103** (3), 259 (1998). (available at http://nvl.nist.gov/pub/nistpubs/jres/jres.htm)
6. D.J. Berkeland, J.D. Miller, J.C. Bergquist, W.M. Itano, and D.J. Wineland, Phys. Rev. Lett. **80**, 2089 (1998).
7. N.F. Ramsey, *Molecular Beams* (Oxford Univ. Press, London, 1963).
8. W. M. Itano, J. C. Bergquist, J. J. Bollinger, J. M. Gilligan, D. J. Heinzen, F. L. Moore, M. G. Raizen, and D. J. Wineland, Phys. Rev. A**47**, 3554 (1993).
9. C.M. Caves, Phys. Rev. D**23**, 1693 (1981).
10. M. Xiao, L. -A. Wu, and H. J. Kimble, Phys. Rev. Lett. **59**, 278 (1987).
11. B. Yurke, Phys. Rev. Lett. **56**, 1515 (1986); B. Yurke, S.L. McCall, and J.R. Klauder, Phys. Rev. A**33**, 4033 (1986).
12. M. Kitagawa and M. Ueda, Phys. Rev. Lett. **67**, 1852 (1991); M. Kitagawa and M. Ueda, Phys. Rev. A**47**, 5138 (1993).
13. D.J. Wineland, J. J. Bollinger, W. M. Itano, F. L. Moore, and D. J. Heinzen, Phys. Rev. A**46**, R6797 (1992); D.J. Wineland, J. J. Bollinger, W. M. Itano, and D. J. Heinzen, Phys. Rev. A**50**, 67 (1994).
14. J.J. Bollinger, D. J. Wineland, W. M. Itano, and D. J. Heinzen, Phys. Rev. A**54**, R4649 (1996).
15. J. Steinbach and C.C. Gerry, quant-ph/9806091.
16. See, for example, S. Haroche, Physics Today **51** (7), 36 (1998).
17. J.J. Bollinger, D. J. Heinzen, W. M. Itano, S. L. Gilbert, and D. J. Wineland, IEEE Trans. on Instrum. and Measurement **40**, 126 (1991).
18. P.T.H. Fisk, M.J. Sellars, M.A. Lawn, C. Coles, A.G. Mann, and D.G. Blair, IEEE Trans. Instrum. Meas. **44**, 113 (1995).
19. J.J. Bollinger, J.D. Prestage, W.M. Itano, and D.J. Wineland, Phys. Rev. Lett. **54**, 1000 (1985).
20. S.F. Huelga, C. Macchiavello, T. Pellizzari, A.K. Ekert, M.B. Plenio, and J.I. Cirac, Phys. Rev. Lett. **79**, 3865 (1997).
21. Q.A. Turchette, C.S. Wood, B.E. King, C.J. Myatt, D. Leibfried, W.M. Itano, C. Monroe, and D.J. Wineland, submitted (quant-ph/9806012).
22. B.E. King, C.S. Wood, C.J. Myatt, Q.A. Turchette, D. Leibfried, W.M. Itano, C. Monroe, and D.J. Wineland, Phys. Rev. Lett. **81**, 1525 (1998).
23. C. Myatt, *et al.*, NIST Ion Storage Group, unpublished.

Quantum Coherence and Decoherence - ISQM - Tokyo '98
Y.A. Ono and K. Fujikawa (Editors)
© 1999 Elsevier Science B.V. All rights reserved.

Atom Interferometry

Fujio Shimizu

Institute for Laser Science and CREST, University of Electrocommunications
1-5-1 Chofugaoka, Chofu, Tokyo 182-8585, Japan

Two types of atom-interferometry with laser-cooled neon atomic beam in the $1s_3$ metastable state are experimentally demonstrated: (1) The two atom temporal correlation experiment that measures the energy spread of the atomic source. (2) The holographic manipulation of the atomic beam, which is the most general case of the interferometric manipulation of a wave.

1. INTRODUCTION

Recent development on the laser cooling of neutral atoms has brought new prospects on studying interferometric phenomena of atomic de Broglie wave. The most spectacular feature of laser-cooled atomic gas is that the kinetic energy spread of the atomic gas is reduced by the factor of as much as 10^9 bringing its frequency less than 1MHz. This enabled us to study temporal interferomtric phenomena with commonly available detectors and electronics. The long de Broglie wave of cold atoms is another feature that opened possibility on spatial interferometry. It enables us to diffract atomic wave at will by artificially made structure. We describe in this paper two atom-interferometric experiments with laser-cooled neon atoms that characterize those features.

2. ATOMIC SOURCE

The flux of an atomic beam is limited by collisions inside the beam and is many orders of magnitude smaller than that of a laser beam with moderate intensity. Therefore, it is important to choose proper atoms that have a small collisional cross section and/or have a high detection efficiency. We use metastable neon atoms in the $1s_3(J = 0)$ state that can be detected with a high quantum efficiency by using a standard ion detector such as a micro-channel plate detector (MCP). Figure 1 shows the schematic configuration of the cold atomic beam source we used in the experiment. Neon has four fine structure states in the first excited state multiplex. Two of them are metastable states. The lowest excited state, $1s_5$, has the angular momentum $J = 2$ and is used to cool the atom by using 640 nm laser light. The atom is then optically pumped into the $1s_3$ state with $J = 0$, producing a downward falling atomic beam.

The $1s_5$ metastable neon atoms were generated (at the left end of the figure) by a weak discharge, and the atoms were extracted through a hole of the anode. The atomic

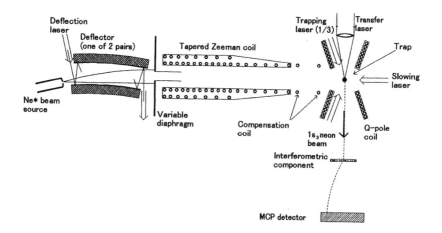

Fig. 1   Schematic diagram of the laser-cooled metastable neon source.

beam passed between two curved mirrors that have a common center of radius. The laser resonant to the cooling transition passed in a zigzag form between the mirrors and collimated the atomic beam perpendicular to the wavefront of the laser beam. The atoms were led into the Zeeman slowing stage and were decelerated by a counter propagating laser beam. At the exit of the Zeeman slowing stage a magneto-optical trap (MOT) was placed. To make the smallest possible atomic cloud the magnetic field gradient of the MOT was set at 300 G/cm. To feed atoms in the Zeeman stage efficiently into the MOT the field in the connecting region was finely adjusted with an auxiliary coil. The diameter of the trapped atomic cloud was typically $50 - 100\,\mu$m. A 598-nm laser that was focused into the trap released the $1s_5$ atoms by optically pumping them into the $1s_3$ state. The $1s_3$ atoms fell vertically and were detected by a micro-channel plate detector (MCP). The flux of the $1s_3$ atoms within the diffraction limited angle was between 10^2 to 10^3 s^{-1}, which depended on the operation condition.

3. TWO-ATOM TEMPORAL CORRELATION SPECTRUM

Quantum statistical nature of a particle beam can be studied only through a multi-particle interferometric phenomenon. The first such experiment was performed with photons by Brown and Twiss,[1] in which they measured the spectral or spatial spread of an optical source by two-photon correlation spectrum. Since the advent of lasers and nonlinear optics, various kind of multi-photon interferometric phenomena have been studied. A similar experiment on particles with mass has never been performed before, because it was impossible to obtain the phase space density that was sufficient to observe multi-particle interferometric effect. Recently we succeeded for the first time in observing the Brown-Twiss type two-atom temporal interferometry by using the $1s_3$ neon beam described in the previous section.[2] Here we used a laser-cooled atomic beam which has the phase space density much higher than a conventional beam due to the narrow energy spread.

The basic scheme of measuring the two-atom correlation function $\Gamma(\tau)$ is straightforward. We intercept the atomic beam with a detector and record the arriving time of all atoms. The probability of finding an another atom at $t + \tau$ when an atom is found at t is calculated as a function of τ from the record. The function should show a peak around $\tau = 0$ that has the width of $\hbar/\Delta\mathcal{E}$, where $\Delta\mathcal{E}$ is the energy spread of the atomic source. When the phase space density is much less than unity, the temporal range of interest is much smaller than the average interval of atom detection. In this case one may simply measure the distribution of arriving time interval of two successive atoms.

In the real experiment, the detector has nonlinear transient responses. When two atoms arrive in a short interval, the quantum efficiency of detecting the second atom is higher than that of the first one. This produces an artificial peak around $\tau = 0$ that interferes with the real signal. To avoid producing the artificial peak, we divided the cross section of the atomic beam into four quadratures and detected atoms hitting each quadrature with an independent detector. The second atomic pulse from a detector that follows within $4\mu s$ after the first pulse was eliminated, and then, pulses from four detectors are summed and fed into the interval counter as the start and stop pulses. When the velocity distribution of the atom at the source is in the Gaussian form,

$$W(v) = \frac{1}{\sqrt{\pi}v_s} \exp\left(-\frac{v^2}{v_s^2}\right),$$

the correlation function should have the form

$$\Gamma(\tau) = 1 + \left\{1 + (\Delta\omega\tau)^2\right\}^{-1/2} \Gamma(\infty), \tag{1}$$

where $\Delta\omega = mv_s^2/(2\hbar)$.

Figure 2(a) shows the correlation spectrum that was obtained by a spatially diffraction limited beam, i.e., coherent atomic beam. Although quantitative comparison with Eq.(1) is not possible, it shows a peak around $\tau = 0$ with the width of approximately 0.15 μs as is expected from the velocity spread 25 cm/s of atoms at the source. To confirm the correlation peak we repeated the same measurement with the atomic beam that contained

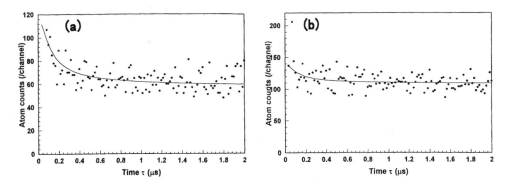

Fig. 2 Two-atom correlation spectrum: (a) the time interval distribution for a coherent atomic beam and (b) the time interval distribution for an incoherent atomic beam.

many spatial modes, i.e., incoherent atomic beam. The result is shown in the spectrum of Fig. 2(b). The peak was much smaller than the case of the diffraction limited beam.

4. GRAY-SCALE ATOM HOLOGRAPHY

Holography is a technique to manipulate wave through thin film called hologram. The hologram diffracts a spatially coherent wave and generates an arbitrary wavefront. The diffraction angle of an atomic wave with the wavelength of λ_{db} that is emitted from a source with the diameter of a is

$$\theta_{diff} \approx \lambda_{db}/d. \tag{2}$$

If the hologram for the atomic wave that has an arbitrary complex transmission function is placed over the diffraction limited cross-section of the beam, one can draw an arbitrary pattern of atoms on screen. The minimum size, d, of the structure of the film required to draw the atomic picture is that the wave diffracted from a hole with the size of d cover the screen,

$$d \approx \lambda_{db}\frac{L}{a_s}, \tag{3}$$

where L is the distance between the film and screen and a_s is the size of the screen. The present micro-fabrication technique allows us to make complex structures of the size $d \sim 100$ nm. For atoms in the magneto-optical trap, $\lambda_{db} \sim 10$ nm, and the atoms are diffracted by a large angle $a_s/L \sim 1$. We have demonstrated the generation of black-and-white pattern of atoms by using a computer-generated binary hologram that was made of SiN$_4$ film with holes.[3, 4] We divided both hologram and screen into 1024×1024 and calculated the transmission amplitude of the hologram $f(j_x, j_y)$ from the amplitude of the atomic wave on the screen $F(m_x, m_y)$ by

$$f(j_x, j_y) = A'e^{-i\frac{md^2}{2\hbar}\left(\frac{1}{t_1}+\frac{1}{t_2}\right)(j_x^2+j_y^2)}$$

$$\sum_{m_x,m_y=1}^{N}\left\{F(m_x, m_y)e^{i\phi_{rand}}e^{-i\frac{mD^2}{2\hbar t_2}(m_x^2+m_y^2)}\right\}e^{i\frac{mdD}{\hbar t_2}(m_x j_x+m_y j_y)}, \tag{4}$$

where t_1 is the transit time of an atom between the source and the hologram and t_2 is the transit time between the hologram and the screen. The d and D are the size of the cell of the hologram and of the screen, respectively. The number of divisions, N, was 1024. We made binary approximation to determine the position of the open cells. The ratio of the open cells with respect to the total cells was approximately 8 %. We accumulated $2-5 \times 10^5$ atoms on the MCP to draw characters, where the resolving power of the entire drawing was approximately 150.

A problem we might encounter when extending this technique to a gray-scale picture is the low flux of the coherent atomic beam. A hologram with $N \times N$ cells has theoretically $N \times N$ resolving power on the screen. Since an atom is sufficient to mark a black point on the screen, $N^2/2$ atoms can draw a black and white pattern of $N \times N$ resolving power. To draw a picture with the gray scale of M the number increases to $MN^2/2$. The number of atoms that hit the detector increases proportional to the area of the hologram. However, if the position and the size of the screen are selected, the size of the cell d is automatically determined by Eq.(3). Therefore, a larger hologram means a larger N. This does not necessarily mean a higher resolving power. If the configuration is optimized, the hologram and the area of the diffraction limited area determined from Eq.(2) should be equal. If the hologram area is further increased, it can be illuminated only partially coherently. Nevertheless, it is worth increasing the area. In this case, one may divide the hologram into smaller sections, and calculate the transmission function of the image independently. The resolving power is determined by the number of cell in a section, however, the calculation time can be reduced considerably.

We made a hologram with $N = 4096$ and divided it into 4×4 sections and calculated the transmission function $f(j_x, j_y)$ independently. To increase further the atomic flux the laser beam at around 630 nm was superimposed on the optical pumping beam forming optical dipole potential for the falling $1s_3$ atoms. The optical dipole potential cooled the transverse momentum of the falling atoms while they passed through the expanding region of the 630 nm beam. This increased the number of atoms hitting the hologram by more than a factor 10. Figure 3 shows the holographic construction obtained with this set up. Figure 3(a) is the original picture, Fig. 3(b), a part of the pattern of the hologram, Fig. 3(c), the atomic pattern that was obtained with the hologarm of Fig. 3(b) after 30 min. of accumulation time.

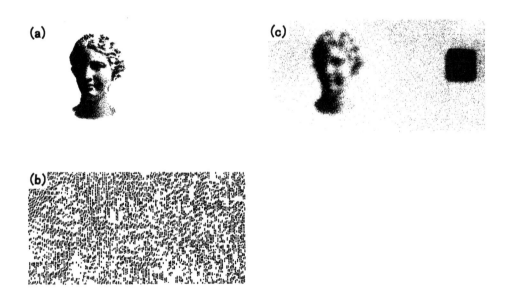

Fig. 3 Holographic construction: (a) original picture, (b) hologram pattern, and (c) obtained atomic pattern.

ACKNOWLEDGMENTS

The correlation experiment was performed in collaboration with M. Yasuda. In the atom holography experiment, the calculation codes of the transmission function were made by M. Morinaga, and the holograms were fabricated by J. Fujita. The data collection was done in collaboration with T. Kishimoto and T. Mitake.

REFERENCES

1. R. H. Brown and R. Q. Twiss, *Proc. Roy. Soc. London* A **242**(1957) 300; **243** (1957) 291.
2. M. Yasuda and F. Shimizu, *Phys. Rev. Lett.* **77** (1996) 3090.
3. J. Fujita, M. Morinaga, T. Kishimoto, M. Yasuda, S. Matsui, and F. Shimizu, *Nature* **380** (1996) 691.
4. M. Morinaga, M. Yasuda, T. Kishimoto, F. Shimizu, J. Fujita, and S. Matsui, *Phys. Rev. Lett.* **77** (1996) 802.

Quantum Coherence and Decoherence - ISQM - Tokyo '98
Y.A. Ono and K. Fujikawa (Editors)
© 1999 Elsevier Science B.V. All rights reserved.

Multiple atomic beam interference in the Ramsey-Bordé atomic interferometer with multiple traveling laser beams

Atsuo Morinaga and Takatoshi Aoki
Department of Physics, Faculty of Science and Technology,
Science University of Tokyo,
2641 Yamazaki, Noda-shi, Chiba 278-8510, Japan

The fringe pattern of the Ramsey-Bordé atomic interferometer with multiple laser beams is discussed. The calculation for a six-zone excitation shows that the fringe pattern strongly depends on the pulse area of the excitation laser beam. In the case of weak excitation, a significant line-narrowing effect occurs on increasing the number of laser beams.

1. INTRODUCTION

Bordé-type atomic interferometers [1], which use coherent interactions of atoms and light as a beamsplitter or a recombiner have several advantage in precision experiments, topological phase investigations and the fundamental test of quantum mechanics. In fact, the Aharanov-Casher effect has been examined within an uncertainty of 2 % using the Ramsey-Bordé atomic interferometer [2].

We have developed atom interferometers with three or four copropagating traveling laser beams and obtained interference fringes with visibilities of 25% or 20%, respectively [3,4]. In those studies we investigated the interference of two partial atomic waves. However, multiple atomic beam interference is interesting, since it produces a very narrow fringe width, similar to a sharply peaked Airy-function for multiple optical beam intereference. It results in an increased relative phase sensitivity between the interfering partial waves. Recently, basic experiments were reported by a few authors using different kinds of atomic interferometers [5,6].

Previously, one of the present authors discussed the optical Ramsey excitation with many separated traveling waves [7]. In the case where n waves propagate unidirectionally (a total of 2n zones), Ramsey fringes with a period of $1/(2mT)$ (T; transit time, $m=1, \cdots n-1$) are generated. In the previous study, the number of sets which generate the Ramsey fringes was counted. It was observed that the fringe pattern was variable depending on the excitation power. The experimental results up to ten zones indicated only a small narrowing effect. Actually, Helmcke observed the Ramsey resonance of the third overtone by an eight-zone excitation at a moderate excitation power [8]. On the other hand, Ertmer's group recently showed a clear narrow spectrum of the Ramsey excitation of a cold Mg atom with more than one hundred pulses

[9]. Therefore, we reexamined the optical Ramsey resonance equations with multiple laser beams, taking into consideration the phase and amplitudes in each Ramsey period. In this paper, we describe the calculation for a six-zone Ramsey interference and present an estimate for multiple-zone interference.

2. SIX-ZONE EXCITATION

Figure 1 shows the interaction geometry of the Ramsey-Bordé interferometer with six laser beams. In the previous paper, we reported 16 atomic trajectories in the case of the six-zone interference and counted the number of sets which generated oscillation periods, regardless of the amplitude and phase [7]. The number of fundamental period of $1/(2T)$ is 32 and that of overtone of $1/(4T)$ is 2. Using these wave functions, we calculate the population probability in the excited state at 5 output ports. In the central port P_3, the six atomic beams interfere and two fringe periods are generated.

For an easier calculation of the interference fringe pattern, we assume that the transit time width is far greater than the Ramsey period and we consider frequencies at near the resonance frequency. In this calculation, we do not integrate the probability over the velocity distribution. However, we assume that the wave functions do not interfere during the five ports, since the transverse velocity distribution smear the fringes after integration over velocity in the case of the thermal atomic beam. All laser beams are assumed to have the same amplitude and the same phase. Hence the parameters which

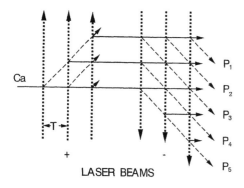

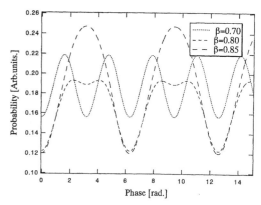

Figure 1. Interaction geometry of six-zone excitation. Dashed line, excited state; solid line, ground state.

Figure 2. Interference fringes of six-zone excitation as a function of excitation.

describe the transition between the states [10] are

$$A^+ = (D^+)^* = (A^-)^* = D^-, \quad |A^+| = \alpha$$
$$B^+ = C^+ = B^- = C^- = i\beta \quad . \tag{1}$$

The probability of the excited state is given by summing up the probabilities of the five ports, as

$$bb^* = \alpha^2\beta^2[3\alpha^8 - 6\alpha^6\beta^2 + 16\alpha^4\beta^4 - 6\alpha^2\beta^6 + 5\beta^8$$
$$-8\alpha^2\beta^2(\alpha^4 - 2\alpha^2\beta^2 + \beta^4)\cos 2\pi\Delta(2T)$$
$$-2\alpha^6\beta^2\cos 2\pi\Delta(4T)] \quad , \tag{2}$$

where Δ is the detuning frequency. This result shows that 32 sets having the fundamental period are composed of different phase and amplitude. We calculated the probability using eq. (2) for several values of α and β. The results are shown in Fig. 2 and summarized as follows. (1) In the case of weak excitation with $\beta < 0.3$, the amplitude ratio of the fundamental period to the overtone is 4. Therefore the pattern is dominated by fundamental fringes, although fringe width becomes narrower. (2) In $\pi/2$ pulse excitation ($\alpha = \beta$ = 0.7), the fundamental period disappears completely, and the fringes of the overtone appear. (3) In the case of $3\alpha^4 - 8\alpha^2\beta^2 + 4\beta^4 = 0$, i.e. $\beta = 0.58$ or 0.78, the amplitudes of both fringes become equal and a line narrowing effect could be discerned. (4) In $\beta > 0.85$, the fundamental period is dominant. Thus, it is ascertained that the interference pattern depends on the excitation power.

3. MULTIPLE-ZONE EXCITATION

We also derived the transition probability for the eight-zone Ramsey resonance. The fringe periods are composed of $1/(2T)$, $1/(4T)$ and $1/(6T)$, and their fringe periods appear by turns, depending on the excitation power. Regrettably, we could not formulate a general equation for the multiple-zone Ramsey interference. However, we can find the condition for obtaining the narrow interference fringes for the case where the excitation power is relatively small ($\beta < 0.3$). The interference fringes are given by

$$I = 2\sum_{m=1}^{n-1}(n-m)^2\cos 2\pi\Delta(2mT) \tag{3}$$

The pattern resembles the Airy function, because the relative ratio of $1/(2T)$ to $1/(4T)$ is nearly equal to 1 for large n. Figure 3 shows the relative interference signal for various n, which is the number of waves propagating in one direction. For $n = 5$, the fringe width is almost half of the fundamental fringe width.

118

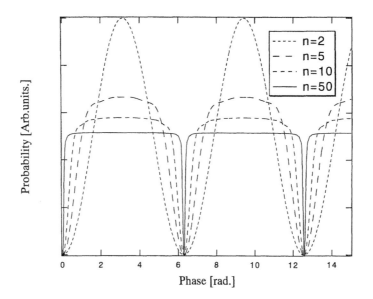

Figure 3. Interference fringes of 2n-zone excitation in the case of weak excitation.

4. SUMMARY

We derived the probability of the excited state for the Ramsey-Bordé interferometer with six traveling laser beams. The interference pattern varies depending on the excitation power. In the case of weak excitation, the ratio of the amplitudes of the fundamental fringes to the overtone approaches a value of one, with increasing number of beams. Consequently, the interference fringe pattern resembles the Airy function, and a strong narrowing effect is expected.

This research was partly supported by the MATSUO FOUNDATION.

REFERENCES

1. Ch. J. Bordé, Phys. Lett A 140 (1989) 10.
2. K. Zeiske et. al., appl. Phys. B 60 (1995) 205.
3. Y. Omi and A. Morinaga, to be published in Appl. Phys. B (1998)
4. S. Yanagimachi, Y. Omi and A. Morinaga, Phys. Rev. A 57 (1998) 3803.
5. M. Weitz, T. Heupel and T. W. Hänsch, Phys. Rev. Lett. 77 (1997) 2356.
6. H. Hinderthür et. al. Phys. Rev. A 56 (1997) 2085.
7. A. Morinaga, Phys. Rev. A 45 (1992) 8019.
8. J. Helmcke, private communication.
9. H. Hinderthür et al: Digest of IQEC' 98 (San Francisco, 1998) p.157.
10. Ch. J. Bordé et. al., Phys. Rev. A 30 (1984) 1836.

Quantum Coherence and Decoherence - ISQM - Tokyo '98
Y.A. Ono and K. Fujikawa (Editors)
© 1999 Elsevier Science B.V. All rights reserved.

Sympathetically Laser-Cooled Mass Spectroscopy using Ba⁺ ions

Izumi Waki, Takashi Baba, and Dongbing Wang

Advanced Research Laboratory, Hitachi, Ltd.
Hatoyama, Saitama 350-0395, Japan

We report a sensitive mass spectroscopy of Xe using laser-cooled ^{138}Ba$^+$ ions. We used fluorescence from laser-cooled ^{138}Ba$^+$ ions as a probe to detect and mass-analyze $^{128\sim136}$Xe$^+$ isotope ions which are trapped in a linear radio-frequency quadrupole trap. A total of 50 trapped Xe$^+$ ions were sufficient to produce the isotope mass spectrum at a signal-to-noise level of better than 5.

1. INTRODUCTION

In the last ISQM conference, we reported the first successful mass analysis of molecules using laser-cooling [1, 2]. In this technique, which we call Sympathetically Laser-Cooled Fluorescence Mass Spectroscopy (or in short, Laser-Cooled MS), we used fluorescence from laser-cooled ^{24}Mg$^+$ ions as a probe to detect molecular ions, such as H_3O^+ and COH$^+$, in a linear RF quadrupole trap [Figure 1]. The molecular ions were cooled indirectly (or, sympathetically [3-5]) through Coulomb interaction with the laser-cooled ^{24}Mg$^+$ ions. With a

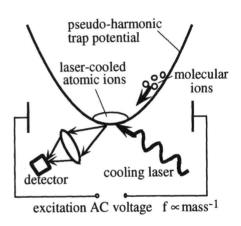

Figure 1. Sympathetically Laser-Cooled Fluorescence Mass Spectroscopy [1,2].

supplementary, weak AC voltage, we excited the secular oscillation of the molecular ions, which disturbed the motion of ^{24}Mg$^+$ through Coulomb collisions. The disturbance was detected by observing the intensity change of ^{24}Mg$^+$ fluorescence. By scanning the frequency of the supplementary AC voltage, we obtained the mass spectrum of a few tens of molecular ions.

Since laser-cooling can detect single atomic ions, we expect that Laser-Cooled MS can ultimately detect cold single molecular ions in situ. The ability to mass-analyze single molecular ions implies that it could enhance the sensitivity of RF ion trap mass spectrometers, which are widely used in organic chemical analysis.

Many organic molecules of analytical interest lie in the mass region above 100. To

120

detect heavier molecules, it is preferable to use heavier laser-cooled ions as the probe, since the efficiency of energy transfer in Coulomb collisions is highest between particles of equal masses [2]. Therefore, extension of Laser-Cooled MS to heavier laser-cooled probe ions is important from analytical point of view.

In this paper, we report Laser-Cooled MS using laser-cooling of $^{138}Ba^+$ ions. In addition to the extension of mass range, use of $^{138}Ba^+$ ions as the probe would facilitate the experimental apparatus for Laser-Cooled MS, because semiconductor diode lasers are commercially available for laser-cooling of Ba^+ ions.

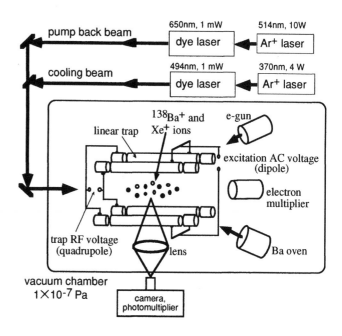

Figure 2. Experimental setup for Laser-Cooled MS using Ba^+ ions

2. PROCEDURE AND RESULTS

The trap was composed of 50 mm long cylindrical electrodes. The facing quadrupole electrodes were separated by 6 mm [Figure 2]. The trap was placed in an UHV chamber, whose base pressure was 2×10^{-7} Pa. The trap quadrupole RF had a frequency of 1.73 MHz and an amplitude of 130 V_{pp} with no static component. Ba and Xe atoms were ionized by electron bombardment. The ions were confined axially by a static voltage of 2 V applied to the end

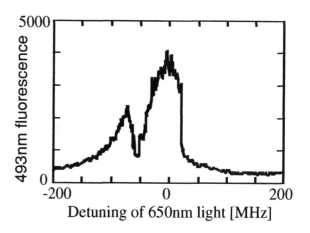

Figure 3. Laser-induced fluorescence excitation spectra of $^{138}Ba^+$.

Figure 4. $^{138}Ba^+$ Wigner crystal. Mean distance between the ions is 20 μm.

electrodes. Secular oscillations were excited by a dipole oscillating voltage with ～ 1mV amplitude.

An electron multiplier was placed on the quadrupole axis, so that the trapped ion number can be calibrated.

$^{138}Ba^+$ was doppler-cooled by exciting $6^2S_{1/2}-6^2P_{1/2}$ transition with 50 μW of 493 nm (blue-green) light. Metastable $5^2D_{3/2}$ state was pumped back to the $6^2P_{1/2}$ state by 200 μW of 650 nm (red) light under magnetic field of 3 Gauss. Both wavelengths were produced by Coherent 899-21 dye lasers (Coumarin 101 dye for the blue-green light, and Kiton Red dye for the red light). The cooling beam entered the trap along the axis with a beam size of 0.2 mm. The 493 nm fluorescence from $^{138}Ba^+$ within a 1 mm region near the center of the trap was collected with a lens for detection.

Figure 3 shows the blue-green fluorescence intensity of the laser-cooled $^{138}Ba^+$ ions as the frequency detuning of the red light is scanned from -200 MHz to +200 MHz. The dip at -50 MHz detuning is produced by the dark state coherence between blue-green and red transitions.

The Ba$^+$ ions phase-changed to a ionic crystalline state near -20 MHz detuning. The photo-image of the ionic Wigner crystal is shown in Figure 4. This image gives us an absolute number of trapped ions. The detection efficiency of the electron multiplier

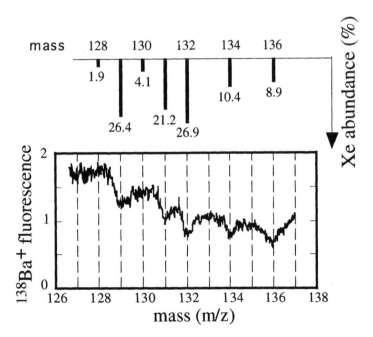

Figure 5. Laser-Cooled Fluorescence Mass Spectrum of Xe using laser-cooled Ba$^+$. The natural abundances of Xe$^+$ isotopes are shown as well. The number of total Xe$^+$ ions is about 50.

was determined using the crystalline image, so that we could calibrate the absolute number of trapped ions other than Ba^+ as well.

Xe gas was introduced into the vacuum chamber through a vacuum leak valve. Figure 5 shows the laser-cooled mass spectrum of Xe isotope ions at a partial pressure of 2×10^{-7} Pa. In this spectrum, the mass was scanned from m=137 to m=126.5. The distribution of the observed line intensities is consistent with the natural abundances of the Xe isotopes, which are also shown in Figure 5. Therefore, we conclude that we have successfully obtained sympathetically laser-cooled mass spectrum of Xe^+ ions using $^{138}Ba^+$ ions as the probe.

3. DISCUSSION

The number of total trapped Xe^+ ions is about 50 in the mass spectrum of Figure 5, which is calibrated using counting data of the electron multiplier and the crystalline image of Ba^+ ions. Therefore, the peak of mass 134 corresponds to about 5 $^{134}Xe^+$ isotope ions, although the number should fluctuate substantially due to Poisson statistics. From Figure 5, we can safely deduce that Laser-Cooled MS can detect a few ions with S/N of better than 5.

The baseline of Figure 5 increases as the mass is scanned from m=137 to m=126.5. We speculate that this increase is induced by partial loss of Xe^+ ions as the analyzing field is scanned. When the analyzing frequency is in resonance with the secular motion of the Xe^+ ions, the Xe^+ ions acquire kinetic energy. The acquired energy must be dissipated through sympathetic cooling with the $^{138}Ba^+$ ions. If the heating rate overcomes the cooling rate, the ions can acquire enough energy to escape from the trap. When the number of trapped Xe^+ ions decrease, the temperature of the $^{138}Ba^+$ ions would decrease due to reduced RF heating. Decrease of the temperature would increase the fluorescence intensity.

The ability to mass-analyze ions in-situ is one advantage of Laser-Cooled MS that was demonstrated with laser-cooled $^{24}Mg^+$ [2]. To retain this feature for Ba^+ Laser-Cooled MS as well, we plan to avoid the loss of ions during the mass scan by finding a right balance condition for cooling and heating.

In summary, we demonstrated Laser-Cooled MS using $^{138}Ba^+$ ions as the probe ions. The mass range of Laser-Cooled MS was extended to 136, with a sensitivity of a few ions. This ability would be valuable for ultra-sensitive mass analysis of organic molecules.

REFERENCES

1. T. Baba and I. Waki, "Quantum Coherence and Decoherence", eds. K. Fujikawa and Y. A. Ono, (Elsevier Science, Amsterdam, 1996) 53.
2. T. Baba and I. Waki, Jpn. J. Appl. Phys. 35 (1996) L1134
3. D. J. Wineland, R. E Drullinger, F. L. and Walls: Phys. Rev. Lett., 40 (1978) 1639.
4. R. E. Drullinger, D. J. Wineland, J. C. Bergquist: Appl. Phys., 22 (1980) 365.
5. D. J. Larson, J. C. Bergquist, J. J. Bollinger, W. M. Itano and D. J. Wineland, Phys. Rev. Lett., 57 (1986) 70.

Quantum Coherence and Decoherence - ISQM - Tokyo '98
Y.A. Ono and K. Fujikawa (Editors)
© 1999 Elsevier Science B.V. All rights reserved.

Measurements of Relative Phase and Quantum Beat Note between Bose-Einstein Condensates

D. S. Hall, M. R. Matthews, C. E. Wieman, and E. A. Cornell*

JILA, National Institute of Standards and Technology
and Department of Physics, University of Colorado, Boulder, Colorado 80309-0440 USA

We use a separated-oscillatory-field condensate interferometer to measure the time-evolution of the relative phase in a two-component Bose-Einstein condensate. The two components are created with a particular relative phase in the $|F = 2, m_f = 1\rangle$ and $|F = 1, m_f = -1\rangle$ states of ^{87}Rb by applying a coupling pulse to a condensate in the $|1, -1\rangle$ state. The components subsequently separate spatially due to their mutual repulsion, damp their relative center-of-mass motion, and come into equilibrium after ~ 45 ms. Meanwhile, the relative phase accumulates at ~ 6.8 GHz and is measured with a second coupling pulse. We find that the coherence initially established between the components is preserved despite the motional damping present in this entangled system.

1. INTRODUCTION

Since its realization in the alkali gases [1–3], Bose-Einstein condensation (BEC) has permitted a number of fascinating glimpses into the macroscopic quantum world. Among the more interesting properties of these condensates is their macroscopic coherence, which has been demonstrated to lead to matter-wave interference [4], reduction in inelastic loss rates [5], and is expected soon to be observed in vortices and analogues of the Josephson junction. Although the absolute phase of a condensate cannot be directly measured, the relative phase between two condensates can. For this reason double condensate systems are especially interesting.

The first double-condensate system was created at JILA in the $|F = 2, m_f = 2\rangle$ and $|F = 1, m_f = -1\rangle$ hyperfine states of ^{87}Rb [6] (Fig. 1). The spatial interference between two condensates of the same atomic hyperfine state (in ^{23}Na) was subsequently demonstrated in a beautiful experiment at MIT [4]. At JILA, we have turned to time-domain interferometry involving the $|F = 1, m_f = -1\rangle$ and $|F = 2, m_f = 1\rangle$ states of Rb [9,10] (henceforth denoted $|1\rangle$ and $|2\rangle$, respectively). The MIT group is also now working with multiple-species condensates in optical traps [7,8].

---

*Quantum Physics Division, National Institute of Standards and Technology.

2. DOUBLE CONDENSATE SYSTEM

Our experiments begin by producing a single condensate in the $|1\rangle$ state. The apparatus and general approach to BEC have been detailed elsewhere [1,11] and will be described here only briefly. We use a double MOT system [12] to load a TOP magnetic trap [13] with up to 10^9 Rb atoms in the $|1\rangle$ state. The hottest atoms are subsequently ejected from the magnetic trap by a technique of forced evaporative cooling. The remaining atoms rethermalize at lower and lower temperatures until they reach a critical temperature, T_c, below which they begin to pile up in the ground state (condensate). We typically continue evaporation until over 75% of the atoms are in the condensate, which ultimately contains about 5×10^5 atoms at a peak density of 10^{14} cm^{-3}.

We create the double condensate system from the single $|1\rangle$ condensate by applying a short (~ 400 μs) pulse of microwave and rf radiation to drive a two-photon hyperfine transition between states $|1\rangle$ and $|2\rangle$ (Fig. 1). By varying the pulse frequency and/or duration, we can effectively create arbitrary superpositions of the two states. One may also think of this process as producing two separate condensates, albeit with a particular relative phase [14], since (i) our imaging system distinguishes between the two states, and (ii) interconversion between the states does not occur with the coupling drive off. We will use both descriptions in this paper.

The data presented here were obtained by turning off the trap and allowing the condensates to expand ballistically for 22 ms. We then selectively image either of the two states (or both at once) by changing the sequence of probing beams. From the absorption images we obtain, we can reconstruct the positions of the condensates as they were in the magnetic trap [9]. The high degree of repeatability of our condensates permits us to follow the evolution of the system by repeating the experiment multiple times, changing only the time between the creation of the double condensate and its subsequent release from the trap.

3. CONDENSATE EVOLUTION

The dynamics of the two condensates are well-described in the mean-field language of coupled Gross-Pitaevskii equations for their order parameters:[15–17]

$$i\hbar\dot{\Phi}_1 = \left(-\frac{\hbar^2}{2m}\nabla^2 + V_1 + u_1|\Phi_1|^2 + u_{12}|\Phi_2|^2\right)\Phi_1 + \frac{\hbar\Omega(t)}{2}e^{i\omega_{rf}t}\Phi_2 \tag{1}$$

and

$$i\hbar\dot{\Phi}_2 = \left(-\frac{\hbar^2}{2m}\nabla^2 + V_2 + V_{hf} + u_2|\Phi_2|^2 + u_{21}|\Phi_1|^2\right)\Phi_2 + \frac{\hbar\Omega(t)}{2}e^{-i\omega_{rf}t}\Phi_1, \tag{2}$$

where m is the mass of the Rb atom, V_{hf} is the magnetic field-dependent hyperfine splitting between the two states in the absence of interactions, V_i is the trapping potential for state i, $u_i = 4\pi\hbar^2 a_i/m$ and $u_{ij} = 4\pi\hbar^2 a_{ij}/m$, $|\Phi_i|^2$ is the condensate density, and the intraspecies and interspecies scattering lengths are a_i and $a_{ij} = a_{ji}$, respectively. The coupling drive is characterized by its frequency ω_{rf} and Rabi frequency $\Omega(t)$, such that

$$\Omega(t) = \begin{cases} 2\pi \cdot 625 \text{ Hz}, & \text{coupling drive on;} \\ 0, & \text{coupling drive off.} \end{cases} \tag{3}$$

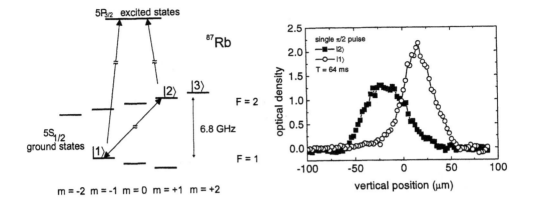

Figure 1. The relevant hyperfine and Zeeman structures of ^{87}Rb in a magnetic field. The atoms are moved coherently from state $|1\rangle$ to state $|2\rangle$ with a microwave pulse. Imaging is accomplished by absorption in optical transitions to the excited states.

Figure 2. Typical vertical cross-section through the condensates after they have come to equilibrium. Note the overlapping region at their boundary.

Once the double condensate is created, the coupling drive is turned off and the final term in Eqs. 1 and 2 vanishes. Moreover, for the number of atoms in our condensates, the kinetic energy terms are insignificant (Thomas-Fermi approximation). If we choose our trapping potentials such that $V_1 = V_2$, the dynamics are determined almost exclusively by the scattering lengths a_i and a_{ij}, which are positive and known at the 1% level to be in the ratio $a_1{:}a_{12}{:}a_2{::}1.03{:}1.00{:}0.97$. The condensates repel one another; in order to control the direction in which the condensates separate we deliberately introduce a vertical offset in the trapping potentials such that $V_1(z) = V_2(z - 0.4~\mu\text{m})$ [9,18]. This offset is much smaller than the vertical extent of the condensates (typically 14 μm). The condensates then separate vertically, driven by this offset as well as by their mutual repulsion.

The observed component separation dynamics is quite complicated, although in each realization of the experiment we observe similar patterns in the condensates' density distributions [9]. After ~ 45 ms, however, the two condensates reach a steady state, in which the $|1\rangle$ condensate sits slightly above the $|2\rangle$ condensate (see Fig. 2). The damping mechanism that leads to equilibrium is not yet understood, but it leaves the condensates with a stable, well-defined overlapping region.

4. MEASUREMENT OF THE RELATIVE PHASE

The atoms in each condensate possess a common phase. This phase evolves at a rate proportional to the condensate chemical potential μ_i. Thus, the relative phase between the two components evolves at the rate $(\mu_2 - \mu_1)$. This phase evolution is, however, unobservable, unless we make the condensates interfere with one another. We accomplish this

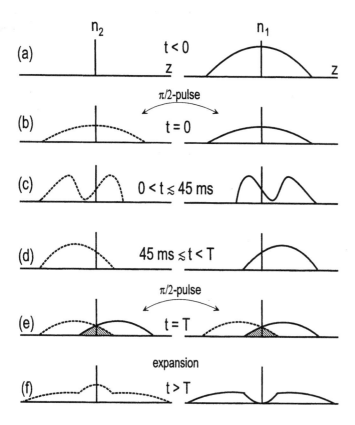

Figure 3. A schematic [10] of the condensate interferometer. (a) We begin with a single $|1\rangle$ condensate in steady-state. (b) After the first $\frac{\pi}{2}$-pulse, the condensate has been split into two components with a well-defined initial relative phase. (c) The components separate in a complicated fashion due to mutual repulsion as well as the vertical offset in the confining potentials. (d) The relative motion between the components damps, leaving the condensates with an overlapping boundary. Relative phase accumulates between the condensates until (e) at time T a second $\frac{\pi}{2}$-pulse remixes the components. Interference occurs in the hatched regions at the boundary. (f) The cloud is released immediately after the second pulse and allowed to expand for imaging. In this example, the second pulse led to destructive interference in the $|1\rangle$ state and constructive interference in the $|2\rangle$ state.

by applying a second $\frac{\pi}{2}$ coupling pulse at a time T after the first $\frac{\pi}{2}$-pulse (Fig.3), thereby implementing a time-domain version of the "Ramsey" method of separated oscillatory fields for the overlapping region. The condensates can interfere with one another through the two possible "paths" by which their atoms can arrive in the detected state (Fig. 3e–f). As a function of T, the number in state $|2\rangle$ in the overlap region varies as

$$N_2 \propto \cos[(\mu_2 - \mu_1 - \omega_{\mathrm{rf}})T + \Delta\varphi_{\mathrm{diff}}], \tag{4}$$

where $\Delta\varphi_{\mathrm{diff}}$ represents phase-diffusing terms arising from interactions between the condensates and their environment. We emphasize that the parts of the condensates that are not spatially overlapping cannot interfere with one another.

We plot the density of the $|2\rangle$ atoms in the overlap region after the second $\frac{\pi}{2}$-pulse in Fig. 4. The individual realizations of the experiment are given by the '+' symbol, and we can clearly see the effect of the interference between the condensates. The ensemble averages (solid circles) reveal that the initial relative phase written into the condensates at time $t = 0$ *is also preserved, despite the damping of the entangled external degree of freedom.*

At the moment the double-condensate system is created, the external (position) degree of freedom becomes entangled with the internal (hyperfine state) degree of freedom. The decoherence times for damped entangled states tend to be much faster than the damping times [19–21], and we there-

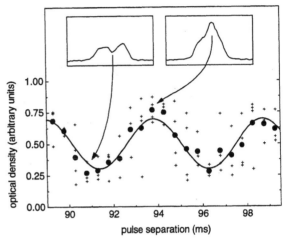

Figure 4: The value of the condensate density in the $|2\rangle$ state is extracted at the center of the overlap region (inset) and plotted as a function of T. Each large point (circles) represents the average of 4 separate realizations (pluses). The line is a sinusoidal fit to the data, from which we extract the angular frequency $\mu_2 - \mu_1 - \omega_{\mathrm{rf}}$.

fore might expect to lose all memory of the initial relative phase established between the condensates at $t = 0$. This situation corresponds to $\Delta\varphi_{\mathrm{diff}} \gg \pi$. From the fringe contrast of the ensemble we see that $\Delta\varphi_{\mathrm{diff}} \sim \frac{\pi}{3}$. The condensates evidently possess a phase-robustness that withstands the effects of such decohering influences.

5. FUTURE

In the future, we plan to study the dependence of the phase preservation as a function of system parameters such as temperature and number of atoms in the condensate. Further, we plan to explore the effect of a weak coupling drive, resonantly tuned to the overlap region between the condensates, to form a kind of nonlinear Josephson junction [23].

We have recently constructed a "nondestructive" imaging system (similar to that in Refs. [24] and [25]) that permits us to examine the double condensates while they are in the magnetic trap. Use of this system is expected to lead to additional interesting physics. For instance, if we can measure the relative number of atoms in each of the two states, then we expect to irrevocably disturb the relative phase, since these are conjugate variables. Experiments of this sort may someday probe the limits of quantum measurement.

The authors would like to acknowledge useful discussions with A. Leggett, F. Sols, K. Burnett, and M. J. Holland. This research is supported by the ONR, NSF, and NIST.

REFERENCES

1. M. H. Anderson *et al.*, Science 269 (1995) 198.
2. K. B. Davis *et al.*, Phys. Rev. Lett. 75 (1995) 3969.
3. C. C. Bradley, C. A. Sackett, J. J. Tollett, and R. G. Hulet, Phys. Rev. Lett. 75 (1995) 1687; *ibid.* 79 (1997) 1170.
4. M. R. Andrews *et al.*, Science 275 (1997) 637.
5. E. A. Burt *et al.*, Phys. Rev. Lett. 79 (1997) 337.
6. C. J. Myatt *et al.*, Phys. Rev. Lett. 78 (1997) 586.
7. D. M. Stamper-Kurn *et al.*, Phys. Rev. Lett. 80 (1998) 2027.
8. W. Ketterle, private communication.
9. D. S. Hall *et al.*, Phys. Rev. Lett. 81 (1998) 1539.
10. D. S. Hall, M. R. Matthews, C. E. Wieman, and E. A. Cornell, Phys. Rev. Lett. 81 (1998) 1543.
11. M. R. Matthews *et al.*, Phys. Rev. Lett. 81 (1998) 243.
12. C. J. Myatt *et al.*, Opt. Lett. 21 (1996) 290.
13. W. Petrich, M. H. Anderson, J. R. Ensher, and E. A. Cornell, Phys. Rev. Lett. 74 (1995) 3352.
14. E. A. Cornell, D. S. Hall, M. R. Matthews, and C. E. Wieman, J. Low Temp. Phys. (in press).
15. T.-L. Ho and V. B. Shenoy, Phys. Rev. Lett. 77 (1996) 3276.
16. B. D. Esry, C. H. Greene, J. P. Burke, Jr., and J. L. Bohn, Phys. Rev. Lett. 78 (1997) 3594.
17. H. Pu and N. P. Bigelow, Phys. Rev. Lett. 80 (1998) 1130.
18. D. S. Hall *et al.*, Proc. SPIE 3270 (1998) 98.
19. A. O. Caldeira and A. J. Leggett, Phys. Rev. A 31 (1985) 1059.
20. D. F. Walls and G. J. Milburn, Phys. Rev. A 31 (1985) 2403.
21. M. Brune *et al.*, Phys. Rev. Lett. 77 (1996) 4887.
22. K. Burnett and M. J. Holland, private communication; F. Sols, private communication.
23. J. Williams *et al.*, e-print cond-mat/9806337.
24. M. R. Andrews *et al.*, Science 273 (1996) 84.
25. M. R. Andrews *et al.*, Phys. Rev. Lett. 79 (1997) 553; *ibid.* 80 (1998) 2967.

Quantum Coherence and Decoherence - ISQM - Tokyo '98
Y.A. Ono and K. Fujikawa (Editors)
© 1999 Elsevier Science B.V. All rights reserved.

Probing quantum statistical mechanics with Bose gases: Non-trivial order parameter topology from a Bose-Einstein quench

J.R. Anglin[a,b] and W.H. Zurek[b]

[a]Institut für Theoretische Physik, Universität Innsbruck, Austria

[b]T-6, MS B288, Los Alamos National Laboratory, Los Alamos, New Mexico 87545

A rapid second order phase transition can have a significant probability of producing a metastable state instead of the equilibrium state. We consider the case of rapid Bose-Einstein condensation in a toroidal trap resulting in a spontaneous superfluid current, and compare the phenomenological time-dependent Ginzburg-Landau theory with quantum kinetic theory. A simple model suggests the effect should be observable.

1. THE ORDER PARAMETER AND THE EFFECTIVE POTENTIAL

A qualitative understanding of a second order phase transition may be had by considering the system to be described by a two-component *order parameter*, consisting of a modulus R and an angle θ[1]. The order parameter is taken to behave as a particle in a two-dimensional effective potential $V(R)$, with some form of dissipation dragging the system towards the bottom of this potential (possibly opposed by some random noise).

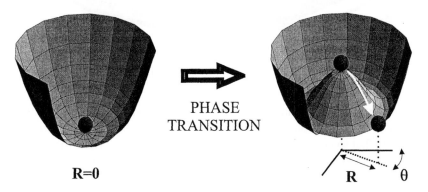

PHASE
TRANSITION

R=0

R θ

Figure 1: Effective potential theory of a second order transition.

In the disordered phase of the system, the minimum of $V(R)$ is at $R = 0$, so that θ is indefinite (or purely random) (Fig. 1, left picture). The phase transition then occurs through a modification of the effective potential, such that $R = 0$ becomes a local maximum, and the new minimum is a circular trough at some finite radius. The system rolls down the central hill to the trough, (Fig. 1, right picture) and θ acquires a definite value (with only small fluctuations due to noise). Which definite value θ acquires, however, is random, determined by small fluctuations while the system is near $R = 0$.

One can also allow R, θ to be functions of position (with $V(R)$ assumed to have the same form everywhere). In the ordered phase, the definite angle $\theta(x)$ may then vary spatially. A qualitative question thus arises, namely the topology of $\theta(x)$. Suppose we let x lie on a circle, for example: what *winding number* $\oint dx\, \partial_x \theta/(2\pi)$ should we typically observe after a transition? If the transition is slow, $\theta(x)$ will have time to organize itself all around the circle into some lowest energy configuration (such as $\theta(x)$ constant); but if it is fast, the random choice of θ around the circle will have some finite correlation length, and non-zero winding number will sometimes result. If it does, it will tend to be stable, since there is a high effective potential barrier for $R = 0$, and without making $R(x) \to 0$ somewhere, it is impossible to alter the winding number. See Fig. 2.

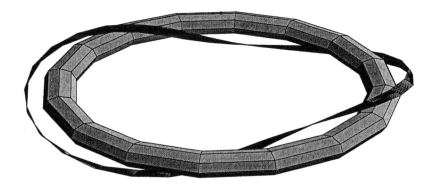

Figure 2: Winding number two; distance of ribbon from torus represents $R(x)$, angle of winding around torus represents $\theta(x)$.

The emergence of non-trivial topology from a rapid (quench-like) phase transition is common to a wide range of physical systems, real and postulated, from vortex loops in superfluid helium[2,3] to cosmic strings in the early universe[4,2]. With the advent of dilute alkali gas Bose condensates[5], we can now investigate this basic statistical mechanical problem in a new arena. The condensate mean field wave function $\psi = Re^{i\theta}$ provides our order parameter; and the current experimental technique of evaporative cooling automatically provides a rapid phase transition (on the Boltzmann scattering time scale). Our example of x lying on a circle can realistically be achieved with a toroidal trap, sufficiently tight for the system to be approximately one-dimensional. (We will consider such a trap of circumference $\sim 100\mu m$, which we understand is a credible prospect for the relatively near future.) In this context, non-zero winding number implies circulation around the trap, since $\nabla\theta$ is the superfluid velocity. (Angular momentum conservation is an issue we will discuss briefly below, only noting for now that the condensate is not isolated.)

2. TIME-DEPENDENT GINZBURG-LANDAU THEORY

To proceed to a quantitative description of our subject, we first consider the phenomenological theory obtained by letting the time evolution of ψ be governed by a first order equation involving a potential of the Ginzburg-Landau form we have sketched above:

$$\tau_0\dot\psi = \beta\Big(\frac{\hbar^2}{2M}\nabla^2 + \mu - \Lambda|\psi|^2\Big)\psi\,, \tag{1}$$

where $\beta = (k_BT)^{-1}$, and τ_0 and $\Lambda > 0$ are phenomenological parameters. The thermodynamical variable μ behaves near the critical point, in the case we consider, as

$$\mu = \frac{3}{2}(T_c - T) + \mathcal{O}(T_c - T)^2\,, \tag{2}$$

where T_c is the critical temperature. The equilibration time for long wavelengths is $\tau = \tau_0 k_B T/|\mu|$. The system's disordered phase is described by $\mu < 0$; the ordered phase appears when $\mu > 0$. Note that this time-dependent Ginzburg-Landau (TDGL) theory also typically assumes some small stochastic forces acting on ψ; we will leave these implicit.

A quench occurs if μ changes with time from negative to positive values. The divergence of the equilibration time τ at the critical point $\mu = 0$ is associated with *critical slowing down*. Because of this critical slowing down, $\frac{d\mu}{dt}/\mu$ must exceed $1/\tau$ in some neighbourhood of the critical point, and so there must be an epoch in which the system is out of equilibrium. What are at the beginning of this epoch mere fluctuations in the disordered phase, in which higher energy modes happen momentarily to be more populated than the lowest mode, can thus pass unsuppressed by equilibration into the ordered phase, to become topologically non-trivial configurations of $\psi(x)$.

The interval within which equilibration is negligible can be identified as the period wherein $|t|/\tau < 1$. If we define the quench time scale τ_Q by letting $\beta\mu = t/\tau_Q$ (choosing $t = 0$ as the moment the system crosses the critical point), this implies that the crucial interval is $-\hat t < t < \hat t$, for $\hat t = \sqrt{\tau_Q\tau_0}$. The correlation length $\hat\xi$ for fluctuations at time $t = -\hat t$ is then given by $\hbar/(2M\hat\xi^2) = \mu(-\hat t)$, which (assuming $T(-\hat t) \doteq T_c$) implies that $\hat\xi = \lambda_{T_c}(\tau_Q/\tau_0)^{1/4}$, for $\lambda_T = \hbar(2Mk_BT)^{-\frac{1}{2}}$ the thermal de Broglie wavelength[2]. Assuming one independently chosen phase within each correlation length $\hat\xi$, then modeling the phase distribution around the torus as a random walk suggests that the net winding number should be of order $\sqrt{L/\hat\xi}$ for L the trap circumference. Since evaporative cooling may be expected to yield $(\tau_Q/\tau_0)^{1/4}$ of order one, and T_c is several hundred nK, we can estimate $\hat\xi \sim 100$ nm, leading us to expect winding numbers in our 100μm torus of order ten. At current experimental densities, this implies a current approaching the Landau critical velocity, in the range of mm/s.

Considering this intriguing possibility raises an obvious question: is TDGL actually relevant to finite samples of dilute gas, far from equilibrium?

3. QUANTUM KINETIC THEORY

Because the condensates now available are dilute enough to be weakly interacting, there are good prospects for answering this question theoretically, by constructing from first principles a quantum kinetic theory to describe the whole process accurately. Using a second-quantized description of the trapped gas, one treats all modes above some judiciously chosen energy level as a reservoir of particles, coupled to the lower modes by two-particle scattering. Tracing over the reservoir modes leads to a master equation for the low energy modes[6], very similar to the equation for a multimode laser. Scattering of particles from the reservoir modes into the low modes, and vice versa, provides gain and loss terms. If the reservoir is described at all times by a grand canonical ensemble (of time-dependent temperature and chemical potential, in general), then the gain and loss processes are related by a type of fluctuation-dissipation relation, in which the reservoir chemical potential and the repulsive self-interaction of the gas combine in precisely the Ginzburg-Landau form. As a result, as long as these gain and loss processes are the only significant scattering channels, and the condensate self-Hamiltonian can be linearized to a good approximation, the system is indeed described quite well by time-dependent Ginzburg-Landau theory. Since these two conditions can be shown to hold near the critical point, all the results of section 2 can be recovered from quantum kinetic theory.

We can also clarify how the onset of a runaway process like Bose-Einstein condensation by evaporation can be consistent with critical slowing down: though the particle numbers in the lowest modes of the trapped gas are growing very fast, the required particle numbers to be in equilibrium with the higher energy modes grow much faster still as the temperature passes through T_c. So even with Bose-enhanced scattering at ever increasing rates, the rate at which the system is able to maintain equilibrium actually decreases.

The problem is that circulating states only become metastable above a threshold density, precisely because nonlinearity is required. So understanding the probability for a circulating condensate to grow into metastability requires extending quantum kinetic theory into the fully nonlinear regime, and without begging the basic question by assuming from the start that some particular mode will end up with the condensate.

3.1. A two-mode model

As a first step towards this goal, we consider a toy model of just two competing modes, such as the torus modes of different winding numbers k_0, k_1. The self-Hamiltonian of this system is taken as

$$\hat{H} = E[\hat{n}_1 + \frac{1}{2N_c}(\hat{n}_0^2 + \hat{n}_1^2 + 4\hat{n}_1\hat{n}_0)] . \tag{3}$$

Since the self-Hamiltonian must conserve both particle number and angular momentum, it can only be a function of $\hat{n}_0$ and $\hat{n}_1$ (which greatly simplifies the derivation of the master equation). Because we have incorporated the Bose enhancement of inter-mode repulsion (the factor of 4 instead of 2 in front of the $\hat{n}_0\hat{n}_1$ term, which is of course the best case value, obtained when the two orbitals overlap completely), we make the state with all particles in the 1 mode a local minimum of the energy for fixed $n_0 + n_1 > (N_c + \frac{1}{4})$. We have thus obtained an simple model which exhibits metastability above a threshold.

To make our results more meaningful, we estimate the experimental ranges of our parameters E and N_c. For our toroidal trap βE would be on the order of 10^{-4} at current experimental temperatures, and proportional to $k_1^2 - k_0^2$. N_c would be of order one for $(k_0, k_1) = (0, 1)$, but higher for higher modes (since E/N_c is actually constant).

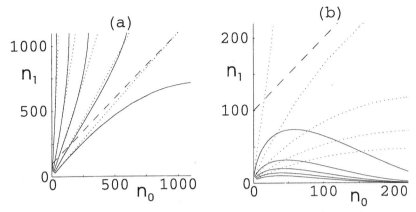

Figure 3: Trajectories from QKT (solid) and TDGL (dotted); heavy dashed line is threshold for metastability of mode 1. Initial times are $\hat{t}$; quench is $\beta\mu = \tanh(t/\tau_Q)$, $\beta = \beta_c e^{\tanh(t/\tau_Q)}$. Parameters are $N_c = 100$, and (a) $\Gamma\tau_Q = 10$, $\beta_c E = 0.01$; (b) $\Gamma\tau_Q = 100$, $\beta_c E = 0.05$. All trajectories shown initially have $n_1 > n_0$.

Tracing out a reservoir of higher modes as described above leads to a master equation for the two modes 0 and 1, which even for this highly simplified model is only directly tractable numerically. Here we will simply consider it as a kind of Fokker-Plank equation for the particle numbers n_0, n_1, and by neglecting the diffusion terms (valid for fast quenches after the early stage, which can be solved separately by assuming linearity), we extract an equation of motion for n_0 and n_1 which we can compare to TDGL:

$$\dot{n}_0 = \Gamma n_0 \left[e^{\beta\mu} - e^{\frac{\beta E}{N_c}(n_0 + 2n_1)} + 2\beta E n_1 e^{-\frac{\beta E}{2N_c}|N_c + n_0 - n_1|} \sinh\frac{\beta E}{2N_c}(N_c + n_0 - n_1) \right]$$

$$\dot{n}_1 = \Gamma n_1 \left[e^{\beta\mu} - e^{\frac{\beta E}{N_c}(N_c + n_1 + 2n_0)} - 2\beta E n_0 e^{-\frac{\beta E}{2N_c}|N_c + n_0 - n_1|} \sinh\frac{\beta E}{2N_c}(N_c + n_0 - n_1) \right].$$

When $n_0 + 2n_1$ and $N_c + n_1 + 2n_0$ are both close to $N_c\mu/E$, or for low enough particle numbers, we may replace $n_j \to |\psi_j|^2$ in the first two terms in each equation of (3) to obtain a TDGL equation, in the sense that $\dot{\psi}_j$ is set equal to the variation of a Ginzburg-Landau effective potential with respect to ψ_j^*. But the last term in each equation is not of Ginzburg-Landau form (it does not even involve μ). These non-GL terms conserve $n_0 + n_1$, and so are neither gain nor loss but another scattering channel hitherto neglected: collisions with reservoir particles shifting particles between our two low modes. These processes are insignificant at low particle numbers, but grow much stronger as the condensate grows larger, because they are doubly Bose-enhanced. And they turn out to have a simple but potentially drastic effect: they allow they system to equilibrate in energy faster than in particle number.

Representative solutions to (3) are shown in Fig. 3, together with the $|\psi_j|^2$ given by the TDGL theory formed by keeping only the first two terms of each equation in (3), expanding the exponentials to first order. It is clear that for sufficiently fast quenches the two theories accord quite well, but that for slower quenches TDGL significantly overestimates the probability of reaching the metastable state, because more rapid equilibration in energy than particle number strongly favours the lowest mode. This is the main new feature we have identified in dilute Bose gases as a GL-like system.

3.2. Angular momentum

Finally, we return to the question of angular momentum. Spontaneous appearance of a circulating state, condensing out of a cold atomic cloud with no net rotation, obviously implies angular momentum concentration in the condensate. Compensating ejection of fast atoms with opposite angular momentum during the evaporation process, in which after all most of the initial atoms are expelled, is a plausible mechanism by which this can proceed. Our assumption that the non-condensate atoms remain in a grand canonical ensemble relies implicitly on some such process. In our two-mode model, constraining the reservoir's mean angular momentum to be always opposite to that of the condensate will effectively just raise E and N_c. Angular momentum transport during evaporative cooling has as yet received no detailed study, however.

4. CONCLUSIONS

Despite the shortcomings of TDGL revealed by our toy model, we emphasize that kinetic theory does show that TDGL is indeed relevant to trapped dilute gases: what TDGL requires is not outright rejection, but corrections. And although the corrections may be substantial, the qualitative features predicted by TDGL are still recovered. While the extension of quantum kinetic theory beyond toy models, to realistic descriptions of spontaneous currents, will obviously require further study, the prospects for experimental realization of spontaneous currents are quite encouraging. Athough in this brief treatment we have neglected noise, diffusive nucleation actually gives a lower bound of typical winding number one for condensation in our hundred micron toroidal trap; with 10^6 atoms, even this has an equilibrium probability of order e^{-10}.

REFERENCES

1. V.L. Ginzburg and L.D. Landau, Zh. Eksperim. i. Teor. Fiz. **20**, 1064 (1950) .
2. W.H. Zurek, Nature **317**, 505 (1985); Acta Physica Polonica B **24**, 1301 (1993); Phys. Rep. **276**, 177 (1996).
3. P.C. Hendry *et al.*, Nature **368**, 315 (1994); V.M.H. Ruutu *et al.*, Nature **382**, 332 (1996); C. Baüerle *et al.*, Nature **382**, 334 (1996).
4. T.W.B. Kibble, J. Phys. **A 9**, 1387 (1976).
5. M. Anderson *et al.*, Science **269**, 198 (1995); K.B. Davis *et al.*, Phys. Rev. Lett. **75**, 3969 (1995); C.C. Bradley *et al.*, Phys. Rev. Lett. **75**, 1687 (1995).
6. C.W. Gardiner and P. Zoller, Phys. Rev. **A55**, 2901 (1997); cond-mat/9712002; D. Jaksch, C.W. Gardiner, and P. Zoller, Phys. Rev. **A56**, 575 (1997); J.R. Anglin, Phys. Rev. Lett. **79**, 6 (1997).

Quantum Coherence and Decoherence - ISQM - Tokyo '98
Y.A. Ono and K. Fujikawa (Editors)
© 1999 Elsevier Science B.V. All rights reserved.

Nonequilibrium Time Evolution of the Condensed State of a Fixed Number of Interacting Bosons

Akira Shimizu, Jun-ichi Inoue, and Takayuki Miyadera

Institute of Physics, University of Tokyo, 3-8-1 Komaba, Tokyo 153-8902, Japan
and Core Research for Evolutional Science and Technology (CREST), JST

We study the condensed state of interacting bosons confined in a box of finite volume V, in the case where the number N of bosons are exactly fixed for time $t < 0$, while a small leakage occurs for $t \geq 0$. It is shown that many properties can be simply described by an operator $\hat{b}_0$, which is a nonlinear function of bare operators. It is found that the gauge symmetry is almost broken after the leakage of only a few bosons.

1. Introduction

Many-body interactions are essential to interesting behaviors of the condensated state of bosons, such as the superfluidity [1]. The condensed state is usually taken as the state that has fluctuations both in the boson number N and the phase ϕ, whose magnitudes are $\langle \delta N^2 \rangle = \langle N \rangle$ and $\langle \delta \phi^2 \rangle \approx 1/4\langle N \rangle$, respectively. [1]. These fluctuations are non-negligible in small systems, such as Helium atoms in a micro bubble [2] and laser-trapped atoms [3], where $\langle N \rangle$ is typically $10^3 - 10^6$. We note that if one fixes N to the accuracy of $3 - 0.1\%$ in such systems, the above state is forbidden because $\langle \delta N^2 \rangle < \langle N \rangle$. The purpose of this paper is to investigate what happens to the interacting bosons in such a case. We stufy the ground state, the time evolution when a small leakage of bosons is possible, the order parameter, its time evolution, the macroscopic limit, and the physical origin of the time evolution.

2. The ground state of a fixed number of interacting bosons

We approximate the standard Hamiltonian for interacting bosons, retaining $\hat{a}_0$ as an operator (in contrast to the conventional treatment that replaces $\hat{a}_0$ with a c-number), as

$$\hat{H} = g(1 + Z_g)\hat{N}^2/2V + \sum_{\mathbf{k} \neq 0} \left(\epsilon_k^{(0)} + g\hat{N}/V \right) \hat{a}_{\mathbf{k}}^\dagger \hat{a}_{\mathbf{k}} + (g/2V)(\hat{a}_0 \hat{a}_0 \sum_{\mathbf{k} \neq 0} \hat{a}_{\mathbf{k}}^\dagger \hat{a}_{-\mathbf{k}}^\dagger + \text{h.c.}). \quad (1)$$

Here, $\epsilon_k^{(0)} \equiv \hbar^2 k^2/2m$, $Z_g \equiv (g/2V)\sum_{\mathbf{k} \neq 0}(1/\epsilon_k^{(0)})$, $g \equiv 4\pi\hbar^2 a/m$, where a is the scattering length, and we have assumed that $0 < na^3 \ll 1$ ($n \equiv \langle N \rangle/V$). Since $[\hat{H}, \hat{N}] = 0$, we can in principle find the ground state for which N is exactly fixed. As its approximate wavefunction, we propose the following form:

$$|N, \mathbf{y}\rangle \equiv e^{iG(\mathbf{y})}(1/\sqrt{N!})(\hat{a}_0^\dagger)^N|0\rangle, \text{ where } \hat{G}(\mathbf{y}) \equiv (-i/2nV)\hat{a}_0^\dagger \hat{a}_0^\dagger \sum_{\mathbf{q} \neq 0} y_q \hat{a}_{\mathbf{q}} \hat{a}_{-\mathbf{q}} + \text{h.c.} \quad (2)$$

We find that the energy expectation value becomes as low as that of the usual wavefunction when the variational parameters $\mathbf{y} \equiv \{y_q\}$ is taken as $\cosh y_q = \{(\epsilon_q + \epsilon_q^{(0)} + gn)/2\epsilon_q\}^{1/2}$, $\sinh y_q = gn/\{2\epsilon_q(\epsilon_q + \epsilon_q^{(0)} + gn)\}^{1/2}$. Equation (2) suggests that we should analyze the interacting bosons in terms of $\hat{b}_k \equiv e^{i\hat{G}(\mathbf{y})}\hat{a}_k e^{-i\hat{G}(\mathbf{y})}$. In fact, we find that $\hat{b}_0$ is a "natural coordinate," by which many physical properties are simply described [4]. For example, $|N,\mathbf{y}\rangle$ is simply the number state of $\hat{b}_0$: $|N,\mathbf{y}\rangle = (1/\sqrt{N!})(\hat{b}_0^\dagger)^N|0,\mathbf{y}\rangle$, where $|0,\mathbf{y}\rangle \equiv e^{i\hat{G}(\mathbf{y})}|0\rangle = |0\rangle$. Since the matrix element $\langle N-1,\mathbf{y}|\,\hat{b}_0|N,\mathbf{y}\rangle = \sqrt{N} = \sqrt{nV}$ diverges as $V \to \infty$, it is appropriate to separate $\hat{b}_0$ from the other terms of the boson field as $\hat{\psi} = (Z^{1/2}/\sqrt{V})\hat{b}_0 + \hat{\psi}'$, where Z is a renormalization constant. From the translational symmetry, we can take $Z^{1/2} = \Xi/\sqrt{n}$, where $\Xi \equiv \langle N-1,\mathbf{y}|\hat{\psi}|N,\mathbf{y}\rangle$. It is then easy to show that $\langle N-\Delta N,\mathbf{y}|\hat{\psi}'|N,\mathbf{y}\rangle = 0$ for all ΔN such that $|\Delta N| \ll N$. We now define two number operators by $\hat{N}' \equiv \int d^3r\,\hat{\psi}'^\dagger(r)\hat{\psi}'(r)$, $\hat{N}_0 \equiv \hat{N} - \hat{N}'$. We then find $\langle N,\mathbf{y}|\hat{N}|N,\mathbf{y}\rangle = V|\Xi|^2 + \langle N,\mathbf{y}|\hat{N}'|N,\mathbf{y}\rangle$. Hence, $\langle N,\mathbf{y}|\hat{N}_0|N,\mathbf{y}\rangle = V|\Xi|^2$, which can be interpreted as the "number of condensate particles." Writing Ξ as $\Xi = \sqrt{n_0}e^{i\varphi}$, where $|\Xi|^2 = \langle N,\mathbf{y}|\hat{N}_0|N,\mathbf{y}\rangle/V \equiv n_0$ denotes the density of the condensate particles, we obtain $\hat{\psi} = e^{i\varphi}\sqrt{n_0/nV}\,\hat{b}_0 + \hat{\psi}'$ [5], which is extremely useful in the following analysis.

3. Nonequilibrium time evolution of interacting bosons in a leaky box

In most real systems, there is a finite probability of exchanging bosons between the box and the environment. Hence, even if one fixes N at some time, N will fluctuate at later times. We simulate this situation by the following gedanken experiment. Suppose that at some time t (<0) one confines *exactly* N bosons in a box of volume V, and that at $t=0$ he makes a small hole in the box, so that a leakage flux J of the bosons is induced. We are interested in the time evolution of $\hat{\rho}(t) \equiv \mathrm{Tr}_{\mathrm{E}}[\hat{\rho}_{total}(t)]$, where $\hat{\rho}_{total}(t)$ denotes the density operator of the total system, and Tr_{E} the trace operation over environment's degrees of freedom. Since it is clear that $\hat{\rho}(t)$ approaches the equilibrium state in the limit of $t \to \infty$, we are mainly interested in an *early time stage* in which $0 \le Jt \ll N$. Assuming that J is small enough such that it does not excite the boson system, we evaluate the time evolution by a method which is equivalent to solving the master equation, as

$$\hat{\rho}(t) = Ke^{-Jt}\sum_{m=0}^{N}\frac{(Jt)^m}{m!}|N-m,\mathbf{y}\rangle\langle N-m,\mathbf{y}|, \quad \text{when } N \gg 1 \text{ and } 0 \le Jt \ll N , \quad (3)$$

where K is the normalization constant ($K \approx 1$). Note that *this formula is quite general* because all details have been absorbed in J. We find that Eq. (3) can be rewritten as

$$\hat{\rho}(t) = \int_{-\pi}^{\pi}\frac{d\phi}{2\pi}|e^{i\phi}\sqrt{Jt},N,\mathbf{y}\rangle\langle e^{i\phi}\sqrt{Jt},N,\mathbf{y}|, \quad |\xi,N,\mathbf{y}\rangle \equiv \sqrt{K}\sum_{m=0}^{N}\frac{(\xi^*)^m(\hat{b}_0^\dagger)^{N-m}}{\sqrt{m!(N-m)!}}|0,\mathbf{y}\rangle \quad (4)$$

where $\xi \equiv e^{i\phi}\sqrt{Jt}$. For the state $|\xi,N,\mathbf{y}\rangle$, we have $\langle\hat{N}\rangle_{\xi,N,\mathbf{y}} = N - |\xi|^2 = N - Jt$ and $\langle\delta\hat{N}^2\rangle_{\xi,N,\mathbf{y}} = |\xi|^2 = Jt$. In comparison, $\langle\hat{N}\rangle_{\alpha,\mathbf{y}} = \langle\delta\hat{N}^2\rangle_{\alpha,\mathbf{y}} = |\alpha|^2 = N$ for the coherent state of $\hat{b}_0$, which is defined by $|\alpha,\mathbf{y}\rangle \equiv e^{-|\alpha|^2/2}\sum_{n=0}^{\infty}[(\alpha\hat{b}_0^\dagger)^n/n!]|0,\mathbf{y}\rangle$, where $\alpha \equiv e^{i\phi}\sqrt{N}$. Hence, $\langle\delta\hat{N}^2\rangle_{\xi,N,\mathbf{y}} \ll \langle\delta\hat{N}^2\rangle_{\alpha,\mathbf{y}}$. The conjugate to N is roughly the phase. More precisely, we must consider the cosine and sine operators. To define them, we must identify a proper "coordinate" among many degrees of freedom. Although this is generally difficult for interacting many-particle systems, we find that $\hat{b}_0$ is the proper coordinate for interacting bosons: We define $\cos\phi \equiv [(\hat{b}_0^\dagger\hat{b}_0 + 1)^{-1/2}\hat{b}_0 + \hat{b}_0^\dagger(\hat{b}_0^\dagger\hat{b}_0 + 1)^{-1/2}]/2$, $\sin\phi \equiv [(\hat{b}_0^\dagger\hat{b}_0 + 1)^{-1/2}\hat{b}_0 - \hat{b}_0^\dagger(\hat{b}_0^\dagger\hat{b}_0 + 1)^{-1/2}]/2i$, as in the case of a single harmonic oscillator [4]. We then find $\langle\phi\rangle_{\xi,N,\mathbf{y}} \approx \phi$, $\langle\delta\phi^2\rangle_{\xi,N,\mathbf{y}} \approx 1/4|\xi|^2 = 1/4Jt$ for $|\xi,N,\mathbf{y}\rangle$, whereas $\langle\phi\rangle_{\alpha,\mathbf{y}} \approx \phi$, $\langle\delta\phi^2\rangle_{\alpha,\mathbf{y}} \approx 1/4|\alpha|^2 = 1/4N$ for $|\alpha,\mathbf{y}\rangle$. Hence, $\langle\delta\phi^2\rangle_{\xi,N,\mathbf{y}} \gg \langle\delta\phi^2\rangle_{\alpha,\mathbf{y}}$. Therefore, $|\xi,N,\mathbf{y}\rangle$ is a sort of a "number-phase squeezed state" (NPSS), which is obtained by squeezing $|\alpha, \mathbf{y}\rangle$ in the N direction, while keeping the uncertainty product $\langle \delta \hat{N}^2 \rangle \langle \delta \phi^2 \rangle$ minimum.

We have obtained two equivalent expressions (3) and (4). In the former expression, the state of the box is one of the number states of $\hat{b}_0$, each of which has a definite (but unknown) N, whereas ϕ is completely indefinite. In the latter, the state is one of the NPSSs of $\hat{b}_0$, each of which has small fluctuation in N, whereas ϕ is almost definite (but unknown) for $Jt \gg 1$ [6]. Depending upon the physical situation, either expression is convenient. For example, Eq. (4) is convenient for the description of an experiment by which information is obtained on ϕ [6]. For example, if the measurement error $\delta\phi_{err} > \sqrt{\langle \delta\phi^2 \rangle_{\xi, N, \mathbf{y}}} = 1/2\sqrt{Jt}$, and if the measurement is of the first kind, then the action of the measurement is just to "find" the "true" value of ϕ, to the accuracy of $\delta\phi_{err}$. Therefore, the density operator immediately after the measurement is generally given by $\int_{-\pi}^{\pi} \frac{d\phi}{2\pi} D(\phi - \bar{\phi}) |e^{i\phi}\sqrt{Jt}, N, \mathbf{y}\rangle\langle e^{i\phi}\sqrt{Jt}, N, \mathbf{y}|$. Here, $\bar{\phi}$ is the value of ϕ obtained by the measurement, $D(\phi - \bar{\phi})$ is a smooth positive function that is finite only for $|\phi - \bar{\phi}| \lesssim \delta\phi_{err}$, and $\int_{-\pi}^{\pi} \frac{d\phi}{2\pi} D(\phi - \bar{\phi}) = 1$. Such a state may be obtained by an interference experiment of *two independent condensates* prepared in two boxes, where the interference pattern of the two fluxes leaking from the boxes is observed, and the fluxes are small enough such that the measurement is approximately of the first kind. It is clear from Eq. (4) that the two fluxes exhibit interference *at each experimental run*, as in the case of non-interacting bosons [7], as the interference between the two NPSSs (one for each box), where $\bar{\phi}$ in this case is the relative phase of the two condensates. Note that these transparent views have been obtained because we have identified $\hat{b}_0$.

4. Order parameter and its time evolution

We now investigate the order parameter according to two typical definitions [1]. The first one utilizes $\Upsilon(\mathbf{r}_1, \mathbf{r}_2) \equiv \text{Tr}[\hat{\rho}(t)\hat{\psi}^\dagger(\mathbf{r}_1)\hat{\psi}(\mathbf{r}_2)]$. The system is said to possess the off-diagonal long-range order (ODLRO) if $\lim_{|\mathbf{r}_1 - \mathbf{r}_2| \to \infty} \Upsilon(\mathbf{r}_1, \mathbf{r}_2) \neq 0$. In this case, it is customary to define the order parameter Ξ by $\Upsilon(\mathbf{r}_1, \mathbf{r}_2) \sim \Xi^*(\mathbf{r}_1)\Xi(\mathbf{r}_2)$. According to this definition, we obtain the same result, $\lim_{|\mathbf{r}_1 - \mathbf{r}_2| \to \infty} \Upsilon(\mathbf{r}_1, \mathbf{r}_2) = n_0$ (hence $\Xi = \sqrt{n_0} e^{i\varphi}$), for *all* of $|N, \mathbf{y}\rangle$, $|\xi, N, \mathbf{y}\rangle$, and $|\alpha, \mathbf{y}\rangle$ (where $\xi = e^{i\phi}\sqrt{Jt}$ and $\alpha = e^{i\phi}\sqrt{N}$). Therefore, neither the ODLRO nor Ξ is able to distinguish between these states. The second definition of the order parameter is the expectation value of $\hat{\psi}$, which we denote by Ψ. According to this definition, $|N, \mathbf{y}\rangle$ does not have a finite order parameter; $\Psi = \langle N, \mathbf{y}|\hat{\psi}|N, \mathbf{y}\rangle = 0$. In our gedanken experiment, $\hat{\rho}(t)$ evolves as Eq. (3), or, equivalently, Eq. (4). For this mixed ensemble, $\Psi = \text{Tr}[\hat{\rho}(t)\hat{\psi}] = 0$. This is the *average over all elements* in the mixed ensemble, and corresponds to the *average over many experimental runs*. On the other hand, Ψ of *each element*, which corresponds to a possible result for *a single experimental run*, is different between the two expressions, Eqs. (3) and (4). When we are interested in the phase-sensitive nature, such as the interference experiment discussed in the previous section, the latter expression is convenient. By evaluating Ψ of each element of Eq. (4) as a function of t, we find that Ψ grows very rapidly, until it attains a constant value, $\Psi = \langle e^{i\phi}\sqrt{Jt}, N, \mathbf{y}|\hat{\psi}|e^{i\phi}\sqrt{Jt}, N, \mathbf{y}\rangle \to e^{i(\phi + \varphi)}\sqrt{n_0}$ [8], for $Jt \gtrsim 2$. This value equals Ψ of the coherent state $|\alpha, \mathbf{y}\rangle$ with $\alpha = e^{i\phi}\sqrt{N}$. That is, *after the leakage of only two or three bosons*, $|e^{i\phi}\sqrt{Jt}, N, \mathbf{y}\rangle$ *acquires the full and stable value of* Ψ, which is equal to that of the coherent state $|\alpha, \mathbf{y}\rangle$, and thus the *gauge symmetry is broken*. Practically, it seems rather difficult to fix N to such high accuracy that $\delta N \lesssim 2$. Therefore, in most cases, δN would be larger than 2 from the beginning, and each element of the mixture has the full and stable value of Ψ from the beginning.

5. Macroscopic limit and physical origin of the time evolution

It is well-known in many fields of physics that coherent states are stable against small perturbations. Note, however, that this statement is useful only when the proper coordinate by which the coherent state is defined is specified, because a coherent state of one coordinate is not a coherent state of another coordinate. Using an input-output formalism, we can show that $\hat{b}_0$ is the proper coordinate (i.e., $|\alpha, \mathbf{y}\rangle$ is stable), and that $\hat{\rho}(t)$ approaches the phase-randomized mixture (PRM) of $|\alpha, \mathbf{y}\rangle$ as $t \to \infty$. Therefore, in terms of $\hat{b}_0$, the system evolves simply from the number state, to the PRM of the NPSSs, and finally to the PRM of the coherent states. (In contrast, these evolutions would be very complicated in terms of bare operators.) This evolution is *not* induced by energy differences among $|N, \mathbf{y}\rangle$, $|\xi, N, \mathbf{y}\rangle$ and $|\alpha, \mathbf{y}\rangle$, because they all have the *same* energy expectation value when they have the same value of $\langle N \rangle$. Therefore, the evolutions is actually induced by some difference in the nature of the wavefunctions. To explore this point, we investigate whether the pure states, $|N, \mathbf{y}\rangle$, $|\xi, N, \mathbf{y}\rangle$ and $|\alpha, \mathbf{y}\rangle$ remain pure as $V \to \infty$. We note that the conventional definition of a pure state (i.e., $\hat{\rho}^2 = \hat{\rho}$) is invalid in an infinite system, and we must use the exact definition [9]. We can show that as $V \to \infty$, $|\alpha, \mathbf{y}\rangle$ remains pure, hence becomes a macroscopic pure state, whereas $|N, \mathbf{y}\rangle$ becomes a mixed state (indicating the formation of a domain structure), and $|\xi, N, \mathbf{y}\rangle$ $(1 \ll |\xi|^2 \ll N)$ approaches a state which is intermediate between the two in the sense that although it is a mixed state it behaves like a pure state for observables which are low-degree polynomials of $\hat{\psi}$, $\hat{\psi}^\dagger$, and their derivatives. These striking differences for $V \to \infty$ should be reflected in natures of these state for finite V. We consider this is the physical origin of the time evolution, $|N, \mathbf{y}\rangle \to |\xi, N, \mathbf{y}\rangle \to |\alpha, \mathbf{y}\rangle$.

[1] See, e.g., A. Griffin et al. (eds.), Bose-Einstein Condensation, Cambridge, New York, 1995.

[2] E. G. Syskakis, F. Pobell and H. Ullmaier, Phys. Rev. Lett. **55** (1985) 2964.

[3] M. H. Anderson et al., Science **269** (1995) 198; C. C. Bradley et al., Phys. Rev. Lett. **75** (1995) 1687; K. B. Davis et al., ibid 3969.

[4] In constrast to the case where the Bogoliubov transformation diagonalizes the Hamiltonian, $\hat{H}$ is *not* diagonal in terms of $\hat{b}_0$, and $\hat{b}_0$ is a *nonlinear* function of $\hat{a}_\mathbf{k}$'s and $\hat{a}_\mathbf{k}^\dagger$'s.

[5] This decomposition was suggested in E.M. Lifshitz and L.P. Pitaevskii, Statistical Physics Part 2, Pergamon, New York, 1980. Here we have *derived* it, and obtained the *explicit* form. Note that the finite renormalization ($|Z| = n_0/n < 1$) has been correctly obtained.

[6] When we talk about the phase, we of course mean the phase *relative to* some reference.

[7] Y. Castin and J. Dalibard, Phys. Rev. A **55** (1997) 4330, and references therein.

[8] This value decreases gradually because $\langle N(t) \rangle$ ($\propto n_0$) decreases as $\langle N(t) \rangle = N - Jt$.

[9] Let us denote a quantum state symbolically by ω and the expectation value of an observable A by $\omega(A)$. (e.g., $\omega = \hat{\rho}$, $\omega(A) = \mathrm{Tr}[\hat{\rho}\hat{A}]$.) A state ω is called *mixed* if there exist states ω_1, ω_2 ($\neq \omega_1$) and a positive number λ ($0 < \lambda < 1$), such that $\omega(A) = \lambda\omega_1(A) + (1 - \lambda)\omega_2(A)$ for *any* observable A. Otherwise, ω is called a *pure* state. Note here that in the field theory observables are restricted to (gauge-invariant) operators which have a finite support in the space-time. See, e.g., R. Haag, Local Quantum Physics, Springer, Berlin, 1992.

Quantum Coherence and Decoherence - ISQM - Tokyo '98
Y.A. Ono and K. Fujikawa (Editors)
© 1999 Elsevier Science B.V. All rights reserved.

Macroscopic Quantum Tunneling of a Bose-Einstein Condensate with Attractive Interaction: Fate of a False Vacuum

Masahito Ueda

Department of Physical Electronics, Hiroshima University, Higashi-Hiroshima 739-8527, Japan, and Core Research for Evolutional Science and Technology (CREST), Japan

Bosons interacting attractively with one another in a uniform system are believed not to undergo Bose-Einstein condensation (BEC) because they tend to collapse upon itself. However, when they are spatially confined, the zero-point kinetic pressure arising from Heisenberg's unceratin principle allows BEC to be formed in a metastable phase. By comparing various competing mechanisms such as collisional decay and fragmentation into pseudo-condensates, we show that near the critical point the decay of BEC will be dominated by macroscopic quantum tunneling.

1. INTRODUCTION

BEC represents a genuinely quantum-statistical phase transition in that it occurs without the help of interaction (Einstein called it "condensation without attraction"). Whether BEC occurs in coordinate space, momentum space, or somewhere inbetween depends on the nature of the interaction and on the boundary conditions. When the system is spatially uniform, the repulsive interaction suppresses density fluctuations such that the condensation occurs in momentum space, while the attractive interaction is believed not to lead to BEC because it causes the system to collapse upon itself. However, it is possible for a limited number of confined bosons with attractive interaction to form BEC, because the zero-point kinetic pressure arising from Heisenberg's uncertainty principle counterbalances the attractive interaction. It is believed that this is the mechanism that has allowed BEC of the ^{7}Li system to be experimentally realized [1]. Both numerical [2] and variational [3] approaches based on the Gross-Pitaevskii equation support this idea with the proviso that the number of bosons is below a certain critical value.

In this paper, I discuss possible senarios for its decay or collapse, including collisional decay, exchange-induced decay, and macroscopic quantum tunneling.

2. METASTABILITY OF BEC WITH ATTRACTIVE INTERACTION

For the metastable BEC to exist the zero-point energy $\sim \hbar\omega$ must exceed the mean-field interaction energy per particle, i.e., $\hbar\omega > N_0|U_0|/V$, where $U_0 = 4\pi a\hbar^2/M$, a is the S-wave scattering length, and M is the atomic mass. This implies that BEC can exist only insofar as the interaction acts perturbatively on the ideal Bose-gas system. The critical number of condensate bosons N_c can therefore be estimated from the condition $\hbar\omega \sim N_c|U_0|/V$, where the volume of the condensate is estimated to be $V \sim 4\pi d^3/3$. This allows us to estimate N_c as $N_c \sim \frac{d}{|a|}$. The more quantitative argument below shows that N_c is indeed proportional to $d/|a|$, and the constant of proportionality is of the order of unity. One might conclude that N_c can be made as large as desired by increasing the value of d. Unfortunately, this is not so, since the corresponding density of particles $n \sim N_c/d^3 \sim 1/|a|d^2$ decreases with increasing d. Since n must be larger than λ_{dB}^{-3} (λ_{dB} is the thermal deBroglie length), d cannot be larger than $\sim \sqrt{\lambda_{dB}^3/|a|}$, and thus we have the following fundamental upper bound for N_c:

$$N_c \leq \left(\frac{\lambda_{dB}}{|a|} \right)^{\frac{3}{2}}. \tag{1}$$

We note that this upper bound can be much larger than ~ 1300, which is the maximum number of condensate bosons observed in recent experiments [1].

3. POSSIBLE SENARIOS FOR THE DECAY OF THE CONDENSATE

3.1. Collisional decay

The collisional decay occurs via two-body dipolar collision and three-body recombination. These inelastic collisions lead to heating and loss of BEC atoms. The total loss rate is given by [4]

$$R(N) = \alpha \int d\mathbf{r} |\Phi(\mathbf{r})|^4 + L \int d\mathbf{r} |\Phi(\mathbf{r})|^6, \tag{2}$$

where $\alpha \simeq 1.2 \times 10^{-14} cm^3 s^{-1}$ is the two-body dipolar loss rate coefficient and $L \simeq 2.6 \times 10^{-28} cm^6 s^{-1}$ is the three-body recombination loss rate. Note that Eq. (2) is different from Eq. (3) of Ref. [4] in that factors N^2 and N^3 are absorbed in the normalization of the wave function. Near the critical point, the total loss rate is about three hundred atoms per second, and therefore the condensate decays in about four to five seconds.

3.2. Exchange-induced decay

The second possibility —exchange-induced decay— is connected with the concept of fragmented pseudo-condensate [5]. It is generally held that the condensate particles accumulate in a single state to form a genuine condensate. This understanding, however, raises the following question. Suppose that the condensate peak extends over a number of pair states having opposite momenta. If the number of pairs is large compared with one but small compared with the total number of bosons N, the distribution should look like a delta function on a macroscopic scale. How then can we tell a genuine condensate from such a fragmented pseudo-condensate? To answer this question, consider the interaction of the form $H_{int} = \sum_{\xi_1 \xi_2 \xi_3 \xi_4} V_{\xi_1 \xi_2 \xi_3 \xi_4} c_{\xi_1}^\dagger c_{\xi_2}^\dagger c_{\xi_3} c_{\xi_4}$. For a genuine condensate in which all particles are in the same quantum state $|N\rangle$, we have only the Hartree energy $\langle N|H_{int}|N\rangle = VN(N-1)$, where we assume that the matrix elements are constant for simplicity. If the condensate is fragmented into two states $|N_1, N_2\rangle$, we have not only the Hartree energy but also the Fock exchange energy, that is, $\langle N_1, N_2|H_{int}|N_1, N_2\rangle = VN(N-1) + 2VN_1 N_2$. For the case of repulsive interaction, fragmentation costs an extensive exchange energy. Genuine condensate is therefore due to the exchange interaction. But the situation drastically changes for the case of attractive interaction.

For the case of attractive interactions, the excitation may be energetically favorable because the interaction energy per particle within the condensate is nU_0, while the interaction energy per particle between the above-condensate particles and the condensate is $2nU_0$. The net energy difference between the genuine condensate and the pseudo-condensate is therefore given by the absolute mean-field energy per particle $n|U_0|$ minus the single-particle excitation energy $\hbar\omega$. Thus if $\hbar\omega$ is larger than $n|U_0|$, the genuiune condensate is metastable. But in the opposite case, fragmentation occurs so that the condensate will eventually be evaporated via exchange-induced excitations [6]. For the present experiments at Rice University, the mean-field energy is 1nK, while the zero-point energy is 7nK. Thus, while the idea of exchange-induced decay is interesting, it does not apply to the present experiments.

3.3. Macroscopic Quantum Tunneling

Collapse of the condensate due to macroscopic quantum tunneling is sometimes asserted to be unlikely to occur at absolute zero. The reasoning is as follows. The tunneling rate is proportional to the square of the overlap integral between the condensate wave function and the collapsed wave function, which is, naively thinking, proportional to $\exp(-\text{const}.N_0)$, and thus negligibly small. This argument relies heavily on a simple one-body model. In actuality, however, the macroscopic quantum tunneling is triggered by density fluctuations of long wavelength, and therefore cannot be described by a simple one-body model. Accordingly, we here apply the instanton method to the GP energy functional to evaluate the rate Γ of macroscopic quantum tunneling. For this purpose, let us expand $f(r)$ around its local minimum $r^{\min}$ up to the third order in $r - r^{\min}$:

$$f(r) \simeq f(r^{\min}) + \frac{3\omega_M^2}{\omega^2}(r - r^{\min})^2 + 20 r_c^{-1}(r - r^{\min})^3, \tag{3}$$

where ω_M is the frequency of the monopole mode, and $r_c = 5^{-1/4}$. Because the condensate undergoes zero-point oscillations around the minimum, it has kinetic energy. Let it be denoted by

$$\text{K.E.} = N\frac{M^* d_0^2}{2}\dot{r}^2, \tag{4}$$

where M^* is the effective mass. Because the frequency of the small oscillations around the local minimum must coincide with ω_M, we should have $M^* = 3M/2$. Thus we obtain the effective Hamiltonian of the collective mode as follows:

$$H^{\text{eff}} = \frac{N\hbar\omega}{4}\left[\left(3\frac{dr}{d(\omega t)}\right)^2 + f(r)\right]. \tag{5}$$

We note that the "potential" $f(r)$ has a turning point r^L such that $f(r^L) = f(r^{\min})$, where

$$r^L = r^{\min} - \frac{3\omega_M^2}{20\omega^2}r_c = r^{\min} - \frac{3}{2}\sqrt{r_c(\gamma_c - \gamma)}. \tag{6}$$

Once the system "tunnels out" at the point $r = r^L$, there is no obstacle to prevent the system from collapsing upon itself. The semiclassical tunneling rate of this process is given as

$$\Gamma = Ae^{-S^B/\hbar}, \tag{7}$$

where S^B is the bounce exponent given by

$$S^B/\hbar = \frac{N}{4}\int_{-\infty}^{\infty} d\tau \left[3(\dot{r}^B)^2 + f(r^B) - f(r^{\min})\right]. \tag{8}$$

Here $r^B(\tau)$ is called the bounce trajectory, which is determined so as to make the bounce exponent stationary, that is,

$$\frac{\delta S^B}{\delta r^B} = \frac{N}{4}\left\{-6\ddot{r}^B + \frac{f''(r^{\min})}{2}\left[2(r^B - r^{\min}) + 3\frac{(r^B - r^{\min})^2}{r^{\min} - r^L}\right]\right\} = 0. \tag{9}$$

Multiplying both sides by $\dot{r}^B$ and integrating over τ gives

$$(\dot{r}^B)^2 = \frac{f''(r^{min})}{6}\left[(r^B - r^{min})^2 + \frac{(r^B - r^{min})^3}{r^{min} - r^L}\right]. \tag{10}$$

Solving this for r^B with the initial condition $r^B(0) = r^L$, we obtain

$$r^B(\tau) = r^{min} - \frac{r^{min} - r^L}{\cosh^2(\sqrt{f''(r^{min})/24}\tau)}, \tag{11}$$

where $f''(r^{min}) = 6(\omega_M/\omega)^2$ and $r^{min} - r^L = 3\omega_M^2 r_c/20\omega^2$. Substituting Eq. (11) into Eq. (8), we obtain

$$S^B/\hbar = \frac{9N}{500\sqrt{5}}\left(\frac{\omega_M}{\omega}\right)^5 \simeq 4.58\left(1 - \frac{N}{N_c}\right)^{\frac{5}{4}}N. \tag{12}$$

The prefactor A in Eq. (7) arises from the Gaussian fluctuations around the bounce trajectory. For the quadratic-plus-cubic potential, the prefactor is given by

$$A = \omega_M\left(\frac{15S^B}{2\pi\hbar}\right)^{\frac{1}{2}}. \tag{13}$$

Substituting Eq. (12) into this equation, we obtain

$$A = \sqrt{\frac{27}{200\pi\sqrt{5}}}\left(\frac{\omega_c}{\omega}\right)^{\frac{7}{2}}N^{\frac{1}{2}} \simeq 11.76\left(1 - \frac{N}{N_c}\right)^{\frac{7}{8}}N^{\frac{1}{2}}. \tag{14}$$

Remarkably, the tunneling exponent is proportional to the five-fourth power of $1 - N_0/N_c$, rather than the unit power. Because of this extra power of $1/4$, and a relatively large attempt frequency, MQT becomes dominant for N_0 very close to N_c [7]. Using the data from Ref. [1], we have $\omega = (\omega_x\omega_y\omega_z)^{\frac{1}{3}} \simeq 908.4$Hz and $N_c = 1250$. It then follows from Eqs. (12) and (14) that

$$\omega_M \simeq 3231\left(1 - \frac{N}{N_c}\right)^{\frac{1}{4}}, \quad A = 10680\left(1 - \frac{N}{N_c}\right)^{\frac{7}{8}}N^{\frac{1}{2}}, \quad S^B/\hbar \simeq 4.58\left(1 - \frac{N}{N_c}\right)^{\frac{5}{4}}N. \tag{15}$$

If the number of condensate bosons is 99.5% of its critical value, the decay rate is about 2/sec, and MQT should therefore dominate over the collisional decay processes.

It must be kept in mind that the above considerations are valid at zero temperature. In the present experiments [1], however, there are abundant above-condensate atoms. Thus the system is at finite temperature. As far as I know, there is no finite-temperature theory that reproduces the results described above at zero temperature. Much work therefore needs to be done to fully understand the present experiments.

REFERENCES

1. C.C. Bradley, C.A. Sackett, and R.G. Hulet, Phys. Rev. Lett. **78**, 985 (1997).
2. P.A. Ruprecht, M. J. Holland, K. Burnett, and M. Edwards, Phys. Rev. A **51**, 4704 (1995).
3. G. Baym and C. Pethick, Phys. Rev. Lett. **76**, 1 (1996).
4. R. J. Dodd, M. Edwards, C. Williams, C.W. Clark, M. Holland, P. Ruprecht, and K. Burnett, Phys. Rev. A **54**, 661 (1996).
5. P. Nozières and D. Saint James, J. Physique **43**, 1133 (1982).
6. Yu. Kagan, G. Shlyapnikov, and J. Walraven, Phys. Rev. Lett. **76**, 2670 (1996).
7. M. Ueda and A. J. Leggett, Phys. Rev. Lett. **81**, 1343 (1998).

Quantum Coherence and Decoherence - ISQM - Tokyo '98
Y.A. Ono and K. Fujikawa (Editors)
© 1999 Elsevier Science B.V. All rights reserved.

Collective Excitations of a Trapped Bose-Einstein Condensate: A Variational Sum-Rule Approach

Takashi Kimura, Hiroki Saito, and Masahito Ueda

Department of Physical Electronics, Hiroshima University, Higashi-Hiroshima 739-8527, Japan and Core Research for Evolutional Science and Technology (CREST), Japan

We propose a new variational method that combines an excitation-energy sum rule and Fetter's variational wave function to analytically find almost exact low-lying collective-mode frequencies of a trapped Bose-Einstein condensate. Our method gives excitation frequencies in excellent agreement with experimental results as well as numerical ones based on the Bogoliubov approximation at zero temperature.

1. INTRODUCTION

Bose-Einstein condensation (BEC) in trapped atomic gases has enabled us to study many-body physics of weakly interacting bosons. Extensive experimental [1,2] and theoretical [3–5] studies on collective excitations of BEC have been reported. Numerical analyses [3] based on the Bogoliubov approximation have so far found the best agreement with experiments at the low-temperature limit [1]. Stringari [4] has obtained analytic expressions of collective-mode frequencies in the Thomas-Fermi (TF) limit [6].

This paper proposes a new variational method that combines an excitation-energy sum rule [4,7] and Fetter's variational wave function [8] to analytically find almost exact frequencies of a trapped Bose system. Our method is presented in Sec. II and results are shown in Sec. III.

2. VARIATIONAL METHOD FOR COLLECTIVE MODES

We consider a Bose system described by the Hamiltonian, $H = T + U + H_{\text{int}}$, where $T = -(\hbar^2/2M) \sum_i \nabla_i^2$, $U = (M/2) \sum_i (\omega_x^2 x_i^2 + \omega_y^2 y_i^2 + \omega_z^2 z_i^2)$, and $H_{\text{int}} = (2\pi\hbar^2 a/M) \sum_{i \neq j} \delta(\mathbf{r}_i - \mathbf{r}_j)$ describe the kinetic energy, the confining potential energy, and the inter-particle interaction energy, respectively. Here M and a denote the atomic mass and the s-wave scattering length, respectively, and i denotes the atom index.

Let $\{|n\rangle\}$ be a complete set of exact eigenstates of H with eigenvalues $\{E_n\}$ (such that $E_n \geq E_m$ for $n > m$), where n represents a complete set of quantum numbers. We are interested in low-lying excitations of many-body states which are excited by an excitation operator F. Let $|0\rangle$ be the ground state and $|1\rangle$ be the lowest energy state excited by F

with excitation energy $\hbar\omega_{10} = E_1 - E_0$. From the following inequality

$$\omega_{10}^2 \leq \omega_{10}^2 \frac{|\langle 1|F|0\rangle|^2 + \sum_{n\neq 1} |\langle n|F|0\rangle|^2 \left(\frac{\omega_{n0}}{\omega_{10}}\right)^3}{|\langle 1|F|0\rangle|^2 + \sum_{n\neq 1} |\langle n|F|0\rangle|^2 \frac{\omega_{n0}}{\omega_{10}}} = \frac{1}{\hbar^2} \frac{\sum_n |\langle n|F|0\rangle|^2 (\hbar\omega_{n0})^3}{\sum_n |\langle n|F|0\rangle|^2 \hbar\omega_{n0}}, \tag{1}$$

we find that an upper bound $\hbar\omega^{\text{upper}}$ of the lowest excitation energy is given by [4,7]

$$\hbar\omega^{\text{upper}} = \sqrt{m_3/m_1}, \tag{2}$$

where $m_p \equiv \sum_n |\langle n|F|0\rangle|^2 (\hbar\omega_{n0})^p$ is the p-th moment of the excitation energy. The virtue of this formula is that m_1 and m_3 can be expressed as expectation values of commutators between F and H with respect to the ground state $|0\rangle$ as $m_1 = \frac{1}{2}\langle 0|[F^\dagger, [H, F]]|0\rangle$ and $m_3 = \frac{1}{2}\langle 0|[[F^\dagger, H], [H, [H, F]]]|0\rangle$, and we are therefore able to find $\hbar\omega^{\text{upper}}$ from the condensate wave function, without the need of finding excited states.

We employ a Fetter's variational wave function [8] for the condensate defined as

$$\Psi(\mathbf{r}) = c_0 \left(1 - \frac{r_\perp^2}{d_\perp^2 R_\perp^2} - \frac{z^2}{d_z^2 R_z^2}\right)^{(1+\eta)/2}, \tag{3}$$

if $\frac{r_\perp^2}{d_\perp^2 R_\perp^2} + \frac{z^2}{d_z^2 R_z^2} \leq 1$ and zero otherwise, where $r_\perp \equiv \sqrt{x^2 + y^2}$, c_0 is the normalization constant, $d_j \equiv \sqrt{\hbar/M\omega_j}$ ($j = \perp, z$) is the oscillator length, and R_j and η are the variational parameters that are determined so as to minimize the total energy $E_0 = \langle T\rangle + \langle U\rangle + \langle H_{\text{int}}\rangle$. Fetter showed that the variational wave function (3) smoothly interpolates between the noninteracting limit ($\Psi(\mathbf{r}) \propto \exp(-r_\perp^2/2d_\perp^2 - z_\perp^2/2d_z^2)$ for $\eta \sim \sqrt{R_\perp}, \sqrt{R_z} \to \infty$) and the TF limit ($\Psi(\mathbf{r}) \sim c_0(1 - r_\perp^2/d_\perp^2 R_\perp^2 - z^2/d_z^2 R_z^2)^{1/2}$ for $\eta \to 0$).

We first consider the collective-mode frequencies with magnetic quantum numbers $m = 0$ and $|m| = 2$ for an axially-symmetric trap ($\omega_x = \omega_y \equiv \omega_\perp$) which are studied in recent experiments [1,2]. In the $|m| = 2$ mode the condensate expands in one direction and simultaneously contracts in the other. This mode can be excited by modulating two radial trap frequencies out of phase but by the same amount $\delta\omega \ll \omega_\perp$. The resulting perturbation on the system defines the excitation operator $F_{|m|=2} = M/2 \sum_i \{[(\omega_\perp + \delta\omega)^2 - \omega_\perp^2] x_i^2 + [(\omega_\perp - \delta\omega)^2 - \omega_\perp^2] y_i^2\} \approx M\omega_\perp \delta\omega \sum_i (x_i^2 - y_i^2) \propto \sum_i r_i^2 \{Y_{2,2}(\mathbf{r}_i) + Y_{2,-2}(\mathbf{r}_i)\}$, where Y_{lm} is the spherical harmonic function. Substituting $F_{|m|=2}$ into Eq. (2) yields

$$\omega^{\text{upper}}(|m| = 2) = \omega_\perp \sqrt{2(1 + \langle T_\perp\rangle/\langle U_\perp\rangle)}, \tag{4}$$

where $T_\perp$ and $U_\perp$ are the radial component of the kinetic energy and that of the trap potential energy, respectively, and $\langle \cdots \rangle$ denotes the expectation value over the wave function of the condensate. We note that $\omega^{\text{upper}}(|m| = 2)$ does not directly depend on the interaction energy $\langle H_{\text{int}}\rangle$ because the volume is conserved. In the absence of the inter-particle interaction we find $\langle T_\perp\rangle = \langle U_\perp\rangle$, so that $\omega^{\text{upper}}(|m| = 2) = 2\omega_\perp$, while in the TF limit we have $\langle T_\perp\rangle = 0$, so that Eq. (4) reduced to the exact result in the TF limit $\omega^{\text{upper}}(|m| = 2) = \sqrt{2}\omega_\perp$ [4].

In the $m = 0$ mode the condensate alternately expands and contracts in the radial direction. This part of the excitation is described by $F = M/2 \sum_i \{[(\omega_\perp + \delta\omega)^2 - \omega_\perp^2] r_{\perp i}^2\} \approx M\omega_\perp \delta\omega \sum_i r_{\perp i}^2$ [1]. Because of repulsive interaction, however, the condensate should also undergo oscillations in the axial direction which must be out of phase with the radial motion. Hence we consider the excitation operator $F_{m=0} = M\omega_\perp \delta\omega \sum_i (r_{\perp i}^2 - \alpha z_i^2)$, where α is another variational parameter. We note that $F_{m=0}$ is a linear combination of two modes $\sum_i r_i^2$ ($n = 1, l = m = 0$) and $\sum_i r_i^2 Y_{2,0}(\mathbf{r}_i)$ ($n = 0, l = 2, m = 0$). Substituting $F_{m=0}$ into Eq. (2) yields

$$\omega^{\text{upper}}(m = 0, \alpha) = \left[2\frac{2(\langle T_\perp \rangle + \langle U_\perp \rangle) + \alpha^2(\langle T_z \rangle + \langle U_z \rangle) + (1 - \alpha/2)^2 \langle H_{\text{int}} \rangle}{2\langle U_\perp \rangle / \omega_\perp^2 + \alpha^2 \langle U_z \rangle / \omega_z^2} \right]^{\frac{1}{2}}, \tag{5}$$

where $\langle T_z \rangle$ and $\langle U_z \rangle$ are the axial component of the kinetic energy and that of the potential energy, respectively. By minimizing $\omega^{\text{upper}}(m = 0, \alpha)$ with respect to α, we find $\omega^{\text{upper}}(m = 0) = 2\omega_\perp$ in the noninteracting limit and $\omega^{\text{upper}}(m = 0) = \omega_\perp(2 + \frac{3}{2}\lambda^2 - \frac{1}{2}\sqrt{9\lambda^4 - 16\lambda^2 + 16})^{\frac{1}{2}}$ ($\lambda \equiv \omega_z/\omega_\perp$) in the TF limit. This result is identical to what has been obtained from hydrodynamic theory that keeps terms up to first order in the density fluctuation $\delta\rho$ [4].

The agreement shows that the excitation operator $F_{m=0}$ that we choose is indeed correct. We also note that if ω^{upper} is maximized with respect to α, we obtain $\omega^{\text{upper}}(m = 0) = \omega_\perp(2 + \frac{3}{2}\lambda^2 + \frac{1}{2}\sqrt{9\lambda^4 - 16\lambda^2 + 16})^{\frac{1}{2}}$, which also agrees with a hydrodynamic result with higher frequency [4]. Because the states excited by $F_{m=0}$ are restricted to states which are constructed by linear combinations of the $n = 1, l = m = 0$ mode and the $n = 0, l = 2, m = 0$ mode, there should be two values of α that makes ω^{upper} extremal, and the corresponding states should describe the two lowest-energy excited states as described above.

Our method can also be applied to the dipole ($l = 1$) modes [4], which correspond to the center-of-mass motion of the condensate, and are therefore not renormalized by the inter-particle interaction. This is known as generalized Kohn's theorem. We consider the excitation operator $F = M/2 \sum_i \omega_\perp^2 [(x_i + \delta)^2 - x_i^2] \propto \sum_i x_i \propto \sum_i r_i (Y_{1,1}(\mathbf{r}_i) - Y_{1,-1}(\mathbf{r}_i))$ or $F = M/2 \sum_i \omega_\perp^2 [(y_i + \delta)^2 - y_i^2] \propto \sum_i r_i (Y_{1,1}(\mathbf{r}_i) + Y_{1,1}(\mathbf{r}_i))$ for $|m| = 1$, and $F = M/2 \sum_i \omega_\perp^2 [(z_i + \delta)^2 - z_i^2] \propto \sum_i z_i \propto \sum_i r_i Y_{10}(\mathbf{r}_i)$ for $m = 0$. Substituting F into Eq. 2, we easily obtain $\omega^{\text{upper}}(l = 1, |m| = 1) = \omega_\perp$ and $\omega^{\text{upper}}(l = 1, m = 0) = \omega_z$. We thus find that the collective-mode frequencies obtained by the present approach coincide with trap frequencies, being independent of the strength of interaction.

3. RESULTS

We first examine the accuracy of Fetter's variational wave function. Table I compares the expectation values of the total, kinetic, potential, and interaction energies obtained by Fetter's variational wave function with those obtained numerically according to the method of Ref. [9]. The agreement is fairly good.

TABLE I. Results for Fetter's variational wave function in an axially-symmetric trap. We take $\sqrt{8}\omega_\perp = \omega_z = 2\pi \times 220$ Hz and $a = 100a_0$. Energies are shown in units of $\hbar\omega_\perp$ and N_0 is the atom number. Numerical results of Ref. [9] are shown in parentheses.

N_0	$\langle E_{\text{tot}} \rangle$	$\langle T \rangle$	$\langle U \rangle$	$\langle H_{\text{int}} \rangle$
1000	3.86 (3.84)	0.75 (0.76)	2.17 (2.15)	0.95 (0.93)
10000	7.83 (7.76)	0.46 (0.45)	4.61 (4.57)	2.76 (2.74)
20000	10.06 (9.98)	0.40 (0.38)	5.95 (5.91)	3.70 (3.68)

146

We calculate the upper bound of the collective-mode frequencies with Fetter's variational wave function. Figure 1 compares our analytical results (solid curves) with the experimental data (dots) taken from Ref. [1] for ^{87}Rb atoms, where we use the same parameters as those of the experiment; $a = 109a_0$ (a_0 is the Bohr radius) and $\omega_z/\sqrt{8} = \omega_\perp = 2\pi \times 132$Hz. We find that our results for both $m = 0$ (with lower frequency) and $|m| = 2$ modes are in excellent agreement with those of the experiment and with those obtained with the Bogoliubov approximation [3]. We have also calculated the upper-bound frequencies using numerically obtained energy expectation values for 4500 atoms and find $\omega^{\text{upper}}(|m| = 2) = 1.454\omega_\perp$ and $\omega^{\text{upper}}(m = 0) = 1.871\omega_\perp$. These results agree excellently with those obtained using Fetter's variational wave function, that is, $\omega^{\text{upper}}(|m| = 2) = 1.450\omega_\perp$ and $\omega^{\text{upper}}(m = 0) = 1.870\omega_\perp$. Such an excellent agreement

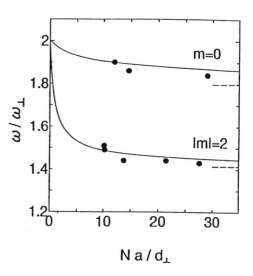

Figure 1. Collective-mode frequencies of the $m = 0$ and the $|m| = 2$ modes in an axially-symmetric trap. The solid curves show our results, the dots show the experimental data of Ref. [1], and the dashed lines show those of the Thomas-Fermi limits [4].

up to three digits is unexpected, considering the fact that expectation values of the kinetic energy, etc. agree only up to two digits.

We also calculate the monopole ($l = 0$) and the quadrupole ($l = 2$) mode excitation frequencies in a spherically-symmetric trap by a similar analysis and the results are also in excellent agreement with numerical results obtained with the Bogoliubov approximation [5].

In conclusion, we have presented a new variational method to analytically evaluate collective-mode frequencies of a trapped Bose gas. The results are in excellent agreement with those of the Bogoliubov approximation [3,5] and those of the experiments [1].

REFERENCES

1. D. S. Jin et al., Phys. Rev. Lett. 77 (1996) 420.
2. M.-O. Mewes et al., Phys. Rev. Lett. 77 (1996) 988.
3. M. Edwards et al., Phys. Rev. Lett. 77 (1996) 1671.
4. S. Stringari, Phys. Rev. Lett. 77 (1996) 2360.
5. D. A. W. Hutchinson, E. Zaremba, and A. Griffin, Phys. Rev. Lett. 78 (1997) 1847.
6. V. V. Goldman, I. F. Silvera, and A. J. Leggett, Phys. Rev. B 24 (1981) 2870.
7. H. Wagner, Z. Physik 195 (1966) 273.
8. A. L. Fetter, J. Low Temp. Phys. 106 (1997) 643.
9. F. Dalfovo and S. Stringari, Phys. Rev. A 53 (1996) 2477.

Quantum Coherence and Decoherence - ISQM - Tokyo '98
Y.A. Ono and K. Fujikawa (Editors)
© 1999 Elsevier Science B.V. All rights reserved.

Quantum Structure of Three Coupled Atomic Bose-Einstein Condensates

K. Nemoto*, W. J. Munro and G. J. Milburn

Centre for Laser Science, Department of Physics,
University of Queensland, QLD 4072, Australia

The simplest model of three coupled Bose-Einstein Condensates (BEC) is investigated using a group theoretical method. The stationary solutions are determined using the SU(3) group under the mean field approximation. This semiclassical model is compared to a full quantum treatment detailing the energy eigenstates of the system. We establish the utility of group theoretic methods.

1 Introduction

The recent creation of neutral atom Bose-Einstein condensates (BEC) [1, 2, 3, 4, 5] stimulated theoretical research aimed at understanding this new state of matter. Models of two coupled BECs in a two mode approximation are considered a tractable system when total particle number is conserved and eigenstates of the two well system are labelled by the particle number difference between the wells. Two coupled BECs in a symmetric double-well potential has been analyzed with the use of the SU(2) group [6, 7]. This model system included tunnelling between the BECs in each potential trap, and also incorporated two-body interactions making the model nonlinear.

However, any extensions to this model significantly increase the nonlinearity while higher dimensional effects increase the complexity of the model structure. These more complex systems are of interest as the richer dynamics and model structure allows us to model quantum states with non-zero currents, for instance. In the limit of large mode number, Bose-Hubbard type approaches are useful using a mean field approximation [8]. However for small $2 < N \approx 3, 4, 5$, systems are complex and models must exploit system symmetries in order to obtain solutions. The symmetries of these groups are the SU(N) group symmetries. In this paper we analyze a three coupled BEC system using the operator algebra of SU(3).

We specify the three coupled BEC system in Section 2, and demonstrate how the group theoretic approach is applied in Section 3. In Section 4, we analyze this system under the semiclassical mean field approach. Finally, in Section 5, we treat this system fully quantum mechanically to estimate the ground state. This treatment also allows us to detail the quantum mechanism of the classical bifurcation mentioned in Ref. [9] which treated the same Hamiltonian as this paper.

*Supported by Australian International Education Foundation (AIEF).

2 Three Coupled BEC Model

We consider the simplest three coupled BEC model with a symmetric triple-well single-particle potential V. For the purpose of simplicity, we treat this model as a one dimensional system. Since we assume that the three potential traps are well separated spatially, the lowest localized states in each trap are approximately orthogonal to each other. We define a mode in each localized trap using particle annihilation and creation operators $\hat{c}_j$, $\hat{c}_j^\dagger$ which commute with every other mode operator to first order in the overlap of the localized states. We also assume that these three lowest localized states are sufficiently well separated from higher energy states, and that interactions between particles do not change this basic property of the system configuration. These assumptions allow us to treat this model system in a three mode approximation.

The many-body Hamiltonian describing atomic BECs [10] can be written in terms of the mode operators as

$$\hat{H} = \omega \sum_{j=1}^{3} \hat{c}_j^\dagger \hat{c}_j + \Omega \sum_{j,k=1,i\neq k}^{3} \hat{c}_j^\dagger \hat{c}_k + \chi \sum_{j=1}^{3} \hat{c}_j^\dagger \hat{c}_j^\dagger \hat{c}_j \hat{c}_j, \tag{1}$$

where χ is the two particle interaction strength and Ω is the tunnelling frequency. This is the Hamiltonian of our model system in this paper.

3 SU(3) Group Approach

This section shows the group theoretic treatment of the system with the Hamiltonian specified in (1). In order to describe this system with SU(3) generators, we extend the Schwinger boson method [11] and we define the eight generators of SU(3) as

$$\begin{cases} \hat{J}_{x_1} = \hat{c}_1^\dagger \hat{c}_1 - \hat{c}_2^\dagger \hat{c}_2 & \hat{J}_{y_k} = i(\hat{c}_k^\dagger \hat{c}_j - \hat{c}_j^\dagger \hat{c}_k), \\ \hat{J}_{x_2} = \frac{1}{3}(\hat{c}_1^\dagger \hat{c}_1 + \hat{c}_2^\dagger \hat{c}_2 - 2\hat{c}_3^\dagger \hat{c}_3), & \hat{J}_{z_k} = \hat{c}_k^\dagger \hat{c}_j + \hat{c}_j^\dagger \hat{c}_k, \end{cases} \tag{2}$$

where $k = 1, 2, 3$ and $j = (k+1) \bmod 3$. We note that the two operators J_{x_i} commute with each other. The most important relation which the generators satisfy is the Casimir invariant of SU(3), $4\hat{N}(\hat{N}/3+1)$. This and other conservation relations play an important role in the stationary analysis of Section 4.

With the use of SU(3) generators, we can represent the Hamiltonian (1) in the form

$$\hat{H} = \Omega(\hat{J}_{z_1} + \hat{J}_{z_2} + \hat{J}_{z_3}) + \frac{\chi}{2}(\hat{J}_{x_1}^2 + 3\hat{J}_{x_2}^2). \tag{3}$$

Here we ignore constant terms involving the total number of particles which do not change the dynamics of the system.

It is evident that this method can readily be applied to the Bose-Hubbard model (BHM) with n potential traps. Defining appropriate SU(n) generators, the Bose-Hubbard Hamiltonian has the form

$$\hat{H} = \omega \sum_{i=1}^{n} J_{z_i} + \sum_{j=1}^{n-1} d_j J_{x_j}. \tag{4}$$

The coefficients d_j depend on the definition of the J_x generators.

4   Mean-field approximation

We treat the three coupled BEC model using the semiclassical mean-field approximation. Ignoring correlations between all operators and taking expectation values converts the eight operator differential equations for the quantum system into eight differential equations for real variables in the semiclassical system. The expectation values of generators are denoted by their subscripts, and the expectation value of the total number operator is N. The equations of motion can be derived from the Heisenberg equations of motion of the Hamiltonian (3). These eight variables are not independent due to conservation relations which constrain the fixed points to be

$$x_1 = x_2 = 0, \tag{5}$$
$$\frac{4}{3}N(\frac{N}{3}+1) = y^2 + z^2 \tag{6}$$

where $y \equiv y_1 = y_2 = y_3$, and $z \equiv z_1 = z_2 = z_3$. This shows that these fixed points lie on a hyper-surface, and y can be non-zero. These properties come from the high dimensional effects of the system structure.

5   Quantum Properties

In this section we consider the energy eigenstates fully quantum mechanically to explore the quantum mechanism underlying the classical bifurcation mentioned in Ref. [9] which treated the same Hamiltonian as this paper. We determine the energy eigenstates in both of the dominant tunnelling limit and the dominant interaction limit, and we discuss the changes in the eigenstate structure as the system moves from one limit to the other through the semi-classical bifurcation point.

In the limit that the tunnelling dominates the interaction $\Omega \gg \chi$ so the ratio $r = \Omega/\chi \to \infty$, an N particle system can be considered a set of N one-particle systems. Each one particle system has a lowest energy ground state separated from two degenerate excited states, with the energy gap being equal to 3Ω. All the energy eigenstates of the N particle system consist of product states of these three single particle states. The ground state of the N particle system is a product state of N of the lowest energy states of the one-particle system. The first excited state of the N particle system is given by a product state of N of the lowest energy single particle states and one excited single particle state, which shows that the first excited state is doubly degenerate with energy gap 3Ω. Hence the energy eigenvalue structure in this limit is that the n-th excited energy eigenstates are (n+1)-fold degenerate with the constant energy gap 3Ω.

On the other hand, when the interaction dominates the tunnelling $r = \Omega/\chi \to 0$, and for attractive two-body forces, it is evident that the three lowest energy eigenvectors $|e_1\rangle, |e_2\rangle, |e_3\rangle$ correspond to having all three atoms localized in the same well, for each of the three wells. The degeneracy in the three lowest energy states comes from the symmetry of the interaction terms in the Hamiltonian which is invariant against permutations of the well index. Then, the ground states must be invariant against permutations of the well index, which implies these states are also eigenstates of the permutation operators. To define three ground states, it is necessary and sufficient that we consider the operator of the symmetry group of rotations of an equilateral triangle. Using this operator, we can determine ground states of the form

$$g_1 = \frac{1}{\sqrt{3}}(|e_1\rangle + |e_2\rangle + |e_3\rangle) \tag{7}$$

$$g_2 = \frac{1}{\sqrt{3}}(|e_1\rangle + e^{i2\pi/3}|e_2\rangle + e^{-i2\pi/3}|e_3\rangle) \tag{8}$$

$$g_3 = \frac{1}{\sqrt{3}}(e^{i2\pi/3}|e_1\rangle + |e_2\rangle + e^{-i2\pi/3}|e_3\rangle). \tag{9}$$

Now we turn to consider the eignestate structure between the above limits $0 < r < 1$. From numerical diagonalization of the Hamiltonian, it is known that the system has a single symmetric ground state and two degenerate excited states between the above two limits. The ground state and the doubly degenerate excited states are almost degenerate for ratio values between $0 \leq r \leq r_b$ where r_b is the classical bifurcation point. Analysis shows that these three states converge to the states (7 ~ 9) in the limit $r \to 0$. As r becomes larger than r_b, the symmetric ground state becomes well separated in energy from the two doubly degenerate excited states.

References

[1] M. H. Anderson, J. R. Ensher, M. R. Matthews, C. E. Wieman, and E. A. Cornell, Science **269**, 198 (1995).

[2] C. C. Bradley, C.A. Sackett, J.J. Tollett, and R. G. Hulet, Phys. Rev. Lett. **75**, 1687 (1995).

[3] K. B. Davies, M. -O. Mewes, M. R. Andrews, N. J. van Druten, D. S. Durfee, D. M. Kurn, and W. Ketterle, Phys. Rev. Lett. **75**, 3969 (1995).

[4] M. -O. Mewes, M. R. Andrews, N. J. van Druten, D. M. Kurn, D. S. Durfee, and W. Ketterle, Phys. Rev. Lett. **77**, 416 (1996).

[5] J. R. Ensher, D. S. Jin, M. R. Matthews, C. E. Wieman, and E. A. Cornell, Phys. Rev. Lett. **77**, 4984 (1996).

[6] G. J. Milburn, J. Corney, E. M. Wright, and D. F. Walls, Phys. Rev. A **55** 4318 (1997).

[7] J. F. Corney and G. J. Milburn, Phys. Rev. A in print.

[8] D. Jaksch, C. Bruder, J. I. Cirac, C. W. Gardiner, and P. Zoller, cond-mat/9805329.

[9] E. Wright, J. C. Eilbeck, M. H. Hays, P. D. Miller and A. C. Scott, Physica D **65** 18 (1993).

[10] For instance, *Bose-Einstein Condensation*, edited by A. Griffin, D. W. Snoke, and S. Stringari (Cambridge University Press, Cambridge, England,1995).

[11] J. Schwinger, *Quantum Theory of Angular Momentum*, edited by L. Biedenharn and H. van Dam, (Academic Press, New York, 1965).

[12] See, for example, B. G. Wybourne, *Classical Groups for Physicists*, (John Wiley and Sons, Inc., New York, 1974).

Quantum Coherence and Decoherence - ISQM - Tokyo '98
Y.A. Ono and K. Fujikawa (Editors)
© 1999 Elsevier Science B.V. All rights reserved.

MACROSCOPIC QUANTUM TUNNELING IN MAGNETIC SYSTEMS

Eugene M. Chudnovsky

Physics Department, CUNY Lehman College, Bronx, New York 10468-1589

The theory of thermally assisted quantum decay of metastable spin configurations in magnetic systems is discussed. It is argued that when a system is being cooled down, the crossover on temperature from the classical overbarrier decay of a metastable state to the quantum tunneling under the barrier resembles a phase transition. It is demonstrated that in spin systems such a transition can be first or second order, depending on the strength of the external field.

All magnets exhibit irreversibility when being magnetized. This is the consequence of metastable spin configurations. Due to thermal fluctuations they decay with time at a rate $\Gamma \propto \exp(-U/T)$ (U being the energy barrier). Although at $T \to 0$ thermal fluctuations die out, experiments have shown that in many magnetic systems the relaxation towards the absolute energy minimum persists down to $T = 0$ [1]. Of course, no experiment should be trusted unless supported by theory. The latter suggests that the non-thermal magnetic relaxation should exist due to quantum tunneling [1]. Examples are quantum superparamagnetism in nanoscale single-domain ferro- and antiferromagnetic particles, quantum depinning of domain walls, quantum nucleation of domains, etc. Each of these processes involves the coherent rotation of thousands of atomic spins. Its study in terms of a microscopic spin Hamiltonian is usually hopeless. Fortunately, at a few nanometer scale and low enough temperatures the only trace of atomic exchange interactions in a ferromagnetic crystal is the existence of the fixed-length local spin density $\mathbf{S}_0(\mathbf{r})$ [2]. This variable provides a convenient macroscopic description of the magnet. The total spin, $\mathbf{S} = \int d^3r \mathbf{S}_0(\mathbf{r})$, within a certain volume, satisfies the quantum commutation relation: $S_i S_k - S_k S_i = i\epsilon_{ikl}S_l$. In a nanometer size volume, however, each of the two terms in the left-hand side of this equation is thousand times greater than the right-hand side of the equation. Consequently, $\mathbf{S}$ can be treated as a classical vector whose components almost commute with each other. It satisfies the Landau-Lifshitz equation [3],

$$\frac{\partial \mathbf{S}_0}{\partial t} = -\mathbf{S}_0 \times \frac{\delta E}{\delta \mathbf{S}_0} + \eta \mathbf{S}_0 \times \left(\mathbf{S}_0 \times \frac{\delta E}{\delta \mathbf{S}_0} \right); \tag{1}$$

where $E(\mathbf{S}_0, \mathbf{H})$ and η are the phenomenological energy density and the dissipation parameter respectively. Without dissipation, Eq.(1) is equivalent to the magnetic Lagrangian:

$$L = \int d^3r \left[S_0 \dot{\phi}(\cos\theta - 1) - E(\theta, \phi) \right] , \tag{2}$$

where $\theta(\mathbf{r}, t)$ and $\phi(\mathbf{r}, t)$ are spherical coordinates of $\mathbf{S_0}$. This simply follows from the fact that $S_z = S\cos\theta$ and ϕ are cannonically conjugate variables [4].

Over the last 50 years equations (1) and (2) have been extraordinarily successful in describing all sorts of micromagnetic phenomena including equilibrium magnetic states, domain walls and their dynamics, spin waves, etc. A less appreciated fact is that the very same equations describe quantum decay of metastable magnetic states. If one switches in Eq.(1) to the imaginary time and neglects dissipation, the equation possesses exact solutions, instantons, connecting different equilibrium spin configurations [5,1]. Trajectories which are close to the instanton dominate the path integral over virtual spin configurations. This approach to the spin-tunneling problem, besides of its mathematical beauty, illustrates the point that nature can be quantized at any macroscopic level, provided that we do not exceed energies at which the microscopic structure of the macroscopic variable (in our case $\mathbf{S}(\mathbf{r}, t)$) becomes apparent. Here we will limit our consideration to the crossover from thermal to quantum decay of a metastable spin state in the absence of dissipation. The effect of dissipation can be treated in the spirit of the Caldeira-Leggett approach [6]. In spin problems dissipation is usually weak, unless the question of quantum coherence is concerned.

The rate of the decay of a metatable state of a macroscopic spin in a thermal bath at temperature T is given by [7] $\Gamma = -2Im(F)$, where $\hbar = 1$, $F = -T\ln(Z)$ is the free energy, and Z is the partition function analytically continued into the complex plane,

$$Z = \oint D\{\mathbf{S}(\mathbf{r}, \tau)\} \exp\left(-\int d\tau L_E\right) . \tag{3}$$

Here $L_E = -L(\tau = it)$ is the Eucledian version of the Lagrangian (2). The integration is performed over $\mathbf{S}(\mathbf{r}, \tau)$ configurations periodic in τ with the period $1/T$. We will illustrate the general point by an example of an uniaxial magnetic particle of total spin $S \gg 1$ in the magnetic field making an arbitrary angle with the anisotropy axis [8,9]. The corresponding spin Hamiltonian,

$$\mathcal{H} = -DS_z^2 - H_z S_z - H_x S_x , \tag{4}$$

(where we have included $g\mu_B$ in the definition of the magnetic field) can be mapped onto a particle problem [10] with the Hamiltonian:

$$\mathcal{H} = -\frac{\nabla^2}{2m} + U(x) , \tag{5}$$

where $m = 1/2D$ and the potential, in terms of the reduced field, $\mathbf{h} = \mathbf{H}/(2S+1)D$, is given by

$$u(x) = \frac{U(x)}{(S+1/2)^2 D} = [h_x \sinh(x) - h_z]^2 - 2h_x \cosh(x) . \tag{6}$$

The equivalence between (4) and (5) means that the low-lying eigenstates of (5) have one-to-one correspondence with the eigenstates of (4). The potential (6) has a double-well shape inside the field range, $h_x^{2/3}+h_z^{2/3} < 1$, allowing the existence of the metastable state.

The temperature dependence of the escape rate can be written in the Arrhenius form,

$$\Gamma = A \exp\left[-\frac{U}{T_{eff}(T)}\right] , \qquad (7)$$

with T_{eff} being the effective escape temperature that describes both thermal fluctuations and quantum tunneling. As has been suggested in the first work cited in Ref.5, Γ is dominated by the classical periodic (with the period $1/T$) trajectories in the inversed potential, thermons [11]. At high temperature the rate of thermal escape is exponentially large compared to the quantum rate and $T_{eff} = T$ with high accuracy. The presence of tunneling results in $T_{eff}\neq 0$ at $T = 0$. For a large system, the crossover from purely thermal activation ($T_{eff} = T$) at high temperature to the thermally assisted tunneling with some slower dependence of T_{eff} on T at lower temperatures occurs in the close vicinity of a certain temperature T_0. In many respects it resembles a phase transition smothened by the finite size of the system [7,12]. The crossover can be sharp (first order) or smooth (second order), depending on the shape of the potential. Note that the difference is qualitative, not just quantitative. The thermally assisted tunneling corresponds to the tunneling from excited levels. Due to the exponential dependence of the escape rate on the height of the excited level, the total rate is dominated by a very narrow group of levels inside the well. As temperature rises, this group of levels moves up. Depending on the shape of the potential, it either does it continuously or jumps discontinuously, at a certain temperature, from the bottom to the top of the well. Which of the two possibilities takes place is determined by whether the period of the oscillations in the inversed potential is a monotonic or a non-monotonic function of energy [11]. For an infinite system, dT_{eff}/dT is continuous at $T = T_0$ in the first case and discontinuous in the second case. To have a non-zero rate, a finite system is needed, however. In that case the discontinuity is smeared but the difference between the first and second order remains pronounced. For commonly used metastable potentials like, e.g., $-x^2 + x^3$ and $-x^2 + x^4$, the transition is always of the second order. This kind of a transition has been intensively studied, including the effect of dissipation [12]. On the contrary, the study of the first-order transition has just begun.

A very interesting feature of spin systems is that they possess both first and second order transitions. One can switch from one type of the transition to another by changing the strength of the external magnetic field. One-half of the potential (6) for $x > 0$ at $h_z = 0$ is shown in Fig.1(left). The whole potential is symmetric about the $x = 0$ axis. It has two wells separated by the energy barrier. Tunneling between the wells corresponds to the tunneling of S_z between positive and negative values. The metastable state exists for $h_x < 1$. One can easily check analytically, as well as numerically, that at $h_x > 1/4$

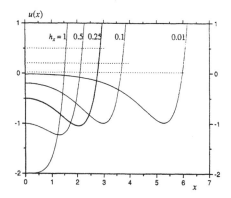 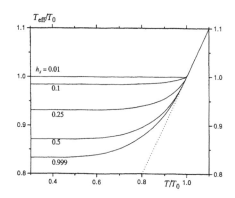

Figure 1: Left: The $x > 0$ half of the reduced effective potential at $h_z = 0$. The dashed lines indicate boundaries between spin levels and unphysical particle levels. Right: Dependence of T_{eff} on T for different values of the transverse field, scaled to the actual temperature, T_0, of the crossover from thermal activation to thermally assisted tunneling.

the period of oscillations inside the well of the inversed potential is a monotonic function of the amplitude. At $h_x < 1/4$ it is not. In that field range the flat shape of the inversed potential near the bottom results in the period being large at small amplitudes compared to the period at higher amplitudes. This flattening of the effective potential at small h_x is easy to understand. In the limit of $h_x \to 0$ the height of the energy barrier is determined by the magnetic anisotropy and h_z. In that limit the tunneling is prohibited because the Hamiltonian (4) commutes with S_z. The only way to achieve that is to send the width of the barrier to infinity at $h_x \to 0$.

The direct numerical computation of the partition function over periodic $\mathbf{S}(\tau)$ configurations [8] gives the dependence of T_{eff} on T at $h_z = 0$ shown in Fig.1(right). Here T_0 is the temperature of the crossover between purely thermal regime and thermally assisted tunneling. The dependence of T_0 on h_x at $h_z \to 0$ is plotted in Fig.2(left). For an arbitrary direction of the field with respect to the anisotropy axis, there is a phase diagram in the (h_x, h_z) plane separating the regions of first and second order transitions [9], Fig.2(right).

The existence of first and second order quantum-classical transitions in the escape rate is a very general feature of spin systems. Besides the uniaxial model considered above, it has been recently demonstrated for the biaxial model in the transversed field [13]. The phase diagram for this model has been also computed for an arbitrary direction of the field [14]. These predictions of the quantum-statistical theory can be tested in experiments on single-domain magnetic particles and on high-spin molecular clusters. The accuracy of experiments on individual single-domain particles [15] is, probably, not

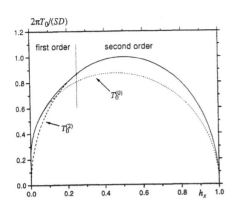
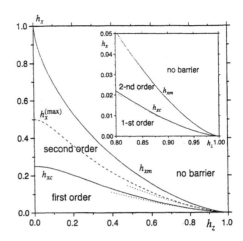

Figure 2: Left: Dependence of the crossover temperature on the transverse field at $h_z = 0$. The dotted lines show results obtained analytically [8]. Right: Phase diagram for the quantum-classical transition in the escape rate. The dashed line corresponds to the maximum of T_0 as a function of h_x. The asymptotes (dotted lines) show analytical results obtained for small barrier [9].

yet at the required level, but high-spin clusters seem very suitable for the above studies [16]. Crystals of Mn-12 (uniaxial anisitropy) and Fe-8 (biaxial anisotropy) contain macroscopic number of identical, weakly interacting spin-10 clusters. They allow to test theoretical suggestions in measurements of the magnetization curve and magnetic relaxation. The required fields are of order 1-3T and the temperatures needed are in the 0.1-1K range.

Highly anisotropic molecular nanomagnets may become the ultimate limit of the miniaturization of elements of magnetic memory. Thus, the study of thermal-quantum crossover in the magnetization reversal of these systems, besides being of fundamental interest, may also be important for applications. The possibility to have a magnet in a superposition of $|\uparrow>$ and $|\downarrow>$ magnetic states opens a remote possibility of using high-spin molecular clusters for elements of quantum computers. Of course, we should learn first how to read and write magnetic information at the nanoscale.

This work has been supported by the U.S. National Science Foundation under the Grant No. DMR-9024250.

REFERENCES

1. E.M.Chudnovsky and J.Tejada, Macroscopic Quantum Tunneling of the Magnetic Moment, Cambridge University Press, 1998.

2. In antiferromagnets and ferrimagnets S carries an isotopic index numerating magnetic sublattices.

3. L.D.Landau and E.M.Lifshitz, Sov. Phys. JETP 8 (1935) 153.

4. The total time derivative term, $S_0\dot\phi$, is relevant only for the quantum problem discussed below. With its addition to the Lagrangian the first term in Eq.(2) acquires a certain geometrical meaning, see, e.g., E.Fradkin, Field Theories of Condensed Matter Systems, Addison-Wesley, 1991.

5. E.M.Chudnovsky, Sov. Phys. JETP 50 (1979) 1035; M.Enz and R.Schilling, J. Phys. C19 (1986) 1765, L711; J.L. van Hemmen and A.Suto, Physica B141 (1986) 37; E.M.Chudnovsky and L.Gunther, Phys. Rev. Lett. 60 (1988) 661.

6. See, e.g., A.J.Leggett et al., Rev. Mod. Phys. 59 (1987) 1.

7. I.Affleck, Phys. Rev. Lett. 46 (1981) 388.

8. E.M.Chudnovsky and D.A.Garanin, Phys. Rev. Lett. 79 (1997) 4469.

9. D.A.Garanin, X.Martinez Hidalgo, and E.M.Chudnovsky, Phys. Rev. B57 (1998) 13639.

10. G.Scharf, Ann. Phys. (N.Y.) 83 (1974) 71; O.B.Zaslavskii et al., Sov. J. Low Temp. Phys. 9 (1983) 259.; G.Sharf, W.F.Wrezinski, and J.L. van Hemmen, J. Phys. A20 (1987) 4309.

11. E.M.Chudnovsky, Phys. Rev. A46 (1992) 8011.

12. A.I.Larkin and Yu.M.Ovchinnikov, Sov. Phys. JETP Lett. 37 (1983) 382; Sov. Phys. JETP 59 (1984) 420; H.Grabert and U.Weiss, Phys. Rev. Lett. 53 (1984) 1787; W.Zwerger, Phys. Rev. A31 (1985) 1745; P.S.Riseborough, P.Hanggi, and E.Freidkin, Phys. Rev. A32 (1985) 489.

13. J.-Q.Liang et al., Phys. Rev. Lett. 81 (1998) 216.

14. D.A.Garanin and E.M.Chudnovsky, to be published.

15. W.Wernsdorfer et al., Phys. Rev. Lett. 79 (1997) 4014.

16. E.M.Chudnovsky, Science 274 (1996) 938.

Quantum Coherence and Decoherence - ISQM - Tokyo '98
Y.A. Ono and K. Fujikawa (Editors)
© 1999 Elsevier Science B.V. All rights reserved.

Quantum Tunneling in Magnetic Molecules and Single Particles

L. Thomas [a], F. Lionti [a], A. Sulpice [b], A. Caneschi [c], W. Wernsdorfer [a], D. Mailly [d], and B. Barbara [a]

a Laboratoire de Magnétisme Louis Néel, CNRS, BP 166, 38042-Grenoble, France.
b CRTBT, CNRS, BP 166, 38042-Grenoble, France.
c Department of Chemistry, University of Firenze, 50144, Italy.
d L2M - CNRS, 196 Av. H. Ravera, 92220 Bagneux, France

The study of Quantum Tunneling of the Magnetization is organized along two main directions : single crystals of magnetic molecules with large spins (say S=10) and individual nanoparticles with huge spins (say $S=10^5$). QTM was clearly observed in the molecular crystals Mn_{12} acetate [2-7]). This system is made of molecules with collective spin S=10. Level quantization plays an important role and in particular QTM occurs when the spin-up and spin-down level schemes are in coincidence (resonant QTM). The same effect was recently found on a similar system, Fe_8 [8], where ground-state tunneling is more easily observable. Strong indications in favor of QTM in insulating nanoparticles of Ba-ferrites with $S=10^5$ are also given.

1. INTRODUCTION

The observation of the quantum behaviour of a macroscopic variable is a challenging problem [1]. In magnetism, Macroscopic Quantum Tunneling (MQT) consists in e.g. the rotation of the magnetization of a single domain small particle by tunneling through its anisotropy energy barrier, or the motion of a small portion of a domain wall through its pinning energy barrier. The first experiments devoted to this question suggested that the total number of spins involved coherently in these MQT effects is of the order of 10^3 to 10^5, depending on the material [9]). Although not at the human scale, this is macroscopic in the framework of quantum mechanics. In order to avoid the complications of energy barrier distributions, two directions of research were more recently taken : the dynamical study of magnetization reversal of (i) arrays of identical magnetic molecules belonging to molecular crystals and (ii) individual nanoparticles (or wires). These approaches were made possible with recent advances achieved in sample elaboration and measurement techniques, and this allowed this field to develop rapidly these last years.

2. RESONANT TUNNELING ON A SINGLE CRYSTAL OF MN$_{12}$-AC

The Mn_{12}-ac crystal, synthetized by Lis [10] in 1980, is built of discrete dodecanuclear $[Mn_{12}(CH_3COO)_{16}(H_2O)_4O_{12}].2CH_3COOH.4H_2O$ molecules (Mn_{12}-ac) with a tetragonal symmetry. Four inner $Mn(1)^{4+}$ (S=3/2) are surrounded by eight Mn^{3+} (S=2). The latter are divided in two crystallographic sites with strong tetragonal $Mn(2)^{3+}$ and orthorhombic $Mn(3)^{3+}$ Jahn-Teller distortions with high crystal field anisotropy. Strong antiferromagnetic interactions between the four $Mn(1)^{4+}$ and the four $Mn(3)^{3+}$ lead to a ferrimagnetic ground state, of spin S=10 [11].

All present interpretations of the magnetic behaviour of Mn12-ac and in particular the phenomenon of resonant tunneling, assume a well defined collective spin S=10. It is therefore important to know the range of validity of this model. Magnetization measurements performed above the superparamagnetic blocking temperature i.e. between 3 and 300 K, along the c-axis and in the basal plane of a single crystal of Mn12-ac show the existence of $S \leq 9$ excited levels at energies between 30 and 90 K [12]. The range of validity of the collective spin S=10, is therefore T<30 K.

Isothermal hysteresis loops of a single crystal of Mn12-ac showed, below 3K, staircase behaviour when the field was applied along the easy axis of magnetization [6, 7] Fig. 1(1). Magnetic relaxation measured at short times showed a series of sharp minima occurring precisely at the fields for which the steps were observed on the hysteresis loops, $H_n = 0.44n$ T, ($H_n= nD/g\mu_B$, with $D/g\mu_B$=0.44 T). This result was interpreted by assuming that the magnetization of the single crystal relaxes much faster when the spin-up and spin-down level schemes coincide, as already suggested in [3,4] (resonant QTM). However, above 2 K, the relaxation times depend on temperature and follow an Arrhenius law : tunneling is thermally assisted. The tunneling probability being much larger

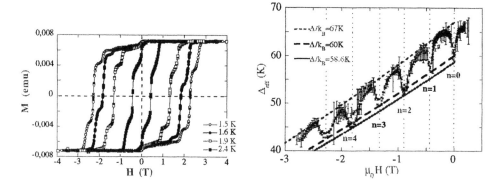

Figure 1. (1) Staircaise hysteresis loops at different temperatures in a single crystal of Mn12-ac. (2) Scaling plot of relaxation times vs applied field. Note a weak but clear parity effect.

on excited states than on the ground state, magnetic paths where tunneling takes place below the top of the barrier are more probable. Nevertheless, relaxation becomes much less dependent on temperature below 2K.

The single-spin Hamiltonian in tetragonal symmetry contains a part (H1) which commutes with S_z and another one (H2) which does not commute with S_z:

$$H_1 = -H_zS_z - DS_z^2 - BS_z^4 + ... \qquad H_2 = -H_xS_x - C(S_x^4 + S_Y^4)...$$

The latter is responsible for tunneling and gives a time-dependent magnetic response. In Mn12-ac experiments [6,7], resonant tunneling was observed with $\Delta m = \pm 1$ showing that each molecule experiences a local transverse field H_x. Dipolar and hyperfine transverse field components (less than 0.05 and 0.01 T respectively) [13,14], as well as fourth order transverse anisotropy have been considered to interpret experimental data. It is important to note that Dzyaloshinsky-Moryia interactions, the possible effect of which on QTM was recognized only very recently [12,15,16], could be the most important term giving off-diagonal matrix elements with $\Delta m = \pm 1$. Even if it is small (which it is not necessarily the case), the term

$$\sum D_{DM}.S_i \times S_j,$$

allowed by symmetry, should be added to   the above equations. Interestingly, it does not conserve S^2, and should mix S=10, with S=9, ...giving a "fine structure" to QTM. Besides, the energy barrier could increase with temperature. The most important role of hyperfine and dipolar interactions is certainly related to their dynamical character. As this is the case for a sweeping field in the Landau-Zener mechanism applied to a magnetic system [17], hyperfine fields (spin-spin and spin-lattice) lead to energy fluctuations of ±m states about a mean value determined by the local field (static) acting on the molecule (sum of dipolar field of other molecules, spin-lattice hyperfine field, D-M field and applied field). When this local field is weak enough, the two levels m and -m are not too far and resonance can be restored by the fluctuations [13,18]. The relaxation times evaluated from this model coincide with the observed ones provided the tunnel splitting is of the oerder of 10^{-8} K in Mn12-ac.

In the presence of a transverse field the magnitudes of the magnetization jumps ΔM increase with the transverse field component $H_T = H\sin\theta$, but the positions of the jumps remain at the same longitudinal field $H_{Ln} = H\cos\theta = nD/g\mu_B$ [7]. The increase of ΔM vs H_T must be due to (a) easier thermal activation resulting from the lowering of the energy barrier ($\propto(1-2H_T/H_A)$), (b) easier QTM resulting from the increase of the tunnel splitting ($\propto(H_T/H_A)^{2m-n}$) of resonant level pairs m and -m+n. A comparison between these two contributions showed that thermally activated tunneling takes place on deeper levels when H_T increases. Ground state tunneling is expected at about 1 K for $H_T \approx 5T$. Not surprisingly, ground state tunneling was recently observed in Fe8, with also a spin S=10, where the ratio H_T/H_A is larger than in Mn12-ac [8]. The study of Mn12-ac is particularly interesting to analyse how the collective spin S=10 behaves in the mesoscopic regime between classical and quantum mechanics. A scaling analysis of magnetic relaxation measured above 2K shows a factor of about one hundred between the relaxation times measured at resonance and out of resonance ($R \approx 100$) (figure 1(2)). When the magnetic states probed by the experiments are closer to the top of the barrier (at shorter time scale and higher temperatures, in a-c susceptibility or EPR experiments), R is smaller and it decreases with the timescale. The classical regime, in which R should be equal to one, is approached asymptotically, but we did not observe it up to 10 K (and 10 kHz). Besides, one can observe in figure 2, that the relaxation minima are deeper for even resonances. This parity effects is a signature of the fourth order transverse anisotropy terms.

3. QTM EFFECTS ON A SINGLE NANOPARTICLE OF BaFeO.

In our investigations for QTM effects at a mesoscopic scale, we also started from "the other side", by looking at relatively large ferromagnetic nanoparticles. In this case the magnetization must be measured on single particles. The first magnetization measurements of individual nanoparticles at low temperature were done using micro-SQUID detectors [19]. For sizes smaller than 20 nm, the particles (Co, FeCu, Ni), are with a single ferromagnetic domain, the probability for non-reversal is exponential ($\approx \exp(-t/\tau)$) and the activation volume is very close to the particle volume [20]. These measurements showed for the first time that the magnetization reversal of a ferromagnetic nanoparticle of good quality can be described by thermal activation over a single-energy barrier as originally proposed by Néel and Brown [21].This agreement with the classical physics constitutes the precondition for the experimental observation of MQT of magnetization in a single particle. Similar single particle measurements were performed on insulating $BaFe_{10.4}Co_{0.8}Ti_{0.8}O_{19}$ nanoparticles (BaFeO) [22]. The agreement with the Néel-Brown model is confirmed above 0.4 K. However, at lower temperature, strong deviation are evidenced which we compare with the MQT theory without dissipation. The BaFeO nanoparticles have a strong uniaxial magnetocrystalline anisotropy depending on the Co-Ti substitution and were made

160

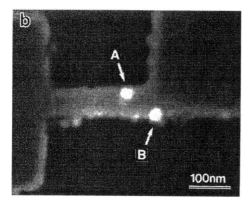

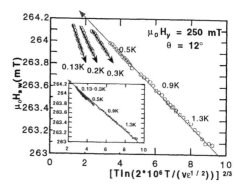

Figure 2: (1) A few nanoparticles were deposited on the micro-bridge of a DC-SQUID. Only the two greatest nanoparticles of about 20 ± 5 nm (white spots) are well coupled with the SQUID-loop. (2) Scaling plots of the mean switching fields H_{sw} for a BaFeO nanoparticle. The field was swept at an angle of $\theta' = 12°$ and a constant perpendicular field of $H_y = 250$ mT was applied. Themeasurements were done at field sweeping rates between 0.01 - 120 mT/s and temperatures between 0.13 - 2.5 K. Inset: Same data of $H_{sw}(v,T)$ but the real temperature T is replaced by an effective temperature T* shown in figure 3(2). The flattening of T* corresponds to a saturation of the escape rate Γ which is a signature of MQT. As measurements at zero temperature are impossible, we investigated the effective temperature at our lowest measuring temperature: T* at $T \approx 0.12$ K.

by Kubo et al. [23].

Between 0.4 K and 6 K, all our measurements on BaFeO nanoparticles were in complete agreement with the Néel-Brown model: (i) exponential probabilities of not-switching, (ii) mean switching fields H_{sw} and rms-values σ following the equations of Kurkijärvi (1 to 5 in ref. 20). Below 0.4 K several of the smallest particles showed strong deviations from the Néel-Brown model. These deviations were (i) a flattening of the thermal dependence of H_{sw} (figure 2(2)), (ii) a higher field sweeping rate (dH/dt = v) dependence of H_{sw} than given by the Néel-Brown model and (iii) a saturation of the width σ of the switching field distribution (figures 3(1)). We investigated these deviations as a function of the field sweeping direction . An example of

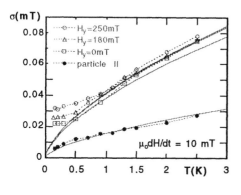

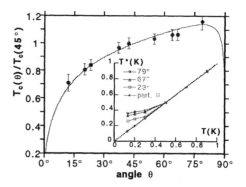

Figure. 3: (1)Temperature dependence of the width of the switching field distribution σ at three different perpendicular applied fields H_y and $\theta' = 12°$. Full points were measured on a particle of about 10^6 μ_B at $H_y = 0$. Lines: prediction of the Kurkijärvi model. (2)-inset : effective temperature T* as a function of the real temperature T at three different applied H_y - fields. Full points were measured on a particle of about 10^6 μ_B.

adjustment of the Néel-Brown model to the measurements of $H_{SW}(v,T)$ and σ is presented in figure 2(2) and 3(1). In all cases for T > 0.4 K, the data of $H_{SW}(v,T)$ can be aligned on the master curve being a straight line. A good agreement between theory and measurements is also found for σ above 0.4K (figure 3 (1)).

The theory predicts an increase of the tunneling rate with the angle θ of the applied field. This increase can also be achieved by applying a constant field perpendicular to the easy axis of magnetization. In the case of magnetocrystalline anisotropy with biaxial symmetry, the expression of the tunneling rate given in [24] allows to determine the crossover temperature $T_c(\theta)$:

$$T_c = \frac{9}{4*6^{1/4}} \frac{E_0}{k\,S} \varepsilon^{1/4} \frac{\sqrt{1 + \frac{K_2}{K_1}\left(1 + |\cot\theta|^{2/3}\right)}}{|\cot\theta|^{1/6}}$$

(defined as the temperature where the measured angle and field dependent Arrhenius law exponent ($E_0\varepsilon^{3/2}/kT$) equals the WKB exponent). S is the total spin of the particle, K_1 and K_2 are the parallel and perpendicular anisotropy constants, respectively. Our measurements are in agreement with these predictions (figure 3(2)). Finally, we found that T_c decreases for bigger particles which can also be explained by the MQT theory as T_c is proportional to $\varepsilon^{1/4}$. For a bigger particle, ε must be smaller than for a smaller one in order to measure a magnetization switching in the same time window, i.e. T_c should be smaller. The level separation between the lowest and the first excited level can be estimated by: $\Delta E \approx DS^2 (1 - (1-1/S)^2) \approx 2E_0/S$. For our particles $\Delta E \approx 1K$, giving for the associated field separation $\Delta H = H_a/2S \approx 0.002$ mT. This is rather small but still measurable, if the level broadening is weak enough. Future measurements will focus on the level quantization of a collective spin state of $S = 10^2$ to 10^4.

4. CONCLUSION

In conclusion, Resonant-QTM was observed in molecules with large collective spins (S=10) : Mn12-ac and more recently Fe8. R-QTM is thermally assisted at high enough temperature or takes place on the ground state. Tunneling on the ground state is more obvious with Fe8 ; it can also be induced in Mn12-ac under the effect of a transverse magnetic field. In this limit of low-temperature, quantum fluctuations dominate magnetic relaxation. In the opposite limit, thermally activated tunneling is dominant and the proportion of thermal activation vs quantum tunneling decreases at higher temperature (and shorter time scales), when the levels probed by the experiment are closer to the top of the barrier. The passage from quantum to classical mechanics appears to be very progressive in Mn12-ac. Finally it is interesting to note that R-QTM could not have been detected with the poor field resolution of conventional SQUID magnetometers, if resonance lines were not importantly broadened. Broadening of magnetic origin plays a major role [13] (hyperfine and dipolar interactions). Diluted clusters with small isotopic fractions of nuclear spins should give narrow transitions, which could be observed with field resolutions of the order of 10^{-4} T. In the next future we should see whether similar effects could be observed with the much larger spins of nanoparticles.

When single particles are small enough and of good quality, micro-SQUID experiments are always in good agreement with the Néel-Brown model [22] of magnetization reversal. This is also the case for a nanocrystal of BaFeO, at temperatures larger than 0.4 K. Below this temperature, strong deviations are observed being quantitatively in agreement with MQT of magnetization. The ultimate proof for MQT in a magnetic nanoparticle would be the observation of level quantization of its collective spin state, as observed in Mn12-ac. This is in principle possible in insulating particles in low dissipation regime.

ACKNOWLEDGEMENTS : we are very pleased to thank D. Gatteschi, R. Sessoli and A. Caneschi from Florence, and as well as O. Kubo from Yokohama, for on-going collaborations.

REFERENCES

1. A.J. Leggett *et al.*, *Rev. Mod. Phys.* **59** (1987) 1 and Lectures in Physics, Les Houches (1986).
2. C. Paulsen, J. G. Park, B. Barbara, R. Sessoli, A. Caneschi, *J. Magn. Magn. Mater.* **140-144** (1995) 379 and 1891.
3. B. Barbara, W. Wernsdorfer, L. Sampaio, J.G. Park, C. Paulsen, M. Novak, R. Ferré, D. Mailly, R. Sessoli, A. Caneschi, K. Hasselbach, A. Benoit, L. Thomas, *J. Magn. Magn. Mater.* **140-144** (1995) 1825.
4. M. Novak and R. Sessoli, in *Quantum Tunneling of the Magnetisation,* eds. L. Gunther and B. Barbara, NATO ASI Series E: Applied Sciences - Vol. 301 (Kluwer, Dordrecht, 1995), p.171.
5. C. Paulsen and J.G. Park, in *Quantum Tunneling of the Magnetisation,* eds. L. Gunther and B. Barbara, NATO ASI Series E: Applied Sciences - Vol. 301 (Kluwer, Dordrecht, 1995), p.189.
6. J. Friedman, M. Sarachik, J. Tejada, J. Maciejewski, and R. Ziolo, *Phys. Rev. Lett.* **76** (1996) 3820 and *J. Appl. Phys.* **81** (1997) 3978; J. Hernandez, X.X. Zhang, F. Luis, J. Barthomomé, J. Tejada, and R. Ziolo, *Euro Phys. Lett.* **35** (1996) 301.
7. L. Thomas, F. Lionti, R. Ballou, D. Gatteschi, R. Sessoli, and B. Barbara, *Nature* **383** (1996) 145; See also *the Proceedings of the Colloque Louis Néel*, January 1996, Le Mont Saint-Odile, F. Lionti, L. Thomas, R. Ballou, D. Gatteschi, R. Sessoli, ana B. Barbara, *J. Appl. Phys.* **81** (1997) 4608.
8. C. Sangregorio, T. Ohm, C. Paulsen, R. Sessoli, and D. Gatteschi, *Phys. Rev. Lett.* **78** (1997) 4654.
9. B. Barbara, *Proc. 2nd Int. Symp. on Anisotropy and Coercivity,* p.137 (1978) and *J. Physique* **34** (1972) 1039; M. Uehara and B. Barbara, *J. Physique* **47** (1985) 235.
10. T. Lis, *Acta Cristallo.* **B36** (1980) 2042.
11. B. Barbara and L. Gunther, *J. Magn. Magn. Mater.* **128** (1993) 35.
12. B. Barbara, L. Thomas, F. Lionti, A. Sulpice, A. Caneschi, *J. Magn. Magn. Mater.* **177-181** (1998) 1324.
13. A.L. Burin *et al.*, *Phys. Rev. Lett.* **76** (1996) 3040; N.V. Prokof'ev and P.C.E. Stamp, *J. Low Temp. Phys.* **104** (1996) 209.
14. F. Hartmann-Boutron, P. Politi, and J. Villain, *Int. J. Mod. Phys.* **B10** (1996) 2577.
15. B. Barbara *et al.*, in *NATO Forum in "Nanoscale Science and Technology"*, Tolede, eds. N. Garcia, M. Nieto-Vesperinas, and H. Rohrer (Kluwer, Dordrecht, 1997)
16. M.I. Katsnelson, V.V. Dobrovitski, and B.N. Harmon, preprint.
17. S. Miyashita, *J. Phys. Soc. Japan* **65** (1996) 2734.
18. N.V. Prokof'ev and P.C.E. Stamp, *Phys. Rev. Lett.* **80** (1998) 5794.
19. W. Wernsdorfer, K. Hasselbach, D. Mailly, B. Barbara, A. Benoit, L. Thomas, *J. Magn. Magn. Mater.* **145** (1995) 33.
20. W. Wernsdorfer, E. Bonet Orozco, K. Hazzelbach, A. Benoît, B. Barbara, N. Domoncy, A. Loiseau, D. Boivin, H. Pascard, D. Mailly, *Phys. Rev. Lett.* **78** (1997) 1791.
21. L Néel, *Ann. Geophys.* **5** (1949) 99; W.F. Brown, *Phys. Rev.* **130** (1963) 1677.
22. W. Werndsorfer, K. Hasselbach, E. Bonet Orozco, A. Benoit, D. Mailly, O. Kubo, and B. Barbara, *Phys. Rev. Lett.* **79** (1997) 4014.
23. O. Kubo, T. Ido, H. Yokoyama, and Y. Koike, *J. Appl. Phys.* **57** (1985) 15; O. Kubo, T. Ido, · and H. Yokoyama, *IEEE Trans. Magn.* **MAG-23** (1987) 3140.
24. M.-C. Miguel and E.M. Chudnovsky, *Phys. Rev.* **B54** (1996) 389; G.-H. Kim and D.S. Hwang, *Phys. Rev.* **B55** (1997) 8918.

Quantum Coherence and Decoherence - ISQM - Tokyo '98
Y.A. Ono and K. Fujikawa (Editors)
© 1999 Elsevier Science B.V. All rights reserved.

Magnetic and Transport Properties of Sub-micron Ferromagnetic Wires

Y. Otani[a], S. G. Kim[a], K. Fukamichi[a], S. Yuasa[b], M. Nyvlt[b], and T. Katayama[b],
O. Kitakami[c] and Y. Shimada[c]

[a]Department of Materials Science, Graduate School of Engineering, Tohoku University, Sendai
980-8579, Japan
[b]Electrotechnical Laboratory, Tsukuba, Ibaraki 305, Japan
[c]Research Institute for Scientific Measurements, Tohoku University, Sendai 980-8577, Japan

The domain wall contribution to the magnetoresistivity was carefully examined by using
two different systems such as polycrystalline Co submicron wires and micron epitaxial (001)
Fe wires. The magnetization reversal takes place via single domain wall propagation in the
Co wires, whereas via formation of regularly spaced domain walls in the epitaxial Fe wires.
In both cases, the domain walls enhance the conductivity at low temperatures. However,
only for the Co wires, the temperature dependence of zero field resistivity exhibits a minimum
at about 30 K, suggesting that the weak localization is responsible for the negative domain
wall contribution to the magnetoresistivity.

1. INTRODUCTION

Micro fabrication techniques now allow the preparation of high quality nano-structured
magnetic wires. Such systems provide an opportunity in exploring the interplay between the
electron transport and well defined magnetic domain structures. A number of low temperature
magnetoresistivity measurements and domain wall observations for polycrystalline submicron
or micron wires of Ni [1], Fe-Ni [2], Co [3] and Fe [4] showed that the magnetic domain wall
nucleation and propagation processes lead to discontinuous jumps in the magnetoresistivity.
Remarkable is that the domain walls enhance the electrical conductivity of the wire. This
effect is quite different from the conventional anisotropy magnetoresistivity due to the spin-orbit
interaction, because the magnetoresistivity depends not on the relative orientation of the
magnetization but on the presence of the domain wall. An independently proposed theoretical
explanation suggests that the domain walls destroy the coherency of the electron wave
functions responsible for weak localization at low temperatures [5]. Recently Ruediger *et al*
have examined the low temperature magnetoresistivity and domain structures of micron scale
epitaxial (110) Fe wires with well defined inplane uniaxial magnetic easy axes. The contribution
of periodically arranged domain walls was evaluated by subtracting the other contributions to
the magnetoresistivity such as the Lorentz and anisotropic magnetoresistivities [6]. They
concluded that the domain walls enhance the wire conductivity below 80 K, which is however
due to the interplay between orbital effects in the modulated internal magnetic field near
domain walls and surface scattering.

We will represent some results obtained for two different types of systems such as
polycrystalline Co submicron wires and micron scale epitaxial (001) bcc Fe wires with four

fold magnetic easy axis symmetry. The magnetization reversal takes place via single domain wall propagation in the Co wires, whereas via formation of regularly spaced domain walls in the epitaxial Fe wires.

2. EXPERIMENTAL

Polycrystalline Co wires 0.6 μm in width and 100 nm in thickness were fabricated on Si substrates by means of high resolution electron beam lithography (JEOL 5000SD) and a lift-off technique. Co was deposited by a dc magnetron sputter method. The base pressure was 5×10^{-8} Torr and Ar pressure during deposition was 3×10^{-3} Torr. X-ray diffraction analyses for a continuous Co film prepared as a reference showed that the film consists of isotropic polycrystallites. The sample structure is shown in Figs. 1 (a) and (b). The wire and current leads consists of Co. Non-magnetic Al voltage leads 1 μm in width are attached to the wire to detect the change in resistivity due to a propagating domain wall without disturbing the magnetic configuration. The distance between the Al voltage leads is 10 μm. Our previous study assures that the magnetization reversal is due to a single domain wall propagation in this configuration [3,4].

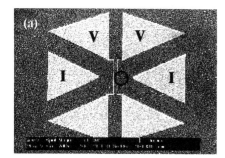 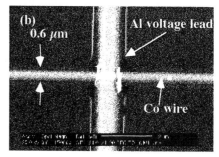

Figure 1 (a) Scanning electron micrograph of a 0.6 μm wide Co wire and (b) the magnified micrograph of the juction area indicated by an open circle.

The 50 nm and 100 nm thick (001) oriented bcc Fe films were grown on (001) MgO substrates at 393 K by means of molecular beam epitaxy. Then 3 nm of (001) Au capping layer was grown on top of the Fe layer to prevent oxidation. The vacuum pressure during deposition was about 10^{-10} Torr. X-ray polar maps showed that the (001) Fe layers grew with their in-plane [110] axis parallel to the [100] axis of the MgO substrate. These (001) Fe films were patterned into a wire structure 2 μm in width with a high resolution electron beam lithographer, followed by an Ar ion etching procedure. The (001) Fe films possess four-fold magnetic easy axes along the <100> directions and hard axes along the <110> directions. The wire structure was cut so as to have its axis parallel to the hard [110] axis.

The magnetoresistivity measurements were performed by an ac 4 terminal method in magnetic fields applied parallel and perpendicular to the wire axis from 5 K to 300 K. The electrical current of 40 μA was applied along the wire axis. The magnetic structure in the remanent state was examined by magnetic force microscopy (MFM SEIKO SPI3800).

3. RESULTS AND DISCUSSION

3.1 Polycrystalline Co wire

Representative longitudinal (i // H) and transverse (i ⊥ H) magnetoresistivity curves measured at 8 K for a 0.6 μm wide polycrystalline Co wire are shown in Figs. 2 (a) and (b). The longitudinal magnetoresistivity curve shows a gradual decrease up to 50 kOe due to the forced magnetoresistivity effect. There are no anomalous behaviors except for the discontinuous jumps observed in low magnetic fields below 1 kOe as shown in Fig. 2 (b). On the other hand, the transverse magnetoresistivity shows an abrupt decrease with increasing magnetic field. This behavior is consistent with the anisotropy magnetoresistivity arising from the magnetization vector rotating out of the wire axis. The magnitude of the anisotropy magnetoresistivity $\Delta\rho_{AMR}$ is about 0.34 μΩ cm.

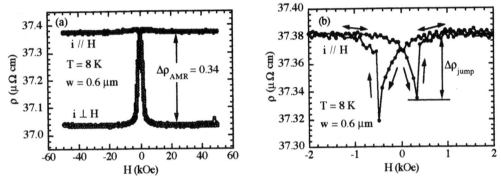

Fig. 2 (a) Longitudinal and transverse magnetoresistivity curves measured at 8 K for a 0.6 μm wide Co wire and (b) a magnified longitudinal magnetoresistivity curve.

The discontinuous jumps observed in the longitudinal magnetoresistivity are caused by the domain wall traversing in between the voltage leads. The change in the longitudinal resistivity $\Delta\rho_{jump}$ amounts ~0.056 μΩ cm. If we assume that $\Delta\rho_{jump}$ is entirely originated from the conventional anisotropy magnetoresistivity, the direction of the magnetization vector averaged out along the wire must be at an angle of ~24° from the wire axis. This is hard to believe because the jumps are pronounced with decreasing temperature below 30 K even though the anisotropy magnetoresistivity $\Delta\rho_{AMR}$ is almost constant as shown in Fig. 3 and the inset. Note that the domain wall width is also almost constant (~10 nm) in this temperature range. Thus the domain wall gives a negative contribution to the magnetoresistivity, which can not be simply explained as the anisotropy magnetoresistivity effect.

Remarked is that the temperature variation of the zero field resistivity ρ_0 exhibits a minimum at 30 K as in Fig. 3. This behavior suggests a possibility of weak localization occurring in this system without domain walls. In other words, the domain wall could suppress the weak localization, which leads to the negative contribution to the magnetoresistivity as suggested by Tatara and Fukuyama [5]. The temperature dependence of ρ_0 is however difficult to be scaled with the power of $1/T$, which is known as a characteristic of the weak localization behavior [7]. Nevertheless the magnitude of the domain wall contribution $\Delta\rho_{jump}/\rho_0$ is proportional to $T^{3/2}$ below 30 K. This variation supports the physical picture where the other decoherence mechanisms such as the electron-electron scattering gradually overtakes the

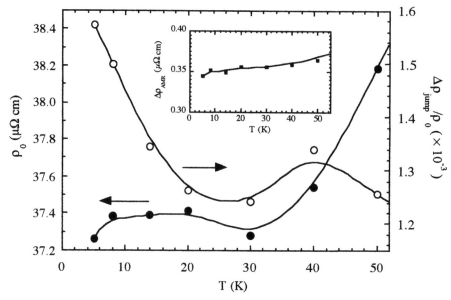

Fig. 3 Zero field resistivity ρ_0 (solid circles) and the magnitude of the domain wall contribution $\Delta\rho_{jump}/\rho_0$ (open circles) as a function of temperature. The inset shows the anisotropy magnetoresistivity $\Delta\rho_{AMR}$ as a function of temperature.

domain wall contribution with increasing temperature.

3.2 Epitaxial Fe wire

Figure 4 shows magnetoresistivity curves measured at 8 K and 200 K in applied fields along [110], [100] and [-110] directions for the 50 nm thick 2 μm wide (001) Fe wire. Hysteresis loops are observed in all curves in the field range where the wire is not saturated towards the applied field directions. Above the saturation field of about 10 kOe, the resistivity increases monotonically as a function of the applied field up to 90 kOe for all curves. The same tendency was observed for the 100 nm thick wire. As seen in the figure, the longitudinal [110] magnetoresistivity is lower than that of the transverse [-110] magnetoresistivity at 8 K, while the relationship between them is completely reversed at 200 K. This is due to temperature dependence of the competing contributions of the Lorentz and anisotropy magnetoresistivities. In general, the Lorentz magnetoresistivity makes the transverse [-110] magnetoresistivity larger than the longitudinal [110] magnetoresistivity, while the anisotropy magnetoresistivity makes them opposite. The crossover was found to take place at 66 K, which coincides well with the value obtained for the (110) Fe wire [6].

We remark that the zero field resistivities ρ_0 coincide well with each other for all the directions. This implies that the remanent magnetic state in the wire is identical. In order to verify this result, MFM images of domain configurations were taken at room temperature after the wire was magnetized in an applied field of 9 T along [110], [100] and [-110] directions. As expected from the resistivity measurements, the domain structure in the remanent

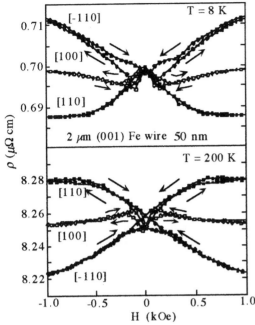

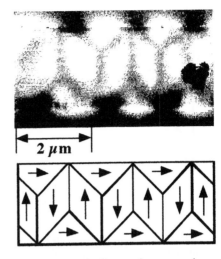

Fig. 5 Magnetic force microscope image of a 50 nm thick (001) Fe wire together with an inferred domain structure.

Fig. 4 Magnetoresistivity curves measured at 8 K and 200 K in applied fields along longitudinal [110], [100] and trasverse [-110] directions for a 50 nm thick 2 μm wide (001) Fe wire.

state is identical, as in Fig. 5, irrespective of the applied field directions. The corresponding inferred structure is given in Fig. 5. There are [1-10] and [-110] oriented magnetic domains separated with periodical 180° domain walls and [110] oriented triangular segments which realize flux closure. The observed period of the 180° walls is about 1 μm. The volume fraction of the [1-10] and [-110] oriented magnetic domains is 0.5. The zero field resistivity ρ_0 can therefore be approximated without taking into account the domain wall contribution as $1/2(\rho_{[-110]} + \rho_{[110]})$, where $\rho_{[-110]}$ is the resistivity of the [1-10] or [-110] oriented domains and $\rho_{[110]}$ is the resistivity of the [110] or [-1-10] oriented domains. The deviation from this value is thus considered as the domain wall contribution.

In order to evaluate the values of $\rho_{[-110]}$ and $\rho_{[110]}$, the magnetoresistivity curves measured at various temperatures were analyzed in terms of Kohler's rule given by $\Delta\rho/\rho(0,T) = [\rho(B,T) - \rho(0,T)]/\rho(0,T) = f(B/\rho(0,T))$, where B is the internal field B (= $\mu_0 H + M$) with the applied field H and the magnetization M, and f is a scaling function characteristic of the material which is independent of the temperature T [8]. This equation means that the magnetoresistivity with different scattering times can be related by rescaling the field with ρ (0, T) since the resistivity in zero internal field ρ (0, T) is proportional to the scattering rate. The quantity B/ρ (0, T) is equivalent to $\omega_c\tau$ where ω_c is the cyclotron frequency and τ is the scattering time. For this analysis, the magnetoresistivity data in the field range from 1 T to 9 T were used, where the wire is saturated along the applied field direction. The value of M is taken as the saturation magnetization M_s of 2.2 T (22 kOe) for Fe. Kohler's rule well scales all the magnetoresistivity curves measured from 5 K to 300 K,

168

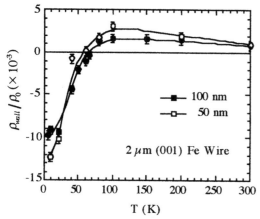

Fig. 6 Domain wall contribution as a function of temperature for 50 nm and 100 nm thick (001) Fe wires.

and the values of $\rho_{[110]}(0,T)$, $\rho_{[110]}(M_s,T)$, $\rho_{[-110]}(0,T)$ and $\rho_{[-110]}(M_s,T)$ were determined. The anisotropy magnetoresistivity given by ($\rho_{[110]}(0,T)$-$\rho_{[-110]}(0,T)$)/ $\rho_{[110]}(0,T)$ is in the range of about 3×10^{-3} from 8 K to 300 K. The values of $\rho_{[110]}(0,T)$ and $\rho_{[-110]}(0,T)$ vary with satisfying the relation αT^2 where $\alpha \sim 2.0 \times 10^{-4}$ $\mu\Omega \mathrm{cm/K}^2$.

Figure 6 shows the domain wall contribution ρ_{wall} as a function of temperature determined as the deviation $\rho_0 - 1/2(\rho_{[-110]}(M_s,T) + \rho_{[110]}(M_s,T))$. The domain wall contribution ρ_{wall} is negative for both the 50 nm and 100 nm thick wires below 66 K where the crossover is observed. The value of ρ_{wall} at 8 K for the 50 nm thick wire amounts -8.7×10^{-9} Ω cm. With increasing temperature, ρ_{wall} abruptly decreases, changing its sign and remains constant up to room temperature. These results are in good agreement with the data obtained Ruediger et al [6] except for the fact that our domain wall contribution is positive above the crossover temperature. These results also support the fact that the presence of the domain walls enhance the conductivity of the wire. However the temperature of 66 K is much higher than the predicted temperature of 10 K from the model based on the weak localization.

This work is partly supported by the new energy and industrial technology development organization (NEDO), RFTF of Japan Society for the Promotion of Science, and the Grant-in-aid for Scientific Research from the Japanese Ministry of Education, Science, Sports and Culture in Japan.

REFERENCES

1. N. Giordano and J. D. Monnier, Physica **B194-196** (1994) 1009; K. Hong and N. Giordano,Phys. Rev. **B51** (1995) 9855.
2. A. O. Adeyeye, J. A. C Bland, C. Daboo, Jaeyong Lee, U. Ebels and H. Ahmed, J. Apple . Phys. **79** (1996) 6120.
3. Y. Otani, S. G. Kim, K. Fukamichi, O. Kitakami and Y. Shimada: IEEE Trans. Magn., **MAG-34** (1998) 1096.
4. Y. Otani, K. Fukamichi, O. Kitakami, Y. Shimada, B. Pannetier, J. P. Nozieres, T. Matsuda and A. Tonomura, Proc. MRS Meeting, (San Francisco) **475** (1998) 215.
5. G. Tatara and H. Fukuyama, Phys. Rev. Lett. **78** (1997) 3773.
6. U. Ruediger, J. Yu, S. Zhang, and A. D. Kent, Phys. Rev. Lett. **80** (1998) 5639.
7. N. Giordano, Phys. Rev. **B22** (1980) 5635.
8. S. G. Kim, Y. Otani, K. Fukamichi, S. Yuasa, M. Nyvlt, and T. Ktayama: to be published in J. Magn. Magn. Mat.

Quantum Coherence and Decoherence - ISQM - Tokyo '98
Y.A. Ono and K. Fujikawa (Editors)
© 1999 Elsevier Science B.V. All rights reserved.

Electronic Transport Properties in Mesoscopic Ferromagnetic Metals

Gen Tatara[a] and Hidetoshi Fukuyama[b]

[a]Graduate School of Science, Osaka University, Toyonaka, Osaka 560-0043, Japan

[b]Graduate School of Science, The University of Tokyo, Hongo, Tokyo 113-0033, Japan

The electronic transport properties in disordered mesoscopic magnets are investigated theoretically. Two cases of a magnetic wire and multilayer are considered. It is indicated that domain walls and induced magnetic moment due to the proximity effect causes dephasing among electrons, which leads to a suppression of localization due to quantum coherence and thus to a negative magnetoresistance at small field.

1. Transport in Mesoscopic Metallic Magnets

In metallic structures which contains magnetic materials the electronic transport properties can strongly be affected by the configuration of the magnetization. Especially close to the coercive field the resistivity will vary by a small magnetic field due to a rearrangement of the magnetization. This effect, called a magnetoresistance (MR) has been observed for a long time in bulk magnets, and recently in magnetic multilayers. The MR in bulk magnets, which is of the order of a few % and shows anisotropy, has been explained by the spin-orbit interaction in d-band[1]. Much bigger MR effects found in multilayers[2] are attributed to the spin-dependent scattering of the electron at the interface.

Recently the magnetoresistance in a mesoscopic magnetic structures such as wires has been studied intensively[3–8]. In the case of non-magnetic metals, the most significant feature of a mesoscopic system would be the effect of the quantum coherence among electrons, which affects substantially the low energy transport properties. For example in the presence of coherence the impurity scattering leads to an interference of the electron wave and thus to a weak localization. In such cases, even a small perturbation can result in a measurable change in the resistivity of the entire sample. One would then naturally expect that the rearrangement of the magnetization in mesoscopic metallic magnets will affect the quantum transport strongly.

In this paper we study the resistivity in a magnetic wire and multilayer based on the linear response theory. We consider the disordered case where the resistivity is dominated by the normal impurities, treating the effect of magnetism perturbatively.

2. Resistivity due to a Domain wall in a wire

We first consider a wire of ferromagnetic metal with z-axis chosen along the wire. We start from the exchange interaction between the electron and the magnetization (M);

$$H_{\text{exch}} = -\Delta \sum_{\mathbf{x}} M(c^\dagger \boldsymbol{\sigma} c). \tag{1}$$

The configuration of M is that of a single domain wall, $M_z = M \tanh \frac{z}{\lambda}$, where λ is the width of the wall (Fig. 1). Classical transport theories tell us that a domain wall reflects the electron and increases the resistivity. Explicit calculation indicates however that the contribution is negligiblly small in $3d$ transition metals[9,8] because walls there are thick compared with the Fermi wavelength, k_F^{-1}, and thus couple to the electron only weakly, except in the case the difference of the lifetime of electrons with opposite spin is large[10].

Such a wall can still affect strongly the quantum correction. To see this it is useful to carry out a local gauge transformation in the electron spin space, so that the wall is expressed as a classical gauge field which flips the electron spin[8]. In the case of strong Zeeman splitting, $\Delta\tau/\hbar \gg 1$ (τ being the lifetime due to impurities), the most dominant process of quantum correction to the conductivity turns is the one shown in Fig. 2. This process represents a dephasing by the wall. In fact summing over the higher

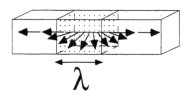

Figure 1. A domain wall in a wire.

Figure 2. Quantum correction due to a domain wall. Shaded squares denote Cooperons and a wavy line represents the interaction with the wall.

order corrections to self-energy of the similar type we find that a particle-particle ladder (Cooperon) acquires an additional mass due to the wall; $1/(Dq^2 + 1/\tau_\varphi) \to 1/(Dq^2 + 1/\tau_\varphi + 1/\tau_w)$. Here $D \equiv \hbar^2 k_F^2/3m^2$ is the diffusion constant and τ_φ denotes the inelastic lifetime due to other sources of dephasing; e.g., phonons. The dephasing time due to the wall, τ_w, is calculated for $\Delta\tau/\hbar \gg 1$ as

$$\frac{1}{\tau_w} = \frac{n_w}{6\lambda k_F^2} \left(\frac{\epsilon_F}{\Delta}\right)^2 \frac{1}{\tau}, \tag{2}$$

n_w being the density of the wall. The correction to the resistivity is then obtained as[8]

$$\frac{\delta\rho}{\rho_0} = -\frac{\tau}{\pi\hbar N(0)} \sum_q \left(\frac{1}{Dq^2\tau + \kappa_\varphi} - \frac{1}{Dq^2\tau + \kappa_w + \kappa_\varphi}\right)$$

$$\simeq -\frac{\sqrt{3}\pi}{(k_F L_\perp)^2}\left(\frac{1}{\sqrt{\kappa_\varphi}} - \frac{1}{\sqrt{\kappa_w + \kappa_\varphi}}\right), \tag{3}$$

where we have evaluated the summation over q with respect to Cooperons as in one dimension, ρ_0 is the resistivity due to impurities, $N(0) \equiv (V m k_F / 2\pi^2 \hbar^2)$ is the density of states $L_\perp$ being the width of the wire. It is interesting that as a quantum correction a domain wall contributes to a decrease of resistivity. The effect can be large for small κ_φ because of a singular behavior of quantum interference.

In the case of wire of the length~ 1000Å, $L_\perp = 300$Å and $k_F^{-1} = 1.5$Å, and if $\lambda \sim$ 150Å and $\Delta/\epsilon_F \sim 0.2$, $\kappa_\varphi = 10^{-4}$ leads to $\delta\rho/\rho_0 \simeq -10^{-4}$, which would be measurable.

Intensive experimental efforts has been put recently in search for a effect of a domain wall in submicron wires of Ni[3], Fe[4,5] and Co[6]. They found one or several discrete jumps of resistivity close to a coercive field, which are argued to be associated with the nucleation, annihilation or a depinning of a domain wall. Most recent results[3–6] suggest a negative contribution of the wall to the resistivity and furthermore the effect grows at lower temperature (below 50K[5] and 20K[6]). These behavior might be due to the quantum effect, although further detailed studies are needed.

3. Resistivity due to a magnetic proximity in multilayers

Next we consider a non-magnetic metallic layer sandwiched between two layers of ferromagnet (Fig. 3). In such systems an effective moment, $M(z)$, is induced by the proximity effect of magnetic layers (z-axis is chosen perpendicular to the plane). We calculate the correction of the conductivity in the plane due to such magnetic moment. We consider the case where the thickness of the layer d is much larger than the elastic mean free path, ℓ. The induced moment, which depends on the configuration of the magnetization of the

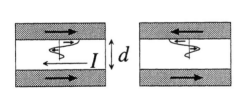

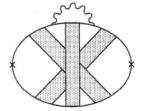

Figure 3. Two configurations of ferromagnets in a multilayer of a nonmagnetic metal and two ferromagnets. The induced magnetic moment is drawn schematically.

Figure 4. Quantum correction due to the induced moment in conduction layer. Wavy line represents the interaction with the induced moment.

two ferromagnetic layers, is treated perturbatively to the second order. In the classical transport theory the contribution is a made up of a self-energy and a vertex correction type processes. It is easy to see that these two processes cancel each other for the current in the plane, which is a natural consequence of a translational invariance in the plane.

The induced moment, however, has a finite effect on the quantum correction of the conductivity, since it modifies the coherence of the electron wave function. The most dominant quantum correction is the process with three Cooperons, shown in Fig. 4. The largest contribution comes from the long range component of $M(Q)$, i.e., $|Q| \lesssim \ell^{-1}$ (Q being the momentum transfer between the induced moment), where all the three Cooperons become important. The quantum correction to the conductivity due to the induced moment is obtained as

$$\delta\sigma = \frac{3\sqrt{3}}{16\pi^2} \frac{e^2}{\hbar} \left(\frac{\Delta_0}{\epsilon_F}\right)^2 k_F^2 \ell \frac{1}{\sqrt{\kappa_\varphi}} \sum_Q |M(Q)|^2 \frac{1}{(Q\ell)^2 + 12\kappa_\varphi}, \tag{4}$$

where Δ_0 is the exchange splitting of the conduction electron at the interface and the moment M is normalized there. Summation over Q is carried out by use of an explicit spatial dependence of the magnetic moment for two cases where the magnetization of the two ferromagnetic layers are parallel or anti-parallel (Fig. 3). The magnitude of the magnetoresistance is measured by the difference of these two cases, $\Delta\rho \equiv \rho^F - \rho^{AF}$ (F and ^{AF} denote the parallel and antiparallel configuration, respectively), which is obtained as[11]

$$\frac{\Delta\rho}{\rho_0} = -5.7 \times \left(\frac{\Delta_0}{\epsilon_F}\right)^2 \frac{1}{k_F^2 d\ell} \frac{1}{\kappa_\varphi} \frac{1}{\sinh \frac{d}{\ell}\sqrt{12\kappa_\varphi}}. \tag{5}$$

For a layer of $k_F d = 450$ and $d/\ell = 10$, and the effective exchange at the interface of $\Delta_0/\epsilon_F \sim 10^{-2}$, $\kappa_\varphi = 10^{-4}$ results in $\Delta\rho/\rho_0 = -7.7 \times 10^{-4}$.

REFERENCES

1. T. R. McGuire and R. I. Potter, IEEE Trans. Magn. **MAG-11** (1975) 1018.
2. M. N. Baibich, J. M. Broto, A. Fert, F. Nguyen Van Dau, F. Petroff, P. Eitenne, G. Creuzet, A. Friederich and J. Chazelas, Phys. Rev. Lett. **61** (1988) 2472.
3. K. Hong and N. Giordano, J. Phys.: Condens. Matter **10** (1998) L401. J. Phys. CM (1997) .
4. Y. Otani, K. Fukamichi, O. Kitakami, Y. Shimada, B. Pannetier, J. P. Nozieres, T. Matsuda and A. Tonomura, Proc. MRS Spring Meeting, San Francisco, 1997.
5. U. Ruediger, J. Yu, S. Zhang, A. D. Kent and S. S. P. Parkin, Phys. Rev. Lett. **80** (1998) 5639.
6. Y. Otani, in preparation.
7. K. Mubu, T. Nagahama, T. Ono and T. Shinjo, submitted to Phys. Rev. **B**.
8. G. Tatara and H. Fukuyama, Phys. Rev. Lett. **78** (1997) 3773.
9. G. G. Cabrera and L. M. Falicov, Phys. Stat. Sol. (b)**61** (1974) 539.
10. P. M. Levy and S. Zhang, Phys. Rev. Lett. **79** (1997) 5110.
11. G. Tatara and H. Fukuyama, in preparation.

Quantum Coherence and Decoherence - ISQM - Tokyo '98
Y.A. Ono and K. Fujikawa (Editors)
© 1999 Elsevier Science B.V. All rights reserved.

Coulomb blockade and enhancement of magnetoresistance change in ultrasmall ferromagnetic double tunnel junctions

Reiko Kitawaki[a], Fujio Wakaya[b] and Shuichi Iwabuchi[a]

[a]Department of Physics, Faculty of Science, Nara Women's University,
Kitauoya-Nishimachi, Nara 630-8506, Japan

[b]Department of Physical Science, Graduate School of Engineering Science,
Osaka University, 1–3 Machikaneyama-cho, Toyonaka, Osaka 560-8531, Japan

A microscopic theory for ultrasmall ferromagnetic *double* tunnel junctions is proposed, which selfconsistently describes the higher order tunneling processes as well as the charged states of the islands for arbitrary environmental impedance. Such a theory is indispensable to understanding of the physics of this system, since the parameter for both stability condition of Coulomb blockade and perturbative treatment of tunneling, $R_q/R_T(H)$ (R_q: quantum resistance, $R_T(H)$: magnetoresistance), strongly depends on applied magnetic field. Tunneling currents and enhancement of negative magnetoresistance due to the Coulomb blockade are briefly discussed.

1. INTRODUCTION

Recently an interesting experimental finding has been reported in ultrasmall ferromagnetic tunnel junctions[1,2]. They found that the change rate of the (negative) magnetoresistance (CRMR) is remarkably enhanced in the Coulomb blockade (CB) regime. This system is a unique system in that the stability condition of CB strongly depends on applied magnetic field in contrast to the CB in the systems ever studied.

In ultrasmall ferromagnetic tunnel junctions, CB is less stable due to quantum fluctuation in the presence of magnetic field for a given charging energy, since the applied magnetic field decreases tunnel resistance (negative magnetoresistance)(Fig. 2). Therefore, the magnetic field tends to cause the sudden breakdown of CB for a certain set of U/k_BT and $R_T(0)/R_q$. To describe the physics of this system the higher order tunneling processes should be taken into account, since perturbative parameter for tunneling is $R_q/R_T(H)$ for a given magnitude of magnetic field (H) and it is not always small enough.

So far, we have already proposed the theory of CB in ultrasmall ferromagnetic *single* tunnel junctions based on the Feynman path integral approach and shown that a remarkable enhancement of CRMR are due to nonlinear current-voltage (I-V) characteristics in Coulomb gap region caused by the sudden breakdown of CB triggered off by the higher order tunneling processes[3,4]. In this theory, however, the electromagnetic environment effect (EMEE) was completely discarded. EMEE is of a main importance in CB problems even in the case of double junctions, since it strongly influences the charged states of the islands, *i.e.*, I-V characteristics[5–8].

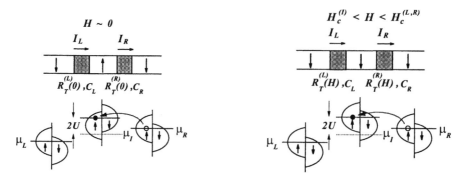

Figure 1. Negative magnetoresistance and CB in UFDJ with tunnel resistance $R_T^{(\alpha)}$ and capacitance C_α. Density of states for each electrode is drawn schematically with chemical potential μ_ν and magnetization (an arrow). $R_T^{(\alpha)}$ is maximum (minimum) for around zero field (for field higher than the coercive magnetic field of the island). For the same charging energy U, Coulomb blockade is less stable for the parallel configuration.

2. HAMILTONIAN OF THE SYSTEM

Let us consider the ultrasmall ferromagnetic *double* junctions (UFDJ) as shown in Fig. 2. The theory we propose is the natural extension of previously proposed one[5–7], which can selfconsistently describe the charged states of the islands for arbitrary environmental impedance. Hamiltonian of UFDJ is taken to have the form

$$\mathcal{H} = \mathcal{H}_0 + \mathcal{H}_T . \tag{1}$$

Here unperturbed Hamiltonian $\mathcal{H}_0$ consists of three parts, $\mathcal{H}_{el}$, $\mathcal{H}_{em}$ and $\mathcal{H}_c$, which respectively describe the electronic states ν [= L(left electrode), R(right electrode), and I (island)], electromagnetic environment effect and charged states of island:

$$\mathcal{H}_0 = \sum_{\nu=L,I,R} \mathcal{H}_{el}^{(\nu)} + \mathcal{H}_{em} + \mathcal{H}_c , \tag{2}$$

where

$$\mathcal{H}_{el}^{(\nu)} = \sum_{k\sigma} \left(\epsilon_k^{(\nu)} - \sigma \frac{\Delta_{ex}^{(\nu)}}{2} \right) a_{k\sigma}^{(\nu)\dagger} a_{k\sigma}^{(\nu)} , \tag{3}$$

$$\mathcal{H}_{em} = \frac{Q^2}{2C} + \frac{\varphi^2}{2L} - QV \quad (C \equiv C_L C_R / C_\Sigma, \ C_\Sigma = C_L + C_R) , \tag{4}$$

$$\mathcal{H}_c = U \left(\frac{q}{e} - n_c \right)^2 \quad (U \equiv \frac{e^2}{2C_\Sigma}) \tag{5}$$

and $\Delta_{ex}^{(\nu)}$ is the exchange splitting energy of the ferromagnetic electrode. Phase variable φ_α, which is canonical conjugate to charge Q_α on junction α, i.e., $[Q_\alpha, \varphi_\alpha] = i\hbar$, is defined as $\varphi_\alpha \equiv c_\alpha \kappa_\alpha \varphi - \psi$ ($c_\alpha = 1(-1)$ for $\alpha = L(R)$, $\kappa_\alpha \equiv C/C_\alpha$). Note that φ and ψ are phases conjugate to $Q \equiv \sum_\alpha \kappa_\alpha Q_\alpha$ (continuos charge) and $q \equiv Q_R - Q_L$ (island charge), respectively, and satisfy $[Q, \varphi] = [q, \psi] = i\hbar$. $\mathcal{H}_T$, on the other hand, describes tunnelings through junctions and is treated as perturbation:

$$\mathcal{H}_T = \sum_{\alpha=L}^{R} \sum_{kk'\sigma} \left[T_{kk'}^{(\alpha)} e^{ie\varphi_\alpha/\hbar} a_{k\sigma}^{(\alpha)\dagger} a_{k'\sigma}^{(I)} + h.c. \right]. \tag{6}$$

3. TUNNELING CURRENT

To take higher order tunneling effect into account, we employ Keldysh formalism. There are two kinds of contributions, one of which is proportional to absolute value of tunneling matrix element, $|T_{kk'}^{(\alpha)}|^2$ and another is the one proportional to combinations such as $T_{kk'}^{(\alpha)} T_{kk''}^{(\alpha)*}$. As for the former type, the current through junction α is expressed as

$$I_\alpha = -c_\alpha \cdot \frac{ie}{2\pi\hbar^2} \sum_{kk'\sigma} |T_{kk'}^{(\alpha)}|^2 \int_{-\infty}^{\infty} \frac{d\omega}{\Lambda_{k'\sigma}(\omega)^2 + \Gamma_{k'\sigma}(\omega)^2} \left\{ \Sigma_{k'\sigma}^{>}(\omega) \mathcal{F}_\alpha^{<}(\omega - \xi_{k\sigma}^{(\alpha)}/\hbar) \, n_F(\xi_{k\sigma}^{(\alpha)}) \right.$$

$$\left. + \Sigma_{k'\sigma}^{<}(\omega) \mathcal{F}_\alpha^{>}(\omega - \xi_{k\sigma}^{(\alpha)}/\hbar) [1 - n_F(\xi_{k\sigma}^{(\alpha)})] \right\} \quad [c_\alpha = 1(-1) \text{ for } \alpha = L(R)] \tag{7}$$

where $\xi_{k\sigma}^{(\alpha)}$ and $n_F(\xi_{k\sigma}^{(\alpha)})$ are the energy of an electron with spin σ in electrode α and the Fermi distribution function, respectively. Using selfenergy of retarded Green's functions for the island electron $G_{k\sigma}^{R}(\omega) = [\omega - \xi_{k\sigma}^{(I)}/\hbar - \Sigma_{k\sigma}^{R}(\omega)]^{-1}$, $\Lambda_{k\sigma}(\omega) \equiv \omega - \xi_{k\sigma}^{(I)}/\hbar - \Re_e\Sigma_{k\sigma}^{R}(\omega)$ and $\Gamma_{k\sigma}(\omega) \equiv -\Im_m\Sigma_{k\sigma}^{R}(\omega)$. Correlation function of phase variable φ_α is defined as $\mathcal{F}_\alpha^{>}(\omega) \equiv \int_{-\infty}^{\infty} dt \ e^{i\omega t} \langle e^{-ie\varphi_\alpha(t)/\hbar} e^{ie\varphi_\alpha(0)/\hbar} \rangle$ and $\mathcal{F}_\alpha^{<}(\omega) \equiv \int_{-\infty}^{\infty} dt \ e^{i\omega t} \langle e^{ie\varphi_\alpha(0)/\hbar} e^{-ie\varphi_\alpha(t)/\hbar} \rangle$. Phase correlation functions with respect to φ and ψ describe EMEE and charged states of the island, respectively[5–7]. Contributions to $\Sigma_{k'\sigma}^{>}(\omega)$, which is topologically similar to $\Sigma_{k'\sigma}^{<}(\omega)$, are diagrammatically shown in Fig. 3. Solid line, dotted line and wavy line denote Green's functions of electrons in the island ($g_I^{\lambda\lambda'}$), in the electrode α ($g_\alpha^{\lambda\lambda'}$) and phase correlation function $\mathcal{F}_\alpha^{\lambda\lambda'}$, respectively. $\lambda = +(-)$ denotes Keldysh's lower (upper) time path. Note that $G^{+-} \equiv G^{>}$ and $G^{-+} \equiv G^{<}$ etc. As easily verified, I_α automatically satisfies current continuity condition, i.e., $I_L = I_R = I$ (current of the system). Note that the wavy lines representing $\mathcal{F}_\alpha^{\lambda\lambda'}$ appears as it were two-body interaction line. Therefore irreducible diagrams appear in any order of $R_q/R_T^{(\alpha)}$ in contrast to the conventional tunneling problem. Even when wave vector dependence of $T_{kk'}^{(\alpha)}$ is neglected, some of them

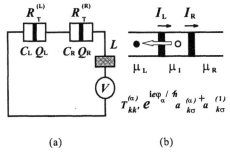

(a)   (b)

Figure 2. (a) Voltage-biased double junction with environmental impedance L. (b) Tunneling electron carries phase φ_α.

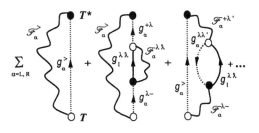

Figure 3. Irreducible selfenergies contributing to $\Sigma_{k\sigma}^{>}(\omega)$.

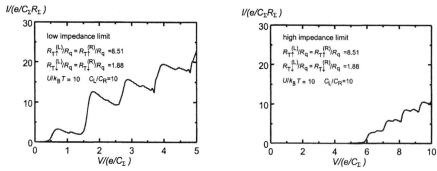

Figure 4. Coulomb staircase. $R_\Sigma/R_q = 3.08$, $U/k_BT = 10$.

yield wave vector dependence of selfenergy. As a first step, retaining the contributions independent of the wave vector up to the order of $(R_q/R_T^{(\alpha)})^2$, I-V characteristics were calculated. Some of the results are shown in Fig. 4 for $R_\Sigma/R_q = 3.08$ ($R_\Sigma \equiv \sum_\alpha R_T^{(\alpha)}$, $R_T^{(\alpha)-1} \equiv 2\pi^2 \sum_\sigma |T^{(\alpha)}|^2 N_\sigma^{(I)}(0) N_\sigma^{(\alpha)}(0)/R_q \equiv \sum_\sigma R_{T\sigma}^{(\alpha)-1}$). Somewhat deformed Coulomb staircases are seen probably due to the higher order tunneling effect. We also evaluated the effective magnetoresistance, $R_{\rm eff}(H)^{-1} \equiv dI(H,V)/dV|_{V=0}$, utilizing prescription $R_T^{(\alpha)} \to R_T^{(\alpha)}(H)$ in Eq. (7) and employing observed values for $R_T^{(\alpha)}(H)$[3]. Within this treatment, the enhancement of $R_{\rm eff}(H)$, $\Delta R_{\rm eff}(H) \equiv [R_{\rm eff}(0) - R_{\rm eff}(H)]/R_{\rm eff}(H)$ is about a few times of $[R_T(0) - R_T(H)]/R_T(H)$ in contrast to the results obtained by path integral approach[3,4]. The wave vector dependence of selfenergy should be treated more carefully. Detailed analyses of I-V characteristics and $\Delta R_{\rm eff}(H)$ of this system including the environmental impedance modulation effect will be reported elsewhere.

4. SUMMARY

We have proposed a microscopic selfconsistent theory for ultrasmall ferromagnetic *double* tunnel junctions based on the Keldysh formalism. In order for the theory to serve detailed study of this system, more careful treatment of the selfenergy effect, which includes the higher order tunneling effect combined in a complicated manner with EMEE, charged states of the island, is necessary. Further study will be reported elsewhere.

ACKNOWLEDGEMENT

This work is supported, in part, by Grant-in-Aid for Scientific Research.

REFERENCES

1. K. Ono, H. Shimada, S. Kobayashi and Y. Ootsuka, J. Phys. Soc. Jpn. **65** (1996) 3449
2. K. Ono, H. Shimada and Y. Ootsuka, Solid State Electronics **42** (1998) 1407
3. S. Iwabuchi, T. Tanamoto and R. Kitawaki, Physica **B249-251** (1998) 276
4. S. Iwabuchi and F. Wakaya, (submitted)
5. S. Iwabuchi, H. Higurashi and Y. Nagaoka, *Proc. 4th Int. Symp. on Foundations of Quantum Mechanics* in Special issue of JJAP: JJAP Series **9** 126 (1993)
6. H. Higurashi, S. Iwabuchi, and Y. Nagaoka, Phys. Rev. **B51** (1995) 2387.
7. S. Iwabuchi, H. Higurashi, F. Wakaya, K. Gamo and Y. Nagaoka, (submitted)
8. F. Wakaya, S. Iwabuchi, H. Higurashi, Y. Nagaoka and K. Gamo, Appl. Phys. Lett. (in press)

Quantum Coherence and Decoherence - ISQM - Tokyo '98
Y.A. Ono and K. Fujikawa (Editors)
© 1999 Elsevier Science B.V. All rights reserved.

Anomalous giant Barkhausen jumps in III-V-based diluted magnetic semiconductor (In,Mn)As at low temperatures

A. Oiwa[a], Y. Hashimoto[a], S. Katsumoto[a*], Y. Iye[a*] and H. Munekata[bt]

[a]Institute for Solid State Physics, University of Tokyo,
7-22-1 Roppongi, Minato-ku, Tokyo 106-8666, Japan

[b]Imaging Science and Engineering Laboratory, Tokyo Institute of Technology,
4259 Nagatsuda, Midori-ku, Yokohama 226-8503, Japan

We have found distinct jumps in the Hall resistance of (In,Mn)As at temperatures well below 1 K. We believe that the stepwise changes are the manifestation of the corresponding changes in the magnetization (giant Barkhausen effect), and are associated with depinning and subsequent propagation of domain walls. The depinning process appears to involve heat generation which is manifested by spike-like changes in the resistivity.

1. INTRODUCTION

Diluted magnetic semiconductors (DMSs) based on III-V compounds[1,2] are a class of new materials that attract significant attention because of their novel magnetic and transport properties. It has been revealed that (Ga,Mn)As and (In,Mn)As exhibit such intriguing phenomena such as carrier-induced magnetism and giant negative magnetoresistance[3,4]. In this paper, we report the observation of jumps in the magnetization curve which are detected through the anomalous Hall effect in the macroscopic-size Hall bar samples of p-(In,Mn)As at dilution-refrigerator-temperature regions and discuss their origin based on their dependence on the temperature, the field sweep rate, as well as the sample size.

2. EXPERIMENTAL

The samples of (In,Mn)As used in this work were prepared by molecular beam epitaxy. An $In_{1-x}Mn_xAs$ layer of thickness about 1.2 μm was grown at 300 °C on top of a 20 nm-thick InAs buffer layer on a GaAs (100) substrate. The details of the growth procedures are described elsewhere[5]. For the present work, we have concentrated on two samples with Mn concentration of $x = 0.004$ (R1255) and $x = 0.007$ (R1265). They were patterned into Hall bars whose width was 100 μm and distance between the voltage probes was 500 μm.

These samples show p-type metallic conduction. The standard way of carrier density estimation from the Hall effect is not straightforwardly applicable to DMSs because the

---

*Also at CREST, Japan Science and Technology Corporation (JST).
†Also at PRESTO, Japan Science and Technology Corporation (JST).

Hall resistivity contains the anomalous Hall term. A rough estimation of the hole density p for these samples from high field data are $\sim 2 \times 10^{19}$ cm^{-3}. These samples undergo a ferromagnetic transition at $T_C \approx 1$ K.

The Hall and magnetoresistance measurements were carried out using a dilution refrigerator and a superconducting magnet. The magnetic field was applied normal to the layer plane. We focus our attention to the behavior of the $x = 0.004$ sample (R1255) at temperatures well below 1 K. Similar results were obtained for the other sample (R1265).

The Hall resistance in this system can be written as follows,

$$\rho_{\text{Hall}} = \frac{R_0}{d} B + \frac{R_S}{d} M. \tag{1}$$

Here, R_0 and R_S are the normal and the anomalous Hall coefficients, respectively, and d is the thickness of the magnetic layer.

In the present p-(In,Mn)As system, The sign of R_0 is positive while that of R_S is negative. At low fields, $\rho_{\text{Hall}}(B)$ is dominated by the anomalous term and hence is proportional to the magnetization $M(B)$. The coefficient of proportionality R_S is related to the longitudinal resistance ρ by $R_S \propto \rho$ (skew scattering) or $R_S \propto \rho^2$ (side jump mechanism) depending on Mn contents x[3,6].

3. RESULTS AND DISCUSSION

The Hall resistance of the $x = 0.004$ sample at different temperatures are shown in Fig. 1. Below 1 K, they exhibit hysteresis reflecting that of the magnetization. As the temperature is decreased, distinct jumps appear in the hysteresis loop. Their number increases with decreasing temperature, but limited to five at the lowest temperature. Figure 2(a) shows the low field part of the hysteresis loop shown in Fig. 1. The longitudinal resistance ρ measured simultaneously is shown in Fig. 2(b). It is seen that the longitudinal resistance also shows jumps at the same fields as the Hall resistance. While the jumps in ρ can affect ρ_{Hall} through R_S in eq. (1), we believe it would be a minor effect. Rather, we believe that the jumps in ρ_{Hall} are basically due to the sudden change in magnetization M, i.e. giant Barkhausen jumps.

Let us next investigate the nature of these giant Barkhausen jumps. The field position of jumps at a given experimental condition (temperature and field sweep rate) are highly reproducible after thermal cycling up to room temperature. Essentially the same phenomenon is observed in the other sample (R1265). In contrast, large area ($\sim 2 \times 2$mm^2) samples cut out from the same wafers showed smooth hysteresis curves without jumps.

When the magnetic field is cycled around a particular jump, ρ_{Hall} traces a minor loop which is a manifestation of the switching between two metastable states. Figure 3 shows ρ_{Hall}-curves measured at different field sweep rates at $T = 112$ mK. With increasing sweep rate, the individual jump tends to shifted toward a higher field.

Despite the macroscopic sample size (~ 100 μm), the Barkhausen jumps are large in magnitude and small in number. This indicates that the change in the magnetic domain configuration and size at each jump occurs with macroscopic scale. Particularly the change at temperatures as low as 40 mK may be reminiscent of macroscopic quantum tunneling. We tentatively infer that the change in magnetic domain is a two-step process consisting of an initial depinning (either thermal or quantum) of the magnetic domain wall and

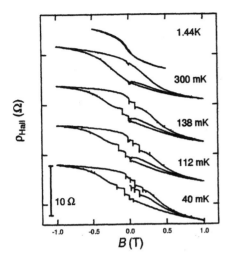

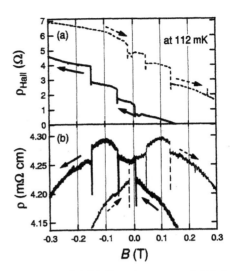

Figure 1. Hall resistance of the $x=0.004$ sample (R1255) at different temperatures at a rate of 0.016 T/min. Each curves are shifted vertically.

Figure 2. (a) Hall resistance and (b) longitudinal resistance of R1255 at 112 mK at a rate of 0.016 T/min.

a subsequent rapid propagation, caused by resultant local heating, of the wall over a macroscopic distance to the next pinning site. This is a mechanism similar to the one called *avalanche effect* proposed by Uehara *et al.*[7] who have explained the staircase-like magnetization curve of $SmCo_{3.5}Cu_{1.5}$ below 4.2 K.

In fact, there is ample evidence in our experimental data that suggests the effect of local heating. It is commonly seen in Figs. 1 and 2 that the stepwise changes in ρ_{Hall} as well as ρ are always accompanied by a sharp spike. The spike-like features in ρ always occur as sudden decrease of ρ and subsequent fast recovery. As far as the spike-like features are concerned, we believe that those in ρ is more fundamental than those in ρ_{Hall}, since the ρ_{Hall} traces in Figs. 1 and 2 most likely comes from the longitudinal resistivity component arising from slight misalignment of Hall probes. In other words, the major part of the spike-like features in ρ_{Hall} are just the shadow of the ρ component. At any rate, unlike the stepwise change in ρ_{Hall}, the spike-like structures should not be attributed to a change in M but to a change in ρ. The change in ρ is caused by sudden increase in sample temperature which is estimated to be $\Delta T \sim 100$ mK above the bath-temperature of 112 mK. Note that $d\rho/dT$ is negative in this temperature range. As seen in Fig. 3, the relaxation after the spike occurs with a time scale on the order of a second. It should be noted that the different width of the relaxation curve are due to the different sweep rates. The fact that the spike-like decrease of ρ_{Hall} always occurs at the leading edge of the stepwise change suggests that the major part of heating is associated with the initial depinning process.

We have argued that the experimentally observed giant Barkhausen jumps can be basically understood in terms of depinning and subsequent rapid propagation of domain walls. On the other hand, the magnetic structure of the present system is known to be rather complex, for example, ferromagnetic cluster might be formed in the real samples[3]. A full account of the phenomena reported here requires further detailed studies.

180

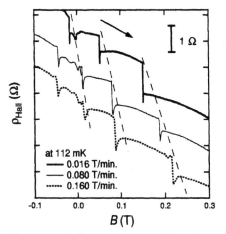

Figure 3. Hall resistance at 112 mK at different sweep rates. Each curves are shifted vertically. Dashed lines are simply guides to the eye.

4. CONCLUSION

We have found for the first time stepwise changes in ρ_{Hall} of the macroscopic-size samples (In,Mn)As, as a manifestation of the corresponding stepwise changes in the magnetization (the giant Barkhausen effect). We have discussed that the stepwise change is due to the initial depinning of the domain wall followed by its rapid propagation. The depinning is accompanied by heat generation, being evident from the spike-like features in ρ.

This work is partly supported by a Grant-in-Aid for the Scientific Research on Priority Area "Spin Controlled Semiconductor Nanostructures" from the Ministry of Education, Science, Sports and Culture, Japan.

REFERENCES

1. H. Munekata, H. Ohno, S. von Molnar, A. Segmüller, L. L. Chang and L. Esaki, Phys. Rev. Lett., 63 (1989) 1849.
2. H. Ohno, A. Shen, F. Matsukura, A. Oiwa, A. Endo, S. Katsumoto and Y. Iye, Appl. Phys. Lett., 69 (1996) 363.
3. H. Ohno, H. Munekata, T. Penny, S. von Molnar and L.L. Chang, Phys. Rev. Lett., 68 (1992) 2664.
4. A. Oiwa, S. Katsumoto, A. Endo, M. Hirasawa, Y. Iye, H. Ohno, F. Matsukura, A. Shen and Y. Sugawara, Solid State Commun., 103 (1997) 209.
5. H. Munekata, H. Ohno, R. R. Ruf, R. J. Gambino and L. L. Chang, J. Crst. Growth, 111 (1991) 1011.
6. A. Oiwa, A. Endo, S. Katsumoto, Y. Iye, H. Ohno and H. Munekata, Phys. Rev. B, to be published.
7. M. Uehara, B. Barbara, B. Dieny and P. C. E. Stamp, Phys. Lett., 114A (1986) 23.

Quantum depinning of a domain wall

Gwang-Hee Kim *

Department of Physics, Sejong University, Seoul 143-747, Republic of Korea

We consider the quantum depinning of a domain wall placed in a magnetic field at an arbitrary angle. Using the classical Landau-Lifshitz equation, we derive a domain wall mass which depends on the magnetic field and find that the WKB exponent and the crossover temperature strongly depend on the direction of the magnetic field.

1. INTRODUCTION

Considerable attention has been devoted to quantum dynamics of a domain wall owing to its importance in understanding macroscopic quantum tunneling in magnetic materials.[1] In general, a domain wall is pinned by an impurity lowering the anisotropy energy locally.[2] At a temperature low enough to neglect the thermal activation,[3] the depinning of a domain wall may occur due to quantum tunneling by applying an external magnetic field. In order for the domain wall to be a good example for quantum tunneling, we need the magnetic field which controls the height and width of the barrier and the effective mass of the system.

Even though several workers[4] suggested the tunneling of a domain wall by using a WKB method, quantum tunneling of a domain wall (QTDW) did not receive wider attention until Stamp[5] investigated QTDW based on the Heisenberg model with a uniaxial anisotropy. Later, Chudnovsky, Iglesias and Stamp (CIS)[6] developed a formulation of the problem which takes the curvature effects of a domain wall into consideration, and confirmed Stamp's work that the quantum tunneling of a domain wall may reveal itself at a macroscopic level. Besides the magnon and phonon studied by Stamp they briefly touched the effects of conduction electrons,[7] photons and the mobiltiy of the domain wall. Since then, QTDW has been the subject of considerable theoretical interest. Among them, recently, Braun, Kyriakidis and Loss (BKL)[8] found that the WKB exponent and the crossover temperature are of different functional forms that found by CIS and the sources for these discrepancies are different soliton mass and functional dependence of the pinning potential on the coercivity.

Up to now the theoretical studies for QTDW have been confined to the condition that the magnetic field is applied in the opposite direction to the initial easy axis. In this work we will extend the previous considerations to a system with a magnetic field applied

*This work was supported in part by the Basic Science Research Institute Program, Ministry of Education, Project No. BSRI-98-2415, and in part by Non-Directed-Research-Fund, Korea Research Foundation 1998.

at some angle to the easy axis of magnetization. We will show that the WKB exponent depends on θ_H via $(1 + \nu \tan \theta_H)^{1/2}$ where ν will be discussed later.

2. DYNAMICS OF A DOMAIN WALL

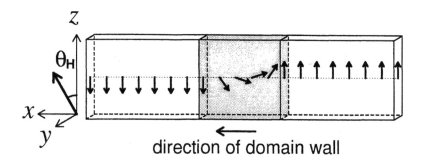

Figure 1. A configuration of magnetization is shown in a thin long slab geometry where a wall plane is parallel to the easy axis(z) and the spin configuratiojn spatially varies along in x direction.

Our considerations will begin by discussing the domain wall of the slab geometry, as is shown in Fig. 1. We work with the systems by assuming that the domain wall thickness λ is sufficiently larger than the lattice constant a between spins in which a continuum approximation for the magnetization $\mathbf{M}$ is valid. Since the phenomena considered occur at temperature far below the Curie temperature, the magnitude of the magnetization M_0 is constant. However, its direction $\hat{M}$ can change depending on the energy which is composed of the magnetic anisotropy energy, the exchange energy, and the demagnetization energy. The dynamics of $\hat{M}$ is determined by the Landau-Lifshitz equation,

$$\frac{d\mathbf{M}}{dt} = -\gamma \mathbf{M} \times \frac{\delta E}{\delta \mathbf{M}}, \tag{1}$$

where $\gamma = g\mu_B/\hbar$ is the gyromagnetic factor and the energy given by

$$\int d^3 r E[\theta_s(\mathbf{r}, t), \phi_s(\mathbf{r}, t)] = \int d^3 r K_{\|} \sin^2 \theta + K_\perp \sin^2 \phi \sin^2 \theta + \frac{1}{2} C[(\nabla \theta)^2 + (\nabla \phi)^2 \sin^2 \theta]$$

$$\simeq 2A_w \sqrt{2CK_{\|}} + \frac{1}{2} M \dot{Q}^2. \tag{2}$$

Here $K_\perp \equiv K_{\perp,a} + 2\pi M_0^2$, $K_{\|}$ and $K_{\perp,a}$ are the parallel and transverse anisotropy constants, C is an exchange constant, the soliton solution $\theta_s(x - Q) = 2 \arctan \exp[(x - Q)/\lambda]$, $M = (A_w M_0^2/\gamma^2 K_\perp)\sqrt{2K_{\|}/C}$ with A_w the cross sectional area of the sample, and Q the center of the domain wall. If defects are present in the samples, they can pin the domain wall. Assuming that the radius R corresponding to the defect volume is much smaller than the wall thickness $\lambda(=\lambda_0/\sqrt{1+k\sin^2\phi_s})$ with $k=K_\perp/K_{||}$, the wall is pinned by a potential form $V_p(Q) = -V_0\mathrm{sech}^2(Q/\lambda_0)$ with V_0 proportional to the volume of the defect, where we have replaced λ by $\lambda_0(=\sqrt{C/2K_{||}})$ and neglected the higher order in $O(V_0/E_0)$ with $E_0 = 2A_w\sqrt{2CK_{||}}$.[5] If we now apply an external magnetic field in the xz plane, the total energy for the wall is given by

$$\int d^3\mathbf{r}(E+E_H) = \frac{1}{2}M_{\mathrm{eff}}\dot{Q}^2 + V_p(Q) - h_zQ - [V_p(Q_0) - h_zQ_0],$$

$$\simeq \frac{1}{2}M_{\mathrm{eff}}\dot{Q}^2 + [V_p'(Q_i) - h_z](q - q_0) + \frac{V_p^{(3)}(Q_i)}{3!}(q^3 - q_0^3), \tag{3}$$

where $M_{\mathrm{eff}} = M + \pi A_w\lambda_0 M_0 H_x(k+1)/(kv_0)^2$, $h_z = 2A_w M_0 H_z$. Here we approximated the potential near the classical depinning field (CDF) and $q = Q - Q_i$, $q_0 = Q_0 - Q_i(<0)$ with Q_i and Q_0 inflection and metastable coordinate of the potential.

Using the standard instanton solution from the least action trajectory obtained by Eqs (2) and (3), we have formulas for the WKB exponent B and the crossover temperature given by

$$B(\theta_H) = B_0\sqrt{1 + v\tan\theta_H}, \quad T_c = \frac{T_c(0)}{\sqrt{1+v\tan\theta_H}}, \tag{4}$$

where

$$B_0 = \frac{12 \times 2^{3/4}}{5}\frac{A_w}{\gamma\hbar}\sqrt{\frac{CM_0^3}{K_{||}K_\perp}}\sqrt{H_z^c}(1-H/H_c)^{5/4}, \quad v = \frac{\pi}{4}M_0 H_z^c\left(\frac{1}{K_{||}} + \frac{1}{K_\perp}\right),$$

$$T_c(0) = \frac{\hbar}{k_B}\frac{(2\epsilon)^{1/4}}{\sqrt{2\pi}}, \quad H_z^c = \frac{2\sqrt{6}}{9}\frac{V_0}{A_w M_0}\sqrt{\frac{K_{||}}{C}}, \quad H_c = \frac{H_z^c}{\cos\theta_H}.$$

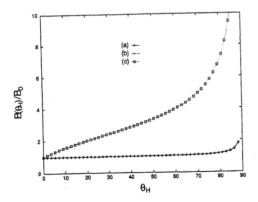
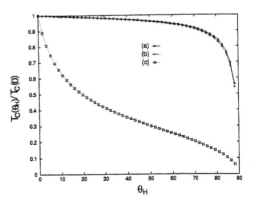

Figure 2. The θ_H dependence of the relative WKB exponent $B(\theta_H)/B_0$ for the samples (a) YIG, (b) Ni, (c) SrRuO$_3$.

Figure 3. The θ_H dependence of the relative crossover temperature $T_c(\theta_H)/T_c(0)$ for the samples (a) YIG, (b) Ni, and (c) SrRuO$_3$.

3. DISCUSSION

The angular dependence of B and T_c are plotted in Figs. 2 and 3 for ferromagnetic samples, Yttrium Iron Garnet (YIG), Ni, and $SrRuO_3$ by using the physical values given in Table 1. In these materials B rises sharply as θ_H approaches $\pi/2$ in accordance with the fact that at $\theta_H = \pi/2$ the potential for Q created by impurities is not deformed by the magnetic field and the position of the domain wall at the pinning center is not metastable any more. In YIG and Ni the behaviors of the ratio $B(\theta_H)/B_0$ and $T_c/T_c(0)$ almost flat for θ_H not close to $\pi/2$. However, in $SrRuO_3$ they have strong dependence on the orientation of the field because of large ν mainly originated from a larger coercivity H_z^c compared with the coercivities for YIG and Ni. It would therefore be interesting to study the angular dependence of the WKB exponent and the crossover temperature in $SrRuO_3$ for future experiments.

Table 1
Saturation magnetization M_0, easy-axis anisotropy constant $K_{\parallel}$, shape anisotropy $K_{\perp} \simeq 2\pi M_0^2$ for a thin film, exchange constant C, wall width λ_0, and coercivity H_z^c taken from Ref. [8] for various materials and the corresponding parameter ν. Also, the WKB exponent B_0 is obtained from a given value of $\epsilon(= 1 - H/H_c)$.

	M_0 [Oe]	$K_{\parallel}$ $[10^5 \frac{erg}{cm^3}]$	$K_{\perp}$ $[10^5 \frac{erg}{cm^3}]$	C $[10^{-6} \frac{erg}{cm}]$	λ_0 $[\mathring{A}]$	H_z^c [Oe]	ν	ϵ $[10^{-3}]$	B_0
YIG	196	0.25	2.4	0.86	414	10	0.068	5	30
Ni	508	8	16	2	112	100	0.075	4	31
$SrRuO_3$	159	20	1.6	0.046	11	10^4	8.43	5	22

REFERENCES

1. L. Gunther and B. Barbara (eds.), Quantum Tunneling of Magnetization-QTM '94, Kluwer Academic, Dordrecht/Boston/London, 1995.
2. A. P. Malozemoff and J. C. Slonczewski, Magnetic Domain Walls in Bubble Materials, Academic Press, New York, 1979.
3. H. A. Kramers, Physica (Utrecht) 7 (1940) 284.
4. T. Egami, Phys. Status Solidi A 20 (1973) 157; B 57 (1973) 211; B. Barbara et al., Solid State Commun. 10 (1972) 1149; J. A. Baldwin and F. Milstein, J. Appl. Phys. 45 (1974) 4006; W. Riehemann and E. Nemback, *ibid.* 55 (1983) 1081;57 (1985) 476.
5. P. C. E. Stamp, Phys. Rev. Lett. 66 (1991) 2802.
6. E. M. Chudnovsky, O. Iglesias and P. C. E. Stamp, Phys. Rev. B 46 (1992) 5392.
7. G. Tatara and H. Fukuyama, Phys. Rev. Lett. 72 (1994) 771.
8. H.-B. Braun, J. Kyriakidis, and D. Loss, Phys. Rev. B 56 (1997) 8129.
9. P. C. E. Stamp, E. M. Chudnovsky and B. Barbara, Int. J. Mod. Phys. B 6 (1992) 1355.

Quantum Coherence and Decoherence - ISQM - Tokyo '98
Y.A. Ono and K. Fujikawa (Editors)
© 1999 Elsevier Science B.V. All rights reserved.

Quantum Dephasing in Mesoscopic Systems

Yoseph Imry

aWeizmann Institute of Science, Department of Condensed Matter Physics,
IL-76100 Rehovot, Israel

After very briefly reviewing some interesting results in mesoscopic physics illustrating nontrivial insights on Quantum Mechanics, we summarize the general principles of dephasing (sometimes called "decoherence") of Quantum-Mechanical interference by coupling to the environment degrees of freedom. A particular recent example of dephasing by a current-carrying (nonequilibrium) system is then discussed in some detail. This system is itself a manifestly Quantum Mechanical one and this is another illustration of detection without the need for "classical observers" etc. Tunneling in a double minimum potential well is revisited. It is shown how it is observable by scattering experiments and how dephasing happens with a large enough scattered flux. The problem of the apparently finite dephasing rate that was recently observed in several systems is discussed. We show that the "standard model" of a conductor with static defects can not have such an effect. However allowing some dynamics of the defects may produce it. It is hoped that these discussions illustrate not only the vitality and interest of mesoscopic physics but also its genuine relevance to fundamental issues in Quantum Mechanics.

1. Introduction

Mesoscopic Physics [1] deals with the realm which is in-between the microscopic (atomic and molecular) scale and the macroscopic one. It can give us fundamental information on the crossover between the microscopic (quantum), and the macroscopic regimes. Macroscopic-type elecrical measurements can be sensitive to nontrivial quantum phenomena in mesoscopic samples at low temperatures. The latter are needed to preserve the phase coherence of the electrons, as will be discussed and reemphsized in this paper. Some outstanding results:

1. Quantization of conductances in the quantum unit of e^2/h, for ballistic orifices and "quantum point contacts" (QPC's).

2. AB conductance oscillations: The A-B effect has been proven to be a convenient way to observe interference effects in mesoscopic samples, because it provides an experimentally straightforward way of shifting the interference pattern. The difference between h/e- and $h/(2e)$ - periodic oscillations gives an indication of the sample-specific vs. ensemble-averaged properties. This has led to the concept of mesoscopic (sample-specific) fluctuations.

3. These mesoscopic fluctuations can have universal magnitudes and reduction factors when symmetries are changed.

4. "Persistent currents" in normal rings and cylinders can flow in the presence of a magnetic flux with no dissipation in the presence of elastic defect scattering.

One of the important ensuing insights is that elastic scattering *does not* destroy phase coherence. It takes *inelastic scattering* – changing the quantum state of other degrees of freedom, to do that.

In section 2, we discuss the general principles of dephasing of Quantum-Mechanical interference by coupling to the degrees of freedom which do not directly participate in the interference. Such degrees of freedom are often referred to as the "environment". The application to linear transport in disordered conductors is presented. The recent Weizmann experiment on dephasing by a current-carrying quantum point contact detector is then discussed in section 3. In section 4 the old problem of tunneling in double-minimum potential is reconsidered. Interesting remarkss are made on the observation of tunneling by scattering experiments. It is then shown how scattering of a strong enough beam can controllably dephase the coherent tunneling oscillation. In section 5 the recent issue of the saturation of the dephasing rate as $T \to 0$ is reviewed. While it does not exist in the usual model, we show that allowing for some motion of the defects may produce this effect. The results from section 4 will be used in this last calculation.

2. General Theory of Quantum-Mechanical Dephasing

We found that it takes *inelastic* scattering [1] of other degrees of freedom (the "environment"), to eliminate phase coherence. This is characterized by the "phase-breaking" time, τ_ϕ, for the electron.

Two descriptions have been used for the dephasing process. The first relies on the electron leaving a trace, in the form of an excitation of the environment, which identifies which path it took. The second description is based on the phase uncertainty induced on the electron by the fluctuating environment. (what counts physically is the uncertainty of the *relative* phases of the paths). In ref.[18] these two descriptions were proven to be equivalent, which turned out to be a useful insight. Some important physical comments:

1. The phase uncertainty remains constant when the interfering wave does not interact with the environment. Thus, if a trace is left by a partial wave on its environment, this trace cannot be wiped out after the interaction is over. The proof of this statement follows from unitarity [18]. Interesting subtleties were discussed in refs.[18,19]. This is relevant for the dephasing problem treated in the next section.

2. If the same environment interacts with the two interfering waves and when both waves emit the same excitation of the medium, each of the partial waves' phases becomes uncertain, but the relative phase is unchanged. A well-known example is that of "coherent inelastic neutron scattering" in crystals (see e.g. ref.[20]). This process is due to the coherent addition of the amplitudes for the neutron exchanging *the same* phonon with *all* scatterers in the crystal.

3. Long-wave excitations (phonons, photons) may not dephase the interference. However, that is *not* because of their low energy but rather because they do not influence the

---

[1] Here the term "inelastic" implies just changing the quantum state of the environment. It is irrelevant how much energy is transferred in this process. This includes zero energy transfer – flipping the environment to a degenerate state.

relative phase of the paths. An equivalent way to state this (see refs.[17,18]) is that, as in the Heisenberg microscope, the radiation with wavelength λ can not resolve the two paths if their separation is smaller than λ.

4. Dephasing may occur by coupling to a discrete or a continuous environment. In the latter case, the excitation *may* move away to infinity and the loss of phase can be regarded as, practically speaking, irreversible. However, in special cases it is possible, even in the continuum case, to have a finite probability to reabsorb the created excitation and thus retain coherence. This happens, for example, in a quantum interference model due to Holstein for the Hall effect in insulators.

In mesoscopic physics the issue is the dephasing of electrons performing diffusive motion due to defects, just above the Fermi energy, and interacting strongly with all the other electrons. This electron-electron interaction provides in many cases the dominant dephasing mechanism. It is easy then to obtain the dephasing rate from the strength of the inelastic scattering of the considered electron by the electron sea, i.e. using the trace left in the environment [18]. A similar result was first obtained in ref.[21] by using the effect of the electromagnetic fluctuations due to the electron gas on the considered electron. The equivalence of these two points of view is guaranteed by the fluctuation-dissipation theorem according to which:

$$\text{Im}\left(\frac{1}{\epsilon(q,\omega)}\right) \sim S(q,\omega)/q^2, \tag{1}$$

where $\epsilon(q,\omega)$ and $S(q,\omega)$ are the dynamic linear response function and structure factor of the same system.

It is easy to manipulate these cited expressions, for a general V_q, into:

$$1/\tau_\phi = \int\int d\mathbf{q}d\omega |V_q|^2 S_p(q,\omega)S_{env}(-q,-\omega). \tag{2}$$

where $S_p(q,\omega)$ is the dynamic structure factor of the diffusing electron (a Lorentzian with width Dq^2) and $S_{env}(-q,-\omega)$ is the same for the environment. Evaluating these integrals is straightforward. We find in 3D:

$$1/\tau_\phi \sim T^{3/2}. \tag{3}$$

For 2D and 1D (thin films and wires) the integrations over the appropriate components of $\mathbf{q}$ are replaced by summations and it is found that the remaining integrations are infrared (small q)-divergent. A careful evaluation of the phase *difference* of two paths shows that this divergence is cured by a cutoff whose physical meaning is exactly that low q excitation can not distinguish paths that are separated in space by less than $1/q$. This is in agreement with the "Heisenberg microscope"-type argument[17,18] mentioned above. The results are:

$$1/\tau_\phi \sim T \ \ (in2D). \qquad (1/\tau_\phi) \sim T^{2/3} \ \ (in1D). \tag{4}$$

These results are in a *quantitative agreement* with experiments.

The case where the "environment" is *far* from equilibrium is of interest since the fluctuations are not given by their well-known equilibrium values and the fluctuation-dissipation theorem is not applicable. Examples are provided in the two next sections.

3. Dephasing by a current-carrying Quantum detector

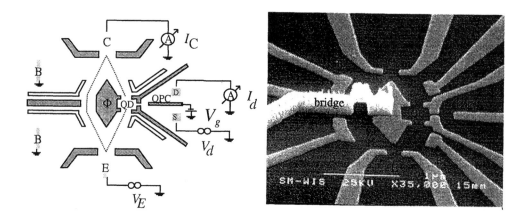

Fig.1: (right) A schematic view of the device used in ref.[24]: Electrons in the 2D gas (B) pass from the emitter (E) to the collector (C), constrained by the reflectors (bordered empty regions). The set of gates (dark regions) deplete the electrons below them and define the AB ring and the quantum dot on its right arm. Near and on the right of the latter, the conductance (I_D/μ_D) of the QPC is measured by the circuit shown.
(left) An SEM micrograph of the device. The gray areas are the gates and reflectors. An "air bridge" biases the central gate which controls the hole of the ring.

The quantum point contact, briefly alluded to in section 1, can serve as a detector[22] sensitive to small changes in parameters, such as the electrostatic field nearby. This sensitivity may be used to detect the presence of an electron in one of the arms of an interferometer, provided the two arms are placed asymmetrically with respect to the QPC. Following discussions by Gurvitz[23], Buks et al.[24] performed measurements confirming "which path" detection by the QPC. The AB oscillations were measured in a ring. On one of its arms the transmission was limited by a "quantum dot" where the electron wave would resonate for a relatively long and controllable (to a degree) dwell time τ_d. A QPC was placed near that arm and the degree of dephasing due to its detecting if the electron is on the quantum dot, could be inferred from the strength of the AB conductance oscillations. Clearly, a necessary and sufficient condition for strong dephasing is that $\tau_\phi \ll \tau_d$. The results were in good agreement with the theory developed by the authors[24], by Levinson[25] and, independently, by Aleiner et al[26]. The new interesting feature of this nonequilibrium dephasing is that a finite current is flowing in the detector and, with increasing time, each electron transmitted there contributes to the decrease of the overlap of the environment wavefunctions. As discussed in the previous section, the reduction in overlap is conserved when further thermalization of the transferred electrons in the downstream reservoir occurs. The alternative picture is that nonequilibrium (shot-noise) fluctuations of the current in the QPC create a phase uncertainty for the electron in the quantum dot. While the equivalence of these two pictures is guaranteed by the discussion of the previous section (ref.[18]) it is interesting and nontrivial to see how it emerges in detail, as was demostrated in ref.[27]. This is all the more interesting, since the former point of view by which the current fluctuations in the QPC cause dephasing, seems superficially to contradict the idea behind eq. 5. According to the latter, dephasing appears to have to *overcome* the shot-noise fluctuations, which therefore may be thought to *oppose* dephasing.

In the model considered, the QPC is taken for simplicity to be single-channel and symmetric and the temperature T is zero (i.e. T is much smaller than the voltage on the QPC). The existence of an electron in the quantum dot is taken to change the transmission coefficient $T = |t|^2$ from T to $T+\Delta T$ and the conductance by $\frac{e^2}{\pi\hbar}\Delta T$. The change in phase of t was neglected. It was later considered by Stodolsky[28]. τ_ϕ can be physically estimated from the condition that the change in the number of electrons, $\langle N \rangle = (I/e)\tau_\phi$, streaming across the QPC within τ_ϕ, $\langle \Delta N \rangle = \frac{e}{\pi\hbar}V\Delta T\tau_\phi$ be larger than the rms fluctuations of N during the same time. For the latter one has the quantun shot-noise result[7] according to which the mean-square fluctuation $\langle(\Delta N)^2\rangle$ is given by $(I/e)\tau_\phi(1-T)$ Thus (the numerical factor follows from more detailed calculations):

$$\frac{1}{\tau_\phi} \sim \frac{e}{8\pi\hbar}\frac{(\Delta T)^2V}{T(1-T)} \tag{5}$$

Several derivations, whose equivalence[27] is guaranteed by the discussion of section 2 have been given of this result.

The experiments of ref.[24] agree better than qualitatively with the above picture. For QPC voltages larger than thermal, the visibility of the AB interference contribution to the conductance of the ring decreased roughly linearly in V and the coefficient was in reasonable agreement with the above. The parameter ΔT was directly measured and the dependence on T was qualitatively observed as well.

4. Tunneling, detection and dephasing

We reconsider in this section the old problem of tunneling in a double mimimum potential well. We shall start by reviewing some old results by the author [29] who calculated the absorption and the Born approximation scattering cross sections from such systems. We shall concentrate here on the latter. The tunnel splitting of the lowest two levels in the double well was taken as Ω and the assymmery between the wells as B. For $B >> \Omega$, the various matrix elements for transitions become very small. Therefore the interesting case is that of $B \sim \Omega$, in which the qualitative behavior is that of a symmetric well, which will be considered, for simplicity, from now on. The elastic and inelastic scattering cross sections are given, up to the same constant, for momentum transfer $\mathbf{q}$ by:

$$\sigma_{el}(q,\omega) \sim cos^2(q.d)\delta(\omega), \qquad \sigma_{inel}(q,\omega) \sim sin^2(q.d)\delta(\omega \pm \Omega). \tag{6}$$

Here **d** is the vector separating the two minima. We remark that the above trigonometric factors (which have been observed experimentally for tunneling of protons) reflect interference of the scattered partial waves from the *same particle being in the two wells!*. The $\pm$ signs in σ_{inel} reflect energy gain and loss of the tunneling energy by the scattered particle, corresponding to situations where the tunneling particle was in the initial symmetric or antisymmetric state. Thus, observing one of these inelastic peaks constitutes a *detection* of in which of these two levels the tunneling particle was!

Consider now the so-called "macroscopic quantum coherence" situation, where the system is prepared at $t = 0$ in the left-hand well (i.e in a wavepacket which is a sum of the symmetric and antisymmetric eigenstates). It is well-known that due to the time-dependent interference between these two states, the wave packet will move back and forth between the two wells, at frequency Ω. We are now going to consider the dephasing of this interference where the "which path" information relates to which of the two eigenstates the system is in. This dephasing will cause the periodic "tunneling" of the wave packet to be damped due to the decrease with time of the overlap of the scattered radiation. This way of looking at the system is different from the usual way of looking at the left- and right- well states and considering how their overlap is modified by coupling to the environment. Imagine now that the flux of x-rays or neutrons probing the system is increased, the damping of the oscillations will be stronger. Eventually, when the integrated inelastic scattered intensity within a time Ω^{-1} in one of the inelastic peaks is of order unity, even the first oscillation will damp out. This will be a "which path observation" very analogous to the one by the QPC in the last section. In fact, it is again possible to control continuously the strength of this dephasing! The delta-function inelastic peaks of σ_{inel} in eq. 6 will gradually broaden and eventually not be resolvable, with increasing flux.

The above results in eq.6 will be used in the next section to evaluate the inelastic scattering (and therefore the dephasing) of electrons in disordered metals at low temperatures.

5. Dephasing when $T \to 0$.

Recently, Mohanty et al[31] have published extensive experimental data indicating that contrary to general theoretical expectations and to eq.4, the dephasing rate in films and wires does not vanish as $T \to 0$. Serious precautions[32] were taken to eliminate experimental artifacts. It was speculated that such a saturation of the dephasing rate when $T \to 0$, might follow from interactions with the zero point motion of the environment. These speculations have received apparent support from calculations in ref.[33]. However, the latter were severely criticized in refs.[34,35] and were in disagreement with experiments in ref.[36]. In fact it is clear that since dephasing must be associated with an excitation of the environment, it cannot happen as $T \to 0$. In that limit neither the electron nor the environment has any energy to exchange. Below, we convert this qualitative argument to a proof. While proving unequivocally that zero point motion does not dephase, our proof does show what *further* physical assumptions can in fact produce a finite dephasing rate for $T \to 0$.

We[30] use eq.2 and apply the very general detailed-balance relationship

$$S(q,\omega) = S(-q,-\omega)e^{-\hbar\omega/k_B T}, \tag{7}$$

to either $S_p(q,\omega)$ or $S_{env}(-q,-\omega)$. It is immediately seen that the integrand of eq.2 is a product of two factors one of which vanishes for $\omega > 0$ and the other for $\omega < 0$, as $T \to 0$. Thus the integral and the dephasing rate vanish in general when $T \to 0$. However, if $S_{env}(-q,-\omega)$ has an approximate delta-function peak at small ω due to an abundance of low-energy excitations, one may get a finite dephasing rate at temperatures higher than the width of that peak. Such near-degeneracies of the ground state are known to exist in disordered, glassy, systems. These follow from the many "mesoscopic" realizations of the disorder configuraton. The system slowly fluctuates among these many states and it may in fact not be in full equilibrium. This may cause [38] the commonly observed low-frequency (often "$1/f$") noise [39].

Let us now estimate the inelastic scattering rate from the set of impurities that are rearranging by tunneling at low temperatures at a rate $1/\tau_0$ satisfying:

$$\hbar/\tau_\phi << \hbar/\tau_0 << k_B T. \tag{8}$$

we denote the fraction of the defects that move on these time scales by $p << 1$. From equation 6 we see that the inelastic component of the scattering from those impurities is smaller by $A \cong (k_F d)^2 << 1$ than the elastic rate. Thus the inelastic scattering rate from these defects is pA/τ, where τ is the elastic scattering time by all defects. Therefore, to get a low temperature dephasing rate of $10^{-4}/\tau$[31], we need, for $A \sim 10^{-2}$, a value of $p \sim 10^{-2}$. The impurity density, for impurities whose scattering length is atomic, is of the order of $1/k_F \ell$. This typically corresponds to total impurity concentrations of the order of 10^{-2}. Thus, we have to assume a ~ 100ppm concentration of *appropriate* low- energy "two-level" defects, to get a low temperature dephasing rate comparable to that of ref. [31]. This does not sound impossible, however the consistency with observed levels of low-temperature $1/f$ noise must be checked. Experimental studies of the possible correlation between the saturation of τ_ϕ and low-frequency noise, would be very valuable.

Thus, while the "standard model" of disordered metals (in which the defects are strictly frozen) gives of course an infinite τ_ϕ at $T = 0$, there may be other physical ingredients that can make τ_ϕ finite at very low temperatures, *without* contradicting any basic law of physics. The TLS model is a particular example and its requirements may or may not be satisfied in the real samples. But, other models with similar dynamics might exist as well. We reemphasize that this does *not* imply dephasing by zero-point fluctuations, which has been repeatedly, and wrongly, claimed in the literature. The failure of the semiclassical approximation used in these considerations was clarified in ref.[30].

6. Acknowledgements

This research was supported by grants from the German-Israel Foundation (GIF) and the Israel Science Foundation, Jerusalem. The author thanks Y. Aharonov, D. Cohen and A. Stern for collaborations on these problems. I. L. Aleiner, N. Argaman, C. W. J. Beenakker, M. Berry, E. Buks, Y. Gefen, B. I. Halperin, M. Heiblum, D.E. Khmelnitskii, R. Landauer, Y. Levinson, Y. Meir, M. Schechter, G. Schön, T.D. Schultz, A. Stern, H.A. Weidenmüller, P. Wölfle and A. Zaikin are thanked for discussions. E. Buks and M. Reznikov are thanked for permission to use their figures.

REFERENCES

1. Y. Imry, *Introduction to Mesoscopic Physics*, Oxford Unversity Press (1997).
2. K., Von Klitzing, G. Dorda, and M. Pepper, Phys. Rev. Lett. **45**, 494 (1980) .
3. R. Landauer, IBM J. Res. Dev. **1**, 223 (1957); R. Landauer, Philosoph. Mag. **21**, 863 (1970) .
4. Y. Imry, Physics of Mesoscopic Systems, in Memorial Volume to S.-k Ma, Grinstein, G. and Mazenko, G., eds. (World Scientific) p. 102 (1986).
5. B.J. van Wees, H. Van Houten, C.W.J. Beenakker, J.G., Williamson, L.P. Kouend-hoven, D. van der Marel and C.T. Foxon, Phys. Rev. Lett. **60**, 848 (1988); D.A. Wharam, T.J. Thornton, R. Newbury, M., Pepper, H. Ahmed, J.E.F. Frost, D.G. Husko, D.C. Peacock, D.A. Ritchie, and G.A.C. Jones J. Phys. **C21**, L209 (1988).
6. M. Reznikov, M. Heiblum, H. Shtrikman and D. Mahalu, Phys. Rev. Lett **75**, 3340 (1995).
7. V.A. Khlus, (1987) JETP **66**, 1243. G.B. Lesovik, JETP Lett., **49**, 592 (1989); Th. Martin and R. Landauer, Phys. Rev. **B45**, 1742 (1992) M. Büttiker, Phys. Rev. **B46**, 12485, (1992).
8. Y. Gefen, Y. Imry, and M.Ya Azbel, Phys. Rev. Lett. **52**, 129 (1984).
9. R.A. Webb, S. Washburn, C.P. Umbach, and R.B. Laibowitz Phys. Rev. Lett. **54**, 2696 (1985).
10. B.L. Altshuler, A.G. Aronov, and B.Z. Spivak, (1981) JETP Lett. **33**, 94.
11. D.Yu Sharvin, and Yu V. Sharvin, JETP Lett. **34**, 272 (1981).
12. Altshuler, B.L. (1985) JETP Lett. **41**, 649 (1985); P.A. Lee and A.D. Stone, Phys. Rev. Lett. **55**, 1622 (1985). P.A. Lee, A.D. Stone and H. Fukuyama, Phys. Rev. **B35**, 1039 (1986).
13. M. Büttiker, Y. Imry and R. Landauer, Phys. Lett. **96A**, 365 (1983) .
14. L.P. Levy, G. Dolan, J. Dunsmuir, and H. Bouchiat, Phys. Rev. Lett. **64**, 2074 (1990); V. Chandrasekhar, R.A. Webb, M.J. Brady, M.B. Ketchen, W.J. Gallagher and A. Kleinsasser Phys. Rev. Lett. **67**, 3578 (1991); D. Mailly, C. Chapelier and A.Benoit, Phys. Rev. Lett. **70**, 2020 (1993).
15. R.P. Feynman and F.L. Vernon, Ann. Phys. NY **24**, 118 (1963).
16. A.O. Caldeira and A.J. Leggett Ann. Phys. **149**, 374 (1983) .
17. R.P. Feynman, R.B. Leighton and M. Sands *The Feyman Lectures on Physics*, Addison Wesley, Reading, MA, Vol. III, pp. 21.14 (1965), contains a beautiful discussion of dephasing.
18. A. Stern, Y. Aharonov, and Y. Imry, (1990) Phys. Rev. **A40**, 3436 and in G. Kramer, ed. *Quantum Coherence in Mesoscopic Systems*, NATO ASI Series no. 254, Plenum., p. 99 (1991).
19. G. Hackenbroich, B. Rosenow and H. A. Weidenmüller, Cond-mat/9807317 and to be published.
20. Kittel, C. *Quantum Theory of Solids*, John Wiley, NY (1963).
21. B.L. Altshuler, A.G. Aronov, and D.E. Khmelnitskii, J. Phys. **C15**, 7367 (1982) .
22. M. Field, C.G. Smith, M. Pepper, D.A. Ritchie, J.E.F. Frost, G.A. Jones and D.G. Hasko, Phys. Rev. Lett **70**, 1311 (1993).
23. S. A. Gurvitz, Phys. Rev. **B56**, 15215 (1997) and Quant-ph/9697029 (1997)

24. E. Buks, R. Schuster, M. Heiblum, D. Mahalu and V. Umansky, Nature **391**, 871 (1998).

25. Y. Levinson, Europhys. Lett. **39**, 299 (1997).

26. I. L. Aleiner, N.S. Wingreen and Y. Meir, Phys. Rev. Lett., **79**, 3740 (1997).

27. Y. Imry, Physica Scripta, in press.

28. L. Stodolsky, quant-ph/9805081.

29. Y. Imry, chapter 35 in *Tunneling in Solids*, proceedings of the 1967 Nato Conference, E. Burstein and S. Lundquist, eds., Plenum Press (N.Y.) 1969.

30. D. Cohen and Y. Imry, Cond-mat/9807038.

31. P. Mohanty, E.M. Jariwala and R.A. Webb, Phys. Rev. Lett. **77**, 3366 (1997).

32. P. Mohanty, E.M. Jariwala and R.A. Webb, these proceedings (1998).

33. D.S. Golubev and A.D.Zaikin, cond-mat/9710079, cond-mat/9712203.

34. I. L. Aleiner, B. L. Altshuler, M. E. Gershenson, cond-mat/9808078, cond-mat/9808053

35. B. L. Altshuler, M. E. Gershenson, I. L. Aleiner, cond-mat/9803125.

36. Yu. B. Khavin, M. E. Gershenson, A. L. Bogdanov, cond-mat/9803067, cond-mat/9805243.

37. H. Fukuyama and Y. Imry, unpublished.

38. S. Feng, P. A. Lee and A. D. Stone, Phys. Rev. Lett. **56**, 1970, 2272(E) (1986).

39. N. O. Birge, B. Golding and W. H. Haemmerle, Phys Rev. **B42**, 2735 (1990)

Quantum Coherence and Decoherence - ISQM - Tokyo '98
Y.A. Ono and K. Fujikawa (Editors)
© 1999 Elsevier Science B.V. All rights reserved.

Decoherence in Mesoscopic Systems*

R.A. Webb, P. Mohanty, E.M.Q. Jariwala, T.R. Stevenson, and A.G. Zharikov

Center For Superconductivity Research, Department of Physics, University of Maryland, College Park, MD 20742

We review previous measurements of the temperature dependence of the electron phase coherence time τ_ϕ in a wide variety of mesoscopic Au 1-D wires. Most of these wires clearly demonstrate that τ_ϕ becomes independent of temperature at low temperatures even though the electrons are in good thermal contact with the temperature reservoir. We present new measurements on the role that the external electromagnetic environment plays in decoherence. We find that up to 26 GHz this external environment does not cause decoherence without a concomitant increase in the energy relaxation rate.

1. INTRODUCTION

One of the most important properties of an electron in any quantum system is the time over which the phase coherence in its wavefunction is maintained. In condensed matter mesoscopic conductors at low temperatures, this coherence time τ_ϕ can be greater than 1 ns, orders of magnitude larger than the elastic scattering time. The large value of τ_ϕ enables the measurement of a number of Aharonov-Bohm effects in transport and magnetic experiments on small samples [1,2]. Theoretically, in a semi-classical picture [3], this coherence time should grow with decreasing temperature becoming infinite at T=0. The main decohering influences are the interactions with the environment. In electron systems, it is the electron-phonon (EP) and the electron-electron (EE) interactions which can cause decoherence, and both alter the coherence time as $\tau_\phi \propto 1/T^p$ where p varies between 0.5 and 3 [4]. Yet in every experiment performed down to very low temperatures, the phase coherence time is universally found to approach a temperature independent value. This saturation temperature depends on the details of the sample studied and varies from 10 K to below 20 mK.

We have recently reported the results of an extensive set of measurements on the temperature dependence of τ_ϕ in 1-D mesoscopic Au wires [5] designed to understand the origin of this saturation. In these experiments, the sample length and classical diffusion constant D were varied by more than a factor of 250, and the width and thickness varied by a

*This work is supported by the National Science Foundation under contract DMR9730577 and by the US Army under contract DAAG559710330.

factor of 7. Some of our data (labeled 1-D Au) is displayed in Fig. 1 along with a representative sample of data from other groups on 1-D Si MOSFET's, 1-D GaAs/GaAlAs quantum wires, and 2-D films. In all these experiments, $L_\phi = \sqrt{D\tau_\phi}$ was determined from fits to weak localization theory of the measured change in resistance as a function of magnetic field near zero field. The measured low temperature value of τ_ϕ in these experiments varies by more than three orders of magnitude but the trend is clear: The lower the coherence time, the higher the temperature at which saturation is observed. As shown by the solid lines in Fig. 1, we discovered that all the 1-D data could be fit to the form $\tau_\phi = \tau_0(\tanh(\alpha\pi^2(\hbar/k_B T\tau_0)^{1/2}))$, where τ_0 is the measured low temperature value of the coherence time and α is a constant on the order of 1 [5]. We suggested [5] that the intrinsic fluctuations of the electromagnetic environment were responsible for the saturation of τ_ϕ.

We showed that the above result could be obtained by including the zero point fluctuations into the spectral density function used in the prior theory [4,7-8] and taking the low temperature limit of the theory. The final step in the calculation requires a summation over the relevant frequency modes of the environment. In a semi-classical picture, the maximum frequency ω is given by $\omega = k_B T/\hbar$, and goes to zero at T = 0 leading to a divergence of τ_ϕ.

We suggested that in the low temperature limit this maximum frequency should be given by the classical energy of the electron [9], $\omega = mv_D^2/2\hbar$, where $v_D = v_F l_e/L$ is the diffusive velocity in a wire and l_e is the mean free path. We also suggested the lowest frequency to be used in the summation should be $2\pi/\tau_\phi$ because wavelengths longer than L_ϕ do not cause dephasing.

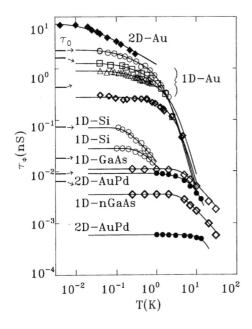

Figure 1. Saturation of decoherence time in various mesoscopic systems.

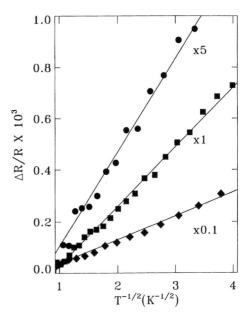

Figure 2. EE interaction correction to the resistance.

Our result leads to the conclusion that the low temperature limiting value of τ_ϕ would be given by $1/\tau_0 = \left(e^2 d^2 R m^* D^{3/2} / 4\pi\hbar^2 L\right)^2$, where d is the classical dimensionality, R is the sample resistance, m^* is the effective mass of the electron, D is the classical diffusion constant and L is the length of the sample. The arrows in Fig.1 represent the values of τ_0 we computed for all the 1-D data displayed. As can be seen, a very respectable agreement with the measurements is obtained. Recently, a more formal zero point fluctuation theory [10] has been used to compute the saturation value of τ_ϕ, giving a reasonable agreement with experiments as well.

Despite its apparent success in producing electron phase decoherence at T = 0, this suggestion of intrinsic fluctuations of the electromagnetic environment has endured considerable theoretical criticism [11]. Earlier arguments used against the validity of the data were that the electron temperature was saturating due to heating by the measurement current or external noise present at the site of the sample. We have convincingly ruled out this possibility by measuring the temperature dependence of the EE interaction [5] in the same samples that τ_ϕ was previously measured. Figure 2 displays the temperature dependence of the change in the normalized sample resistance measured at a low magnetic field that was sufficiently strong to suppress the weak localization contribution to the temperature dependence. If the electrons are in good thermal contact with the temperature reservoir, the change in the resistance due to the EE interaction is given by $\Delta R = \left(2e^2 R^2 / hL\right)\left(\hbar D / k_B T\right)^{1/2}$. The straight lines drawn through the data are the theoretically expected behavior. There are no indications that the electrons are out of thermal equilibrium with the temperature reservoir thereby ruling out heating. The second argument used against the validity of the data in Fig. 1 is that magnetic impurities are causing the saturation of τ_ϕ. For over 15 years, the standard lore in theoretical mesoscopic physics has been that magnetic impurities will cause decoherence and produce a temperature independent τ_ϕ. But since it is believed that both Si MOSFET's and GaAs/GaAlAs heterostructures do not contain magnetic impurities in the region where the 2-D gas is formed, this cannot be responsible for the observed saturation of τ_ϕ in these samples. After measuring τ_ϕ in our Au samples we ion implanted 2-10 ppm of Fe impurities in them. τ_ϕ decreased by more than an order of magnitude after adding these impurities but remained temperature dependent down to our lowest measurement temperatures [5,12]. Thus we have ruled out magnetic impurities as the cause of the τ_ϕ saturation in all the experiments and still conclude that the origin of the saturation is an intrinsic quantum effect.

More recently, new arguments have been given as to why the data in Fig. 1 are flawed [11]. First, the authors of Ref. [11] do not believe that intrinsic quantum fluctuations can cause decoherence in a Fermi Liquid at any temperature. In addition they suggest that external high frequency electromagnetic radiation is present in all the experimental cryostats. If the external noise contains frequencies in the range of $1/\tau_\phi$ at high enough power levels, decoherence could occur without necessarily causing enhanced energy relaxation or real heating of the electrons. In our opinion, this is an extremely important concept for the field of mesoscopics because it clearly places on a different footing the ideas of decoherence and energy exchange (inelastic processes) [8,13] in a condensed matter quantum system.

2. NEW EXPERIMENTS ON HIGH FREQUENCY DECOHERENCE

We fabricated new 1-D Au wires in which the parameters of the samples were carefully selected such that the decoherence due to our suggested zero point mechanism would not occur until very low temperatures. Figure 3 displays the temperature dependence of τ_ϕ for three of these samples measured in the same cryostat using the same electronics as for the 1-D Au data in Fig. 1. As can been seen, there is no saturation of τ_ϕ down to at least 50 mK. The open diamonds shown in Fig. 3 were obtained after winding a single turn coil, radius = 0.77 cm, and placing it 2.8 mm below the plane of the samples. All the samples were placed at r = 0.53 cm from the center of the coil. The ends of the coil were connected to high-quality high-frequency 50 Ω transmission line, both of which were connected to our room temperature high frequency S-parameter measurement electronics. These two transmission lines increased the heat leak to our dilution refrigerator so that the minimum temperature increased to 120 mK. If there was a significant amount of external electromagnetic radiation in our environment, the amount of ac power coupled to our sample should have increased after the insertion of these transmission lines. In fact, τ_ϕ did not change! At 120 mK, τ_ϕ for the three samples was between 0.8-2 ns; the same as before adding the new transmission lines.

The power coupled into our samples was estimated using two techniques. First, standard electromagnetic theory [14] was used to estimate the electric field E_φ at the site of our sample from knowledge of the coil current and geometry. The current in the coil was determined using the four S-parameter transmission and reflection coefficients and from the measurement of the rf power into the transmission lines. This frequency dependent current was measured from 45 MHz to 26.5 GHz. The power coupled into the sample is then $(E_\varphi L_{eff})^2/R_T$ (assumes an open circuit beyond L_{eff}), where $L_{eff} \approx 1$ mm is the effective sample length including the

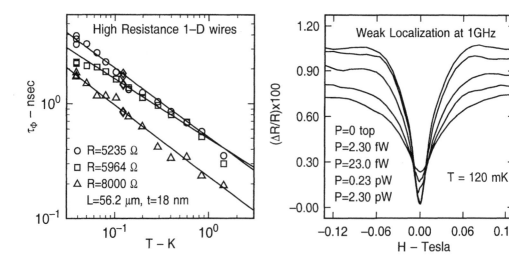

Figure 3. τ_ϕ as a function of temperature for 3 samples. Open diamonds are after insertion of two copper transmission lines.

Figure 4. Effect of rf power at 1 GHz on weak localization for the sample with R=5235 Ω. Signal decreases with increasing rf Power.

connecting leads, and R_T is the total resistance. The second technique is to assume that beyond some distance the sample is effectively shorted because of the high lead capacitance. The power coupled to the sample is I^2R, where $I=\omega I_C M/Z$ and I_C is the current in the high frequency coil, M is the mutual inductance between the coil and the sample circuit, and Z is the total impedance of the sample loop. With in a few orders of magnitude, both estimates give similar results. All powers quoted in this paper assume the larger of the two estimates.

Figure 4 displays the weak localization data obtained at 1 GHz for 5 different rf power levels coupled to the sample. As the power level increases above 1 fW, the size of the weak localization signal decreases indicating a decrease of τ_ϕ. The zero of $\Delta R/R$ was not shifted for these data. Note that the resistance for H > 0.06 T decreases faster than the resistance at H=0 increases for increasing rf power to the sample. This is exactly what one might expect if the effective temperature of the electrons was increasing. $\Delta R/R$ decreases as $1/T^{1/2}$ for the EE interaction with increasing T. $\Delta R/R$ for weak localization in the presence of strong spin orbit scattering increases with increasing temperature as a weaker power of T.

Figure 5 displays the temperature dependence of the EE interaction for one of our samples at six different power levels for a 1 GHz frequency. The power calibration is -20 dBm=0.23 pW and -30 dBm=23 fW. Note that the EE interaction begins to show a saturation above 2.3 fW of power and is consistent with the power at which the weak localization shown in Fig. 4 begins to show a change. Fig. 6 displays the rf power dependence of τ_ϕ at 120 mK for 2 different samples at 3 different frequencies. For all frequencies, the power level at which τ_ϕ begins to decrease is between 2-10 fW and is the same power at which the EE interaction begins to show saturation effects. At high powers we find that $\tau_\phi \propto P^{-1/5}$, in agreement

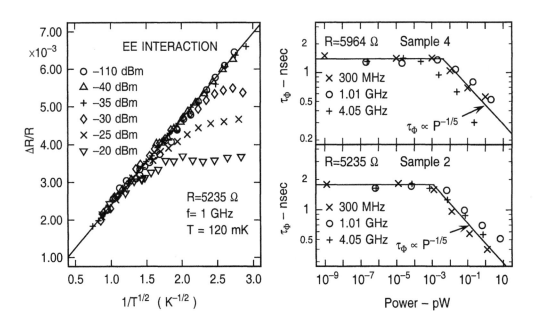

Figure 5. T dependence of the EE correction to the resistance at H=0.9 T in an rf field.

Figure 6. τ_ϕ as a function of rf power coupled into two samples at T=120 mK.

with previous high frequency experiments [15-16] and theory [17].

3. SUMMARY

The new high frequency results reported here clearly demonstrate that at 120 mK, rf powers at the level of a few fW in the frequency range of $1/\tau_\phi$ cause τ_ϕ to decrease. This decrease, however, appears to be entirely due to an enhanced energy relaxation process and not a strict decoherence process. We therefore conclude that the saturation of τ_ϕ displayed in Fig. 1 is not due to external high frequency electromagnetic noise coupled into our samples since in the earlier experiments the EE interaction was found to be temperature dependent. It seems that the proposed mechanism by Altshuler *et al.* [11] is not the source of the saturation in most experiments on the temperature dependence of τ_ϕ. We believe that in order to make theoretical progress on the intrinsic dephasing problem, the standard semi-classical perturbation approach must be abandoned (as was required for the Kondo and Luttinger Liquid problems). A full quantum treatment of the environment is needed [18,19]. Such an approach has recently been attempted [20] with the preliminary conclusion that zero point fluctuations can cause intrinsic decoherence.

REFERENCES

1. S. Washburn and R.A. Webb, Rep. Prog. Phys. **55**, 1311 (1992).
2. B.L. Altshuler, P.A. Lee, R.A. Webb, eds., *Mesoscopic Phenomena in Solids*, [Elsevier Science Publishers, Amsterdam (1991)], vol. 30.
3. B.L. Altshuler and A.G. Aronov, in *Electron-Electron Interactions in Disordered Systems*, edited by A.L. Efros and M. Pollak (Elsevier, Amsterdam, 1985).
4. B.L. Altshuler *et al.*, in *Soviet Scientific Reviews. Section A: Physics Reviews*, edited by I.M. Khalatnikov (Harwood Academic, New York, 1987), Vol. 9.
5. P. Mohanty, E.M.Q. Jariwala, and R.A. Webb, Phys. Rev. Lett. **78**, 3366 (1997).
6. P. Mohanty and R.A. Webb, Phys. Rev. **B55,** R13452 (1997).
7. S. Chakravarty and A. Schmid, Phys. Rep. **140**, 193 (1986).
8. A. Stern, Y. Aharonov, and Y. Imry, Phys. Rev. A **41**, 3436 (1990).
9. W.G. Unruh and W.H. Zurek, Phys. Rev. D **40**, 1071 (1989).
10. D.S. Golubev and A.D. Zaikin, Phys. Rev. Lett. **81**, 1074 (1998); and cond-mat/9804156.
11. B.L. Altshuler, M.E. Gershenson, and I.L. Aleiner, cond-mat/9803125 and cond-mat/9808053.
12. P. Mohanty and R.A. Webb, preprint.
13. D. Loss and K. Mullen, Phys. Rev. B 43, 13252 (1991).
14. R.W.P. King and C.W. Harrison Jr., *Antennas and Waves: a Modern Approach* (M.I.T. Press, 1969).
15. S.A. Vitkalov, G.M. Gusev, Z.D. Kvon, and G.I. Leviev, JETP Lett. **43**, 185 (1986).
16. S. Wang and P.E. Lindelof, Phys. Rev. Lett. **59**, 1156 (1987).
17. B.L. Altshuler, A.G. Aronov, and D.E. Khmelnitskii, Solid State Commun. **39**, 619 (1981).
18. R. Glauber, Phys. Rev. **130**, 2529 (1963).
19. L. Mandel, Phys. Rev. **152**, 438 (1966).
20. P. Mohanty, Ph.D. Thesis, University of Maryland 1998.

Quantum Coherence and Decoherence - ISQM - Tokyo '98
Y.A. Ono and K. Fujikawa (Editors)
© 1999 Elsevier Science B.V. All rights reserved.

Quantum Eraser and Quantum Zeno effect in mesoscopic physics[*]

G. Hackenbroich, B. Rosenow, and H. A. Weidenmüller

MPI für Kernphysik, D–69029 Heidelberg, Germany

[*]Dedicated to H. Horner on the occasion of his sixtieth birthday.

We discuss two novel applications of basic principles of quantum mechanics in mesoscopic physics: The quantum Eraser, and the Quantum Zeno Effect. For the Quantum Eraser, we consider electron transport through an Aharonov–Bohm (AB) interferometer with a quantum dot in one of its arms. The quantum dot is coupled to a which–path quantum detector. Tracing over the detector yields dephasing and a reduction of the AB interference amplitude. For a detector with a finite number of states the interference can be restored by a suitable measurement on the detector.

For the Quantum Zeno effect, we consider electron tunneling between two coupled quantum dots with one of the dots coupled to a quantum point contact detector. The interaction with the detector leads to decoherence and to the suppression of tunneling. Driving the detector with an ac voltage results in parametric resonance and a suppression of decoherence. We propose a new transport experiment with the two quantum dots in a parallel circuit connected to external leads. The experiment allows for the observation of both the quantum Zeno effect and parametric resonance.

1. Purpose

Recent progress in quantum and atom optics has made it possible to test basic tenets of quantum physics. It is the purpose of this work to show that similar tests are possible in the realm of mesoscopic physics. We study two examples: The Quantum Eraser and the Quantum Zeno effect. Aside from extending the domain of tests of quantum theory, such a study is of interest because in contrast to quantum optics, mesoscopic probes are inevitably coupled to macroscopic bodies (leads etc.), opening the possibility to address quantitatively the issue of quantum decoherence. This paper is a short summary of work contained in Refs. [1,2]. Our work is motivated by the first demonstration of controlled dephasing in a which–path semiconductor device by Buks *et al.* [3].

2. Quantum Eraser

Recent experiments in quantum optics [4] and with atomic beams [5] not only confirmed the destruction of multiple–path interference due to a which–path measurement. More importantly, they also demonstrated that the loss of interference need not be irreversible if the which–path detector is itself a quantum system. It was shown that the interference can be restored [6] by erasing the which–path information from the detector in a subsequent measurement. Here, we propose a semiconductor micostructure which can act as a quantum eraser.

We consider an Aharanov–Bohm (AB) interferometer with a quantum dot (QD) embedded in one of its arms. The quantum dot is coupled to a quantum detector with a *finite* number of states. Such a detector can be realized e.g. by utilizing the spin of the electron passing through the QD, or by a pair of quantum dots coupled capacitively to the QD in the AB interferometer. An electron passing through the QD in the AB interferometer changes the quantum state of and leaves which–path information in the detector. Analyzing various possible measurements we show that the detector with the AB interferometer can be used as a quantum eraser.

The quantum detector is taken to be an N–state quantum system. The states of the detector are labelled k, k' with $k, k' = 1, \ldots, N$ and have energies ϵ_k. We assume that the interaction V between electron and detector vanishes unless the electron is located on the QD. With $c_k^\dagger$ and $b^\dagger$ the creation operators for the states in the detector and for the state on the QD, respectively, the Hamiltonian for detector plus coupling is given by

$$H = \sum_{k=1}^{N} \epsilon_k c_k^\dagger c_k + \sum_{k,k'=1}^{N} V_{kk'} b^\dagger b c_k^\dagger c_{k'} . \tag{1}$$

We assume that the QD is in the Coulomb blockade regime and sufficiently close to a resonance so that only a single dot state need be considered. Energy and width of this resonance in the absence of any coupling to the quantum detector are denoted by E_0 and Γ, respectively. Let α, α' denote the leads coupled to the AB interferometer and p, p' the transverse modes in the leads. (For simplicity, we consider only one such mode per lead). Then, $c \equiv \{p\alpha\}$ denotes the channels. For an electron passing through the AB interferometer coupled to the detector, we find the scattering matrix

$$S_{cc',kk'} = S_{cc'}^{(0)} \delta_{kk'} - 2\pi i \gamma_c \gamma_{c'} G_{kk'}. \tag{2}$$

The term $S^{(0)}$ describes an energy– and flux–independent background due to scattering through that arm of the AB interferometer which does not contain the QD. The two–particle Green function G for transitions through the dot and interaction with the detector can be found from the matrix elements of its inverse,

$$[G^{-1}]_{kk'} = (E - \epsilon_k - E_0 + i\Gamma/2)\delta_{kk'} - V_{kk'} . \tag{3}$$

Here, E denotes the total energy. With ϵ_c the kinetic energy in channel c, energy conservation requires that $\epsilon_c + \epsilon_k = \epsilon_{c'} + \epsilon_{k'}$. Both the partial width amplitudes γ_c and the total width Γ depend on the magnetic flux through the AB ring. A non–vanishing V leads to a *splitting of the single Breit–Wigner resonance into (generically) N resonances*. The positions of these resonances are found by diagonalizing the matrix $\epsilon_k \delta_{kk'} + V_{k'k}$.

We calculate the influence of V and of a measurement at the detector on the interference pattern in the AB device. Let $\rho^{(0)} = \rho_{\text{ring}}^{(0)} \rho_{\text{det}}^{(0)}$ be the density matrix of the total system prior to the passage of the electron through the AB interferometer. Then, $\rho = S\rho^{(0)}S^\dagger$ is the density matrix after passage through the AB device. Let $A = A_{\text{ring}} A_{\text{det}}$ be the operator of an observable connected to electron transmission. The expectation value of A is given by $\langle A \rangle = \text{Tr}[\rho A]$. The trace is taken over the states of detector and leads. We focus attention on the interference term involving $S^{(0)}$ and the amplitude through the

QD. We keep only the lowest harmonic in the flux Φ. With $\alpha = 2\pi\Phi/\Phi_0$ and Φ_0 denoting the elementary flux quantum, we find

$$\langle A \rangle = A^{(0)} + \text{Re}\left\{|t|e^{i(\alpha - \alpha_0)}\langle G A_{\text{det}}\rangle_N\right\}. \tag{4}$$

Here, t and the phase α_0 depend on properties of the AB device. We focus attention on the factor $\langle G A_{\text{det}}\rangle_N = \text{Tr}_N[G A_{\text{det}}]$ which describes the loss of interference induced by the coupling of the AB device to the detector. This factor not only depends on the interaction between the interferometer and the detector but also on the actual form of the measurement A_{det} performed on the detector. This fact illustrates the fundamental difference between *detector–induced dephasing* and a true *which–path measurement*. Detector–induced dephasing amounts to putting $A_{\text{det}} = 1$ and tracing out the degrees of freedom of the detector. This procedure reduces the magnitude of the interference term in Eq. (4). By tracing out the detector, however, no information is obtained about the actual path of the electron through the AB device. To get this information one necessarily has to perform a measurement, i.e., use $A_{\text{det}} \neq 1$.

For simplicity, we assume that the detector has $N = 2$ states and use a spin 1/2–terminology although our results are not confined to this case. Prior to the pasage of the electron, the detector (the electron spin) is assumed to be polarized in the +x–direction, i.e., $\rho_{\text{det}}^{(0)} = |x, +\rangle\langle x, +|$. In the AB interferometer, the amplitude of the incoming spin–polarized electron is split into two parts $|\Psi_j\rangle = |\psi_j\rangle \otimes |x, +\rangle$, $j = 1, 2$, each part passing through one arm of the interferometer. Passage through the arm containing the QD causes a transition $|x, +\rangle \to |\chi\rangle = |x, -\rangle$ in the spin degree of freedom of $|\Psi_1\rangle$, say, while the spin part of $|\Psi_2\rangle$ remains unchanged. The flux–dependent interference term is proportional to $|\langle\chi|A_{\text{det}}|x, +\rangle|$. If A_{det} projects onto the $\pm x$–direction, the interference signal is completely wiped out. This fact reflects the complete which–path information encoded by the spin. If, however, A_{det} measures the z–component of the spin, the which–path information is irretrievably lost (erased) and the spin overlap in $|\langle\chi|A_{\text{det}}|x, +\rangle|$ is finite.

Let us consider the case $\epsilon_1 = \epsilon_2 = 0$ and model the interaction V in the form $\Delta\sigma_z$ where σ_z is a Pauli spin matrix. The interference term has the spin dependence

$$\Delta I_{AB} \sim \left|<x,+|\frac{\Gamma/2}{E - E_0 - \Delta\sigma_z/2 + i\Gamma/2} A_{\text{det}}|x,+>\right|. \tag{5}$$

We consider the case $\Delta \gg \Gamma$. The QD can be operated in two possible modes: (i) As a device for rotating the spin orientation within the $x - y$–plane or (ii) as a Stern–Gerlach filter. Case (i) is realized for $E = E_0$, and the angle of spin precession is $\pi - 2\arctan[\Gamma/\Delta]$ resulting in a spin state nearly orthogonal to $|x, +\rangle$. The total current is obtained by tracing out the detector ($A_{\text{det}} = 1$) and shows a strongly suppressed interference term $\Delta I_{AB} \sim \Gamma^2/(\Delta^2 + \Gamma^2)$. Interference can partly be restored by projecting onto the $\pm z$–spin direction. Then $\Delta I_{AB} \sim (\Gamma/2)/(\Delta^2 + \Gamma^2)^{1/2}$ which amounts to an enhancement of $|\langle\chi|A_{\text{det}}|x, +\rangle|$ by a factor of order $\Delta/\Gamma \gg 1$. In case (ii) we choose $E = E_0 + \Delta/2$ with $\Delta \gg \Gamma$. The QD blocks the $(-z)$–component of the spin. Taking the trace over spin orientations (with $A_{\text{det}} = 1$), one finds that the interference term has magnitude $\sim 1/2$. By projecting onto the $(-x)$–direction one performs a true position measurement and only detects the amplitude passing the QD. The interference term vanishes completely. A projection onto the vector bisecting the $x-$ and the z–directions reduces the degree of position information encoded in the spin and improves the contrast as compared to a spin–independent measurement by a factor $1 + 1/\sqrt{2}$.

3. Quantum Zeno Effect and Parametric Resonance

The frequent repetition of a decohering measurement leads to a striking phenomenon known as the quantum Zeno effect [7]: The suppression of transitions between quantum states. The standard example is a two–level system with a tunneling transition between the two levels. For small times, the probability to tunnel out of one of the two levels is $\sim t^2$. With a device that projects the system onto that same level, N repeated measurements yield the reduced probability $\sim N(t/N)^2$. The suppression of tunneling in bound systems has its parallel in systems with unstable states. Here, the quantum Zeno effect predicts the suppression of decay and the enhancement of the life time of unstable states. We study the quantum Zeno effect theoretically in an arrangment of coupled quantum dots one of which interacts with a quantum point contact (QPC). This arrangement is within reach of present–day experimental techniques.

The quantum Zeno effect in mesoscopic physics was first studied by Gurvitz [8]. We go considerably beyond this work and identify several novel aspects. First, we show that the application of an AC voltage with frequency ω across the QPC leads to parametric resonance and to a strong reduction of decoherence. The resonance occurs when ω equals twice the frequency ω_0 of the internal charge oscillations in the double–dot system. Second, the power spectrum of the QPC displays a peak which is a clear signal for the quantum Zeno effect. Third, we propose a new transport experiment with the two quantum dots in a parallel circuit and each dot coupled to external leads. With current flowing into the lower dot (the one coupled to the QPC), we calculate the branching ratio of the current transmitted through the upper dot and that through the lower dot. The coupling to the QPC modifies the branching ratio. The correction is proportional to the decoherence rate. A measurement of the branching ratio provides a direct signature of dephasing, and of the quantum Zeno effect in the two coupled quantum dots.

We first consider a simple system: Two coupled quantum dots without coupling to external leads and occupied by a single (excess) electron. We consider only one energy level in each dot and assume that both levels are degenerate and have energy E_0, while $\Omega_0/2$ with $\Omega_0 = \hbar\omega_0$ is the coupling matrix element between the two dots. The lower dot interacts with a QPC. The QPC is modelled as a single–channel device which is symmetric with respect to the lower dot. As before, we calculate the total density matrix of dots plus QPC, using a scattering approach, and trace out the QPC–variables. We do not give any details of the calculation here. Suffice it to say that for weak coupling between QD and QPC the density matrix ρ_{dot} of the two–dot system can be shown [9] to obey the following master equation.

$$\frac{d\rho_{\text{dot}}}{dt} = -\frac{(\Delta\mathcal{T})^2}{8\mathcal{T}(1-\mathcal{T})} \frac{e\mu}{\pi\hbar} \begin{pmatrix} 0 & \rho_{\text{dot},12}^{(0)} \\ \rho_{\text{dot},21}^{(0)} & 0 \end{pmatrix} - \frac{i\Omega_0}{2\hbar}[\sigma_x, \rho_{\text{dot}}]. \tag{6}$$

We have introduced the transmission coefficients $\mathcal{T}_l, \mathcal{T}_u$ [3] for the lower and upper dot, respectively, and $\Delta\mathcal{T} = \mathcal{T}_u - \mathcal{T}_l$ and $\mathcal{T} = (\mathcal{T}_u + \mathcal{T}_l)/2$ while μ is the voltage drop across the QPC.

Since $\rho_{\text{dot}} = \rho_{\text{dot}}^{\dagger}$ and $\text{Tr}\rho = 1$, we can parameterize ρ_{dot} by the two quantities $a = 1/2 - \rho_{\text{dot},11}$, $b = \rho_{\text{dot},12}$. Substitution into Eq. (6) yields the equations of motion $(da)/(dt) = \omega_0 \text{Im} b$ and

$$\frac{d^2 a}{dt^2} + \frac{(\Delta \mathcal{T})^2}{8\mathcal{T}(1 - \mathcal{T})} \frac{e\mu}{\pi\hbar} \frac{da}{dt} + \omega_0^2 a = 0 . \tag{7}$$

For a time–independent voltage drop $\mu = \mu_0$ both a and b display exponentially damped oscillations. The damping constant κ is equal to one half times the factor multiplying da/dt in Eq. (7). This is a clear demonstration of the quantum Zeno effect. The charge oscillations in the double–dot system modulate the current in the QPC and, hence, modify its power spectrum. They cause a peak with FWHM 2κ centered at the shifted frequency $\omega_0\sqrt{1 - \kappa^2/\omega_0^2}$. This peak is present in addition to standard shot noise. Both location and width of this peak also are clear signatures of the quantum Zeno effect. The experimental investigation of these features would amount to a time–resolved study of the quantum–mechanical measurement process and would be of considerable interest.

Interesting new physical aspects arise if μ has an AC-component, $\mu(t) = \mu_0 - \mu_1 \sin \omega t$ where $\mu_0, \mu_1 \geq 0$ and $\mu_1 \leq \mu_0$. According to Eq. (7) this corresponds to a harmonic oscillator with an *oscillatory* damping constant. One is led to an equation of the Mathieu type which is known to display parametric resonance. The resonance is most pronounced for $\omega \approx 2\omega_0$. The damping near the resonance is strongly reduced, $\kappa = \frac{(\Delta \mathcal{T})^2}{8\mathcal{T}(1-\mathcal{T})} \frac{e}{2\pi\hbar}(\mu_0 - \frac{1}{2}\mu_1)$. The resonance condition $\omega = 2\omega_0$ is interpreted as follows: The position of the electron is not measured when μ is close to zero. The electron uses this time to tunnel from one dot to the other.

We turn to the transport experiment and consider an arrangement with the two dots in parallel and each dot coupled to two external leads. For simplicity we consider only one transverse channel in each lead. The quantum dots are assumed to be in the resonant tunneling regime close to a Coulomb blockade resonance so that it is sufficient to consider only a single level in each dot. Both levels are assumed to have the same energy E_0 and width Γ. The QPC detector is described in terms of plane waves with energy ϵ_k and mean density ρ_F which are scattered from a spatially local potential with Fourier components $U_{kk'}$. To model the capacitive coupling of the QPC with the lower dot, we use the Hamiltonian of Eq. (1) where k now labels plane wave states and $b^{\dagger}$ creates an electron on the lower QD, and where we add to H the term $\sum U_{kk'} c_k^{\dagger} c_{k'}$. We note that the interaction vanishes for an electron on the upper dot.

We restrict ourselves to constant scattering potentials $U_{kk'} \equiv U$ and $V_{kk'} \equiv V$ and calculate the Green function G by expanding to all orders in V and resumming the resulting series. We find two types of contributions to G. The first contribution is independent of V and describes independent elastic scattering through the QPC and the dots. The second contribution represents the connected part and involves energy exchange Ω between the dots and the QPC. It is a complicated function of the energy E but the essential features are displayed at $E = E_0 + \epsilon_k$,

$$G_{kk'}^{\text{conn}}(\Omega) = \frac{4}{\Gamma^2 + \Omega_0^2} \frac{-V}{F_U F_{U+V}} \frac{\delta_{\epsilon_k, \epsilon_{k'} + \Omega}}{(2\Omega + i\Gamma)^2 - \Omega_0^2} \begin{pmatrix} \Omega_0^2 & -i\Omega_0\Gamma \\ -\Omega_0(2\Omega + i\Gamma) & i\Gamma(2\Omega + i\Gamma) \end{pmatrix} . \tag{8}$$

Here, $F_U = 1 + 2\pi i U \rho_F$. The Ω–dependent prefactor effectively limits inelastic processes to an interval of width Γ around $E_0 + \epsilon_k$. The energy exchange allows for a position measurement of the dot electron without violating the Heisenberg uncertainty relation.

We extend this treatment to the simultaneous scattering of $2e\mu\rho_F$ particles in different *longitudinal* QPC modes and calculate the ratio of the transmission coefficients in the upper and lower leads. We find that measurements with the QPC detector have a twofold effect: (i) They universally suppress tunneling from the feeding lead into the lower dot and (ii) they universally suppress tunneling from the lower into the upper dot. Observation (ii) follows from the *decrease* of the branching ratio

$$\frac{T_u}{T_l} = \frac{\Omega_0^2}{\Gamma^2}\left[1 - \frac{e\mu}{\pi\Gamma}\frac{(\Delta\mathcal{T})^2}{4\mathcal{T}(1-\mathcal{T})}\right] . \tag{9}$$

(We recall that without coupling to the QPC, we would have $\Delta\mathcal{T} = 0$). Both effects (i), (ii) have an obvious interpretation as manifestations of the quantum Zeno effect. We note that the second term in the square bracket is up to a factor $\Gamma/(4\hbar)$ the damping constant found for the isolated double–dot system. The appearance of the damping constant in the branching ratio shows that the parametric resonance discussed above for isolated dots can also be observed in a transport experiment.

4. Summary

In summary, we have investigated both a mesoscopic Quantum Eraser and a mesoscopic realization of the Quantum Zeno effect. We have shown that in both systems, it is possible to study generic aspects of quantum mechanics and of the measurement process. The information attainable in this way would complement that from other physical systems. Studies on mesoscopic devices are within reach of present–day experimental possibilities [1,2].

Acknowledgment. We thank Y. Imry for a valuable discussion.

REFERENCES

1. G. Hackenbroich, B. Rosenow, and H. A. Weidenmüller, submitted to Europhys. Lett.
2. G. Hackenbroich, B. Rosenow, and H. A. Weidenmüller, submitted to Phys. Rev. Lett.
3. E. Buks, R. Schuster, M. Heiblum, D. Mahalu, and V. Umansky, Nature (London) **391** (1998) 871.
4. T.J. Herzog, P.G. Kwiat, H. Weinfurter, and A. Zeilinger, Phys.Rev.Lett. **75** (1995) 3034.
5. J. Kunze, K. Dieckmann, and G. Rempe, Phys. Rev. Lett. **78** (1997) 2038.
6. M.O. Scully and Kai Drühl, Phys.Rev. A **25** (1982) 2208. M.O. Scully, B.-G. Englert, and H. Walther, Nature **351** (1991) 111.
7. W. M. Itano, D. J. Heinzen, J. J. Bollinger, and D. J. Wineland, Phys. Rev. A **41** (1990) 2295 and references therein.
8. S. A. Gurvitz, Phys. Rev. B **56** (1997) 15215.
9. S. A. Gurvitz, quant-ph/9607029.

Quantum Coherence and Decoherence - ISQM - Tokyo '98
Y.A. Ono and K. Fujikawa (Editors)
© 1999 Elsevier Science B.V. All rights reserved.

NEW "COHERENT" BILAYER QUANTUM HALL STATES

A. Sawada[1], Z.F. Ezawa[1], H. Ohno[2], Y. Horikoshi[3], Y. Ohno[2],
S. Kishimoto[2], F. Matsukura[2], A. Urayama[1], N. Kumada[1]

[1]Department of Physics, Tohoku University, Sendai 980-8578, Japan
[2]Research Institute of Electrical Commun., Tohoku University, Sendai 980-8577, Japan
[3]School of Science and Engineering, Waseda University, Tokyo 169-8555, Japan

The Hall-plateau width and the activation energy in the typical bilayer quantum Hall
(BQH) states at filling factor $\nu = 2/3$, 1 and 2 have been measured by changing the total
electron density as well as the density ratio. The stability of the QH states are found
remarkably different from one to another. The $\nu = 1$ state is stable over all measured
range of the density difference, and hence it is identified as "coherent" state. The $\nu = 2$
state shows an unexpected phase transition between these two types of the states as the
electron density is changed. The $\nu = 2$ state for the larger density is stable only around
the balanced point, and that for the lower density is stable over all range of the density
difference similarly to the $\nu = 1$ state.

1. Introduction

The quantum Hall (QH) effect in double quantum wells has recently attracted much
attention, where the structure introduces additional degrees of freedom in the third direction. Various bilayer QH states are realized by controlling system parameters such as
the strengths of the interlayer and intralayer Coulomb interaction, the tunneling interaction and Zeeman effect. It has been also pointed out[1] that a novel interlayer quantum
coherence (IQC) may develop spontaneously in the $\nu = 1/m$ state with m an odd integer.
Here, ν is the total filling factor. Murphy et al.[2] have observed an anomalous activation
energy dependence in the bilayer $\nu = 1$ QH state on the tilted magnetic field, which is
probably one of the signals[3, 4] of the IQC. Another unique feature of this IQC[5] is that
the QH state is stable at any electron density difference $n_f - n_b$, where n_f (n_b) is the
electron density in the front (back) quantum well.

The Hall-plateau width and the activation energy in the typical bilayer quantum Hall
(BQH) states have been measured by changing the total electron density as well as the
density ratio.

2. Experimental results

Our sample consists of two modulation doped GaAs quantum wells of width 200 Å,

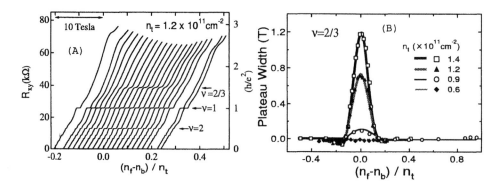

Figure 1: (A) Hall resistance versus magnetic field at a various density difference at a fixed total electron density n_t. The origins of the magnetic field axis are shifted in correspondence with the normalized density difference $(n_f - n_b)/n_t$. (B) The Hall-plateau width of $\nu = 2/3$ state at 50 mK as a function of the electron density difference at several fixed total electron densities. The lines are guides to the eye.

separated by an $Al_{0.3}Ga_{0.7}As$ barrier of thickness 31 Å, and has front and back side gates for controlling the electron density of each well independently. The measured tunneling gap was 6.8 K.

We measured the activation energy in the typical BQH states at filling factor $\nu = 1$ and 2 by changing the total electron density $n_t = n_f + n_b$ as well as the density difference $n_f - n_b$. The activation energy Δ is derived from the temperature dependence of the magnetoresistance: $R_{xx} = R_0 \exp(-\Delta/T)$.

The $\nu = 1$ state is stable over all measured range of the density difference for the total electron density $n_t \leq 1.5 \times 10^{11}$ cm^{-2} and has the minimum stability at the equal density ratio (see Fig. 2). The $\nu = 2$ state shows an unexpected phase transition between these two types of the states as the electron density is changed. The $\nu = 2$ state for the larger density is stable only around the balanced point, and that for the lower density is stable over all range of the density difference similarly to the $\nu = 1$ state (see Fig. 3).

3. Discussion

The $\nu = 2$ state in higher total and equal layer electron density have clearly all the properties of a compound state. The compound state is superposition of the independent QH states in each layer. First, they are sharply enhanced in the balanced configuration as in Fig. 2. Second, these states become unstable as the total electron density decreases (or equivalently d/ℓ_B decreases).

Next, we discuss the case where the interlayer Coulomb interaction is dominant. In general, the bilayer QH state is described by the extended Laughlin wave function $\Psi_{m_f m_b m}$ at

$$\nu = \frac{m_f + m_b - 2m}{m_f m_b - m^2} \leq 1, \tag{1}$$

where odd integers m_f and m_b represent the intralayer electron correlations, while integer

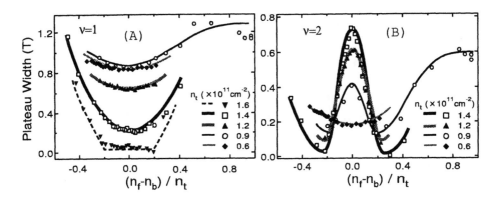

Figure 2: (A) and (B) are the Hall-plateau width of the $\nu = 1$ and 2 states respectivly at $50\,\mathrm{mK}$ as as a function of the electron density difference at several fixed total electron densities.

m represents the interlayer correlation induced by the interlayer Coulomb interaction. (The compound state is obtained as a special limit $m = 0$.) The density ratio is fixed as

$$\frac{n_{\mathrm{f}}}{n_{\mathrm{b}}} = \frac{m_{\mathrm{b}} - m}{m_{\mathrm{f}} - m}. \tag{2}$$

A strong interlayer correlation supports the "coherent" state with $m_f = m_b = m$, where the density ratio (2) becomes undetermined. It is a characteristic feature of this state[1, 3, 5] that it is stable at any density ratio and that the IQC may develop spontaneously.

The $\nu = 1$ state in our data can be identified as this "coherent" state, since the state continues to exist over all measured range of the density difference, as in Fig. 2. As the total density increases, or equivalently as d/ℓ_B increases, the stability decreases as in Fig. 3, because the interlayer Coulomb interaction becomes weaker. The state may be regarded sufficiently stable for $n_t \leq 1.0 \times 10^{11}\mathrm{cm}^{-2}$, where the activation energy is almost constant. The QH state breaks down at and above the critical density of $1.5 \times 10^{11}\ \mathrm{cm}^{-2}$ $(d/\ell_B \simeq 2.2)$.

As found in Fig. 2, the "coherent" state is least stable in the balanced configuration. This is also interpreted as an effect due to an interlayer Coulomb correlation. The activation energy in Fig. 2 is understood if the excitation gap is dominated by the charging energy proportional to $(n_f - n_b)^2$.

The $\nu = 2$ QH state undergoes a phase transition from a compound state to a "coherent" state as the total density decreases or the densities are unbalanced, i.e., as the interlayer Coulomb interaction is increased over the intralayer Coulomb interaction. When the total density becomes sufficiently small at $n_t = 0.6 \times 10^{11}\ \mathrm{cm}^{-2}$, we observe in Fig. 2 that the activation energy of the $\nu = 2$ QH state behave as those of the $\nu = 1$ state which we identify as the "coherent" state. The reason why we have a "coherent" state at $\nu = 2$ can be explained by considering the spin degree of freedom. In the $\nu = 1$ state all electrons are in the *spin-up* states. It is natural to expect another "coherent" state with

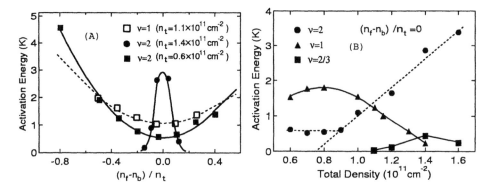

Figure 3: (A) Activation energy of the $\nu = 1$ and 2 QH states as a function of the density difference. The total density is fixed at a constant value. The curves with square data points are fitted by assuming that the activation energy depends on $(n_f - n_b)^2$. (B) Activation energy as a function of the total electro density at the balanced density point. The lines are guides to the eye.

spin-down polarization above the $\nu = 1$ state at $\nu = 2$. In this $\nu = 2$ "coherent" state the total system is spin-unpolarized.

References

[1] Z.F. Ezawa and A. Iwazaki, Int. J. Mod. Phys. B **6**, 3205 (1992).

[2] S.Q. Murphy, J.P. Eisenstein, G.S. Boebinger, L.N. Pfeifer and K.W. West, Phys. Rev. Lett. **72**, 728 (1994).

[3] Z.F. Ezawa, Phys. Rev. B **55** 7771 (1997) ; B **51**, 11152 (1995).

[4] K. Moon, H. Mori, K. Yang, S.M. Girvin, A.H. MacDonald, L. Zheng, D. Yoshioka and S.C. Zhang, Phys. Rev B **51** 5138 (1995).

[5] A. Sawada, Z.F. Ezawa, H. Ohno, Y. Horikoshi, O. Sugie, S.Kishimoto, F. Matsukura, Y. Ohno and M. Yasumoto, Solis State Commun.**103**, 447 (1997).

[6] A. Sawada, Z.F. Ezawa, H. Ohno, Y. Horikoshi, Y. Ohno, S. Kishimoto, F. Matsukura, M. Yasumoto and A. Urayama, Phys. Rev. Lett. **62**, 183 (1998).

Quantum Coherence and Decoherence - ISQM - Tokyo '98
Y.A. Ono and K. Fujikawa (Editors)
© 1999 Elsevier Science B.V. All rights reserved.

Electron-electron scattering in two-dimensional electron gas under spatially modulated magnetic field

Mayumi Kato, Akira Endo, Shingo Katsumoto* and Yasuhiro Iye*

Institute for Solid State Physics, University of Tokyo, Roppongi, Minato-ku, Tokyo, 106-8666 Japan

We have found that two-dimensional electron gas at a GaAs/AlGaAs heterointerface subjected to a spatially alternating magnetic field exhibits an excess resistivity $\Delta\rho = AT^2 + C$ at low temperatures. The T^2-term is a manifestation of the electron-electron umklapp scattering, while the constant term C represents the mass enhancement due to miniband formation and/or electron-electron umklapp process. We have also observed a similar excess resistivity when electrons are heated by a higher bias current, which affirms the electronic origin of the excess resistivity and presents a new method of hot electron thermometry.

1. INTRODUCTION

Electon-electron scattering in a metallic system is known to produce a T^2-term in resistivity at low temperature, which has been observed in a variety of materials ranging from alkali metals to the so-called strong correlation systems such as heavy fermion metals, organic conductors and transition metal oxides. The role of the electron-electron scattering in transport is subtle. As is well known, the momentum conservation precludes electron-electron scattering to contribute to resistance if the system has continuous translational symmetry (Galilean invariance). It can only do so when the Galilean invariance of the system is somehow broken[1–3]. In the case of lattice (periodic) system, this leads to the so-called umklapp process. In this work, we investigate the electron-electron scattering process in one of the simplest and cleanest experimental systems, namely, two-dimensional electron gas (2DEG) at the GaAs/AlGaAs heterointerface. The Galilean invariance is broken in a controlled way by imposing spatially modulated magnetic field. The spatial modulation of magnetic field is created by a stripe-patterned gate electrode made of ferromagnetic metals (Co or Ni) as shown in Fig. 1[4–7]. An advantage of using magnetic field modulation rather than more conventional electrostatic potential modulation[8] is that the modulation amplitude (and hence the degree of translational symmetry breaking) can be varied without affecting the electron density.

---

*Also at CREST, Japan Science and Technology Corporation (JST)

2. EXPERIMENTAL METHOD

The samples used in this study were fabricated from GaAs/AlGaAs single heterostructure with electron density $n_e = 2.9 \times 10^{15} \text{m}^{-2}$ and mobility $\mu = 60 \text{m}^2/\text{Vs}$ at 4.2 K. The 2DEG plane resides at a depth 90nm from the surface. A standard Hall-bar pattern was mesa-etched with the current path along the [100] orientation, and an array of Co (or Ni) stripes with periodicity $a = 500$nm ($a/2$ wide and $a/2$ apart) was fabricated on top. Resistivity masurements were carried out using a standard low-frequency a.c. technique. A cross-coil superconducting magnet system consisting of a 6T split-coil and a homemade 1T solenoid was used to control the horizontal and vertical field components independently. A rotating sample holder was used to align the 2DEG plane horizontally. A horizontal field of 5 T was applied to saturate the magnetization of the ferromagnetic stripes. Magnetoresistance measurements were then done by sweeping the vertical field. The condition of zero normal field component can be realized with precision by fine tuning of the vertical field.

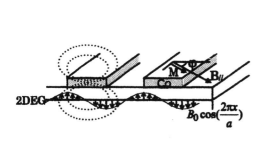

Figure 1. Schematic illustration of the sample. The definition of azimuthal angle φ is also shown.

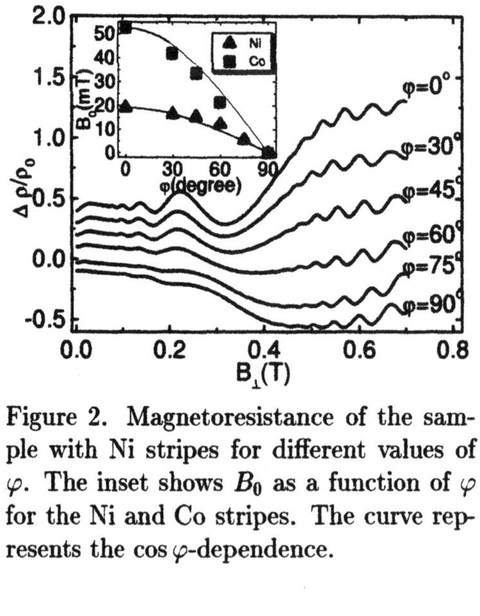

Figure 2. Magnetoresistance of the sample with Ni stripes for different values of φ. The inset shows B_0 as a function of φ for the Ni and Co stripes. The curve represents the $\cos \varphi$-dependence.

3. RESULTS AND DISCUSSION

Figure 2 shows the magnetoresistance traces with $B_\| = 5$T applied at different azimuthal angles φ. The oscillatory behavior in the low field range ($B_\perp \le 0.3$T) is the magnetic Weiss (commensurability) oscillation[5–7]. At higher $B_\perp$, the Shubnikov-de Haas (SdH) oscillations are observed. The amplitude of the magnetic field modulation B_0 at the 2DEG plane is estimated from the analysis of the Weiss oscillation[9,10]. As seen in the inset, B_0 is proportional to $\cos \varphi$ as expected, since only the magnetization component parallel to the direction of modulation is effective in producing B_0.[11] The maximum value of the modulation amplitude (at $\varphi = 0°$) turned out to be $B_0 = 52$mT for Co stripes and 19mT for Ni stripes.

Figure 3 shows the temperature dependence of the resistivity under a parallel magnetic field $B_\parallel = 5T$ applied at $\varphi = 0°$ ($B_0 = 52mT$) and $90°$ ($B_0 = 0$), respectively. The approximately T-linear resistivity for $B_0 = 0$ is governed by the acoustic phonon scattering. We define the excess resistivity $\Delta\rho$ as the increment of resistivity at a finite B_0 from that at $B_0 = 0$. The main panel of Fig. 4 shows the temperature dependence of $\Delta\rho$ for $B_0 = 52mT$. The inset of Fig. 4 demonstrates that $\Delta\rho$ is proportional to $\cos^2\varphi$, as expected, since the relevant scattering matrix element is proportional to $B_0 \propto \cos\varphi$ as seen in Fig. 2[11].

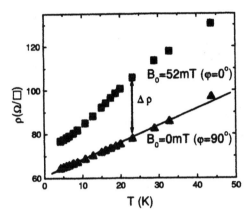

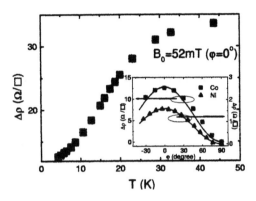

Figure 3. T-dependence of resistivity of the sample with Co stripes with $B_\parallel = 5T$ applied at $\varphi = 0°$ and $90°$.

Figure 4. T-dependence of $\Delta\rho$ for $B_0 = 52mT$. Inset shows the φ-dependence of $\Delta\rho$. The curve represents the $\cos^2\varphi$-dependence.

Figure 5 shows $\Delta\rho$ plotted against T^2 for different values of B_0 produced by different settings of φ. The excess resistivity can be expressed as $\Delta\rho = AT^2 + C$, and both A and C are proportional to B_0^2. We note that the measurement on another sample with the stripe pattern $90°$ rotated with respect to the current channel yielded null result, $i.e.$ no resistivity enhancement associated with the magnetic field modulation. The term AT^2 represents the electron-electron umklapp scattering process, which is brought into play by the imposed artificial superperiodicity. The constant term C which represents the resistivity enhancement in the $T \to 0$ reflects the mass renormalization due to miniband formation and/or electron-electron umklapp process.

We next turn to the behavior at higher current bias. The inset of Fig. 6 shows the current density dependence of the resistivity for $B_0 = 52mT$ and $B_0 = 0$ with the sample at $T = 1.25$ K. The J-independence of ρ for $B_0 = 0$ ensures that the lattice temperature remains unchanged. The J-dependence of ρ for finite B_0, therefore, reflects the dependence on the electron temperature. The electron temperature T_e can be estimated from the SdH oscillation amplitude. The main panel of Fig. 6 show the T_e-dependence of ρ with solid triangles. It agrees well with the T-dependence at low current bias (open squares). This result furnishes a strong piece of evidence that the excess resistivity is associated with the electron-electron scattering. We also comment that it forms the basis of a new method to determine T_e of hot electrons. The method has a merit that it can be used in a high

214

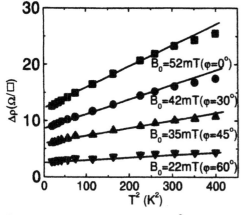

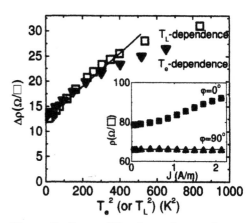

Figure 5. $\Delta\rho$ plotted against T^2 for different values of B_0.

Figure 6. Inset: The dependence of ρ on J for $\varphi = 0°$ ($B_0 = 52$mT) and $\varphi = 90°$ ($B_0 = 0$) $\Delta\rho$ plotted against T_e^2 (triangles) in comparison with the T^2 (open squares)-dependence at low current bias.

T_e region and for low mobility samples where the conventional method based on the SdH effect may not be applicable.

4. SUMMARY

We have found that the 2DEG at a GaAs/AlGaAs heterointerface exhibits an excess resistivity $\Delta\rho = AT^2 + C$ when subjected to a spatially alternating periodic magnetic field. The term AT^2 is a manifestation of the electron-electron umklapp scattering associated with the artificial periodicity, while the term C reflects the mass renormalization due to the miniband formation and/or the electron-electron umklapp process. Similar behavior is observed when the electron temperature is raised by higher bias current, which strongly affirms the electronic origin of the excess resistivity, and suggests a new method of hot electron thermometry.

REFERENCES

1. J.M. Ziman, *Electrons and Phonons* (Clarendon Press, Oxford, 1960).
2. M.Kaveh and N.Wiser, Adv. Phys. **33** (1984) 257.
3. K. Yamada and K. Yosida, Prog. Theor. Phys. **76** (1986) 621.
4. R.Yagi and Y.Iye, J. Phys. Soc. Jpn. **62** (1993) 1279.
5. S.Izawa, S.Katsumoto, A.Endo, and Y.Iye, J.Phys.Soc.Jpn. **64** (1995) 706.
6. P.D.Ye, *et al.*, Phys. Rev. Lett. **74** (1995) 3013.
7. H.A.Carmona, *et al.*, Phys. Rev. Lett. **74** (1995) 3009.
8. A.Messica, *et al.*, Phys. Rev. Lett. **78** (1997) 705.
9. F.M.Peeters and P.Vasilopoulos, Phys. Rev. B**46** (1992) 4667; *ibid* **47** (1993) 1466.
10. A. Endo, S. Izawa, S. Katsumoto, and Y. Iye, Surf. Sci.**361/362**, 333 (1996).
11. M.Kato, A.Endo, S.Katsumoto and Y.Iye, to appear in Phys. Rev. B.

Quantum Coherence and Decoherence - ISQM - Tokyo '98
Y.A. Ono and K. Fujikawa (Editors)
© 1999 Elsevier Science B.V. All rights reserved.

Low-energy transport through a 1D Mott-Hubbard insulator

V. V. Ponomarenko* and N. Nagaosa

Department of Applied Physics, University of Tokyo, Bunkyo-ku, Tokyo 113, Japan
* On leave from A.F.Ioffe Physical Technical Institute, 194021, St. Petersburg, Russia

There is a low energy crossover [1] from a Fermi liquid to a Mott-Hubbard insulator in 1D transport through a one channel wire of finite length L near half-filling. It is examined in the presence of impurity pinning assuming that the Hubbard gap $2M$ is large enough: $M > T_L \equiv v_c/L$ (v_c: charge velocity in the wire). We show that the conductance vs. voltage/temperature displays a zero-energy resonance if the impurity backscattering rate is weak, $\Gamma_1 \ll T_L$.

Recent developments in the nano-fabrication technique have allowed to produce relatively clean one channel wires of $1-10\mu m$ length. In these wires, a 1D Tomonaga-Luttinger Liquid (TLL) behavior [2] has been observed in transport measurements [3,4]. Then it was suggested that the correlated insulating behavior may be tuned with this setup [5]. This behavior of the 1D electron systems is expected [6] at half-filling where any Umklapp scattering has to open a Hubbard gap $2M$ in the charge mode spectrum of the infinite wire if the forward scattering is repulsive. If the wire is connected to reservoirs a crossover is expected [1] from a Fermi liquid transport carried by the reservoirs electrons at low energy to that of a Mott-Hubbard insulator involving soliton excitations of the condensate, which emerges in the one channel wire of long enough length L: $M > T_L \equiv v_c/L$ (v_c: charge velocity in the wire). The Mott-Hubbard insulator, which is probably the easiest for the experimental observation, features two marginal quantities [7]: the charge of the soliton is unchanged e ($e = \hbar = 1$ below) and the exponent of the electron-soliton transition $1/2$ brings about similarity with the free electron tunneling through a resonant level.

The transport through a 1D channel of a wire confined between two leads could be modeled by a 1D system of electrons whose pairwise interaction is local and switched off outside the finite length of the wire. Applying bosonization and spin-charge separation we can describe the charge and spin density fluctuations $\rho_b(x,t) = (\partial_x \phi_b(x,t))/(\sqrt{2\pi})$, $b = c,s$, respectively, with (charge and spin) bosonic fields $\phi_{c,s}$. Without impurities their Lagrangian symmetrical under the spin rotations reads

$$\mathcal{L} = \int dx \sum_{b=c,s} [\frac{v_b}{2g_b}\{\frac{1}{v_b^2}\left(\frac{\partial_t \phi_b(t,x)}{\sqrt{4\pi}}\right)^2 - \left(\frac{\partial_x \phi_b(t,x)}{\sqrt{4\pi}}\right)^2\} - \frac{E_F^2 U_b}{\pi v_F}\bar{\varphi}(x)\cos(\frac{4\mu_b g_b}{v_b}x + \sqrt{2}\phi_c)] \ (1)$$

where $\bar{\varphi}(x) = \theta(x)\theta(L-x)$ specifies a one channel wire of the length L adiabatically attached to the leads $x > L, x < 0$, and $v_F(E_F)$ denotes the Fermi velocity(energy) in the channel. The parameter $\mu_c \equiv \mu$ varies the chemical potential inside the wire from its zero value at half-filling and $\mu_s = 0$. The constants of the forward scattering differ inside the wire $g_b(x) = g_b$ for $x \in [0, L]$ from those in the leads $g_b(x) = 1$, and an Umklapp scattering (backscattering) of strength $U_c(U_s)$ is introduced inside the wire. The velocities $v_{c,s}(x)$ change from v_F outside the wire to some constants $v_{c,s}$ inside it. We can eliminate them rescaling the spacial coordinate x_{old} in the charge and spin Lagrangians of (1) into $x_{new} \equiv \int_0^{x_{old}} dy/v_{c,s}(y)$. As a result, the new coordinate will have an inverse energy dimension and the length of the wire becomes different for the charge mode $(L \to 1/T_L)$ and spin mode $(L \to 1/T'_L)$. Applying renormalization-group results of the uniform sin-Gordon model at energies larger than T_L or T'_L we come to renormalized values of the parameters in (1). For repulsive interaction when initially $g_s > 1 > g_c$, the constant U_s of backscattering flows to zero and g_s to 1, bringing the spin mode into the regime of the free TLL . The constant U_c of Umklapp process increases reaching v_F/v_c at the energy cut-off corresponding to the mass of the soliton M if the chemical potential μ is less than M. Meanwhile, g_c flows to its free fermion value $g_c = 1/2$ [1]. A weak backscattering on a point impurity potential inside the wire $0 < x_0 < L$ will be accounted for with the additional part to the Lagrangian (1):

$$\mathcal{L}_{imp} = -\frac{2V_{imp}}{\pi\alpha}\cos(\frac{\phi_c(t, x_c)}{\sqrt{2}} + \varphi_0)\cos(\frac{\phi_s(t, x_s)}{\sqrt{2}}) \qquad (2)$$

where $x_{c,s} = x_0/v_{c,s}$, $\varphi_0 \equiv \varphi + 2\mu g_c x_c$ includes a phase of the scatterer φ. The amplitude of the potential V_{imp} specifies the free electron transmittance coefficient [8] as $1/(1+V_{imp}^2)$, and $\alpha \simeq 1/E_F$ is momentum cut-off assumed to be determined by the Fermi energy.

Duality Transformation - The finite length of the wire entails an exponentially small but non-zero probability for tunneling between the infinite set of the degenerate vacua of the massive charge mode characterized by the quantized values of $\sqrt{2}\phi_c(\tau, x) + 4g_c\mu x = 2\pi m$, m is integer. In the absence of an impurity, this tunneling has been considered [7] by instanton techniques in imaginary time. Under the assumption that $E_F V_{imp} \ll M$, a crucial modification to that consideration stems from the shift of the m-vacuum energy produced by the impurity. Since it is equal to $(-1)^m \frac{2V_{imp}}{\pi\alpha}\cos\varphi\cos(\frac{\phi_s(\tau, x_s)}{\sqrt{2}})$ the neighbor vacua become non-degenerate. This can be accounted for by ascribing opposite values of the pseudospin variable $\sigma = \pm 1$ to the neighbor vacua which are the eigenvalues of the third component of the Pauli matrix σ_3. The energy splitting becomes an operator $\sigma_3 \frac{2V_{imp}}{\pi\alpha}\cos\varphi\cos(\frac{\phi_s(\tau, x_0)}{\sqrt{2}})$ acting on the pseudospins, and every (anti-)instanton tunneling rotates a σ_3-value into its opposite with the Pauli matrix σ_1. The partition function associated to the Lagrangian (1) plus (2) then can be written as

$$\mathcal{Z} \propto \sum_{N=0}^{\infty}\sum_{e_j=\pm}\int D\phi_s \frac{e^{-\mathcal{S}[\phi_s]}}{N!} Tr_\sigma \left[T \left\{ \int \left(\prod_{i=1}^{N} d\tau_i Pe^{-s_0/T_L}\sigma_1(\tau_i) \right) \right. \right.$$

$$\left. \left. \times \exp[\sum_{i,j}\frac{e_i e_j}{2}F(\tau_i - \tau_j) + \frac{2\cos\varphi V_{imp}}{\pi\alpha}\int d\tau \sigma_3(\tau)\cos(\frac{\phi_s(\tau, x_0)}{\sqrt{2}})] \right\} \right]. \qquad (3)$$

Here $\mathcal{S}[\phi_s] = \int_0^\beta d\tau \int dx\{(\partial_\tau\phi_s(\tau, x))^2 + (\partial_x\phi_s(\tau, x))^2\}/(8\pi)$ is the free TLL Eucleadian action. All τ-integrals run from 0 to the inverse temperature $\beta = 1/T$ and the sum over e_j is subject to the neutrality condition $\sum_j e_j = 0$. To have all $\sigma_{1,3}$-matrices time-ordered under the sign T, we attributed each of them to a corresponding time t assuming that their time evolution is trivial. The exponential part s_0 came from the instanton action $s_0 = \sqrt{M^2 - \mu^2} = M \sin \varpi$ and the pre-exponent P has been calculated [7] as: $P = C\sqrt{D'}(\sin^3 \varpi M T_L)^{1/4}$ with the constant C of the order of 1. The parameter D' is a high-energy cut-off to the long-time asymptotics of the kink-kink interaction: $F(\tau) = \ln\{\sqrt{\tau^2 + 1/D'^2}\}$. It varies with μ from $D' \simeq \sqrt{MT_L}$ at $\mu = 0$ to $D' \simeq (M/\mu)T_L$ if $\mu > T_L$. Noticing that the interaction F coincides with the pair correlator of some bosonic field θ_c whose evolution is described with free TLL action $\mathcal{S}[\theta_c]$, we can, following Schmid [9], sum up the expansion in (3) ascribing a factor $\exp(\mp\theta_c(\tau_j,0)/\sqrt{2})$ to the τ_j (anti-)instanton, respectively. The result reduces to a standard Hamiltonian form $\mathcal{Z} \propto \times Tr\{e^{-\beta\mathcal{H}}\}$ with

$$\mathcal{H} = \mathcal{H}_0[\phi_s(x)] + \mathcal{H}_0[\theta_c(x)] - 2Pe^{\frac{-s_0}{T_L}}\sigma_1 \cos(\theta_c(0)/\sqrt{2}) - \frac{2V_{imp}}{\pi\alpha}\cos(\varphi)\sigma_3 \cos(\frac{\phi_s(x_s)}{\sqrt{2}}) \quad (4)$$

Here $\phi_s(x)$ and $\theta_c(x)$ are Schrödinger's bosonic operators related to the variables $\phi_s(\tau, x)$ and $\theta_c(\tau, x)$ of the functional integration in (3). The operator $\mathcal{H}_0[\phi_s(x)]$ ($\mathcal{H}_0[\theta_c(x)]$), a function of the field $\phi_s(x)$ ($\theta_c(x)$) and its conjugated, is a free TLL Hamiltonian ($g = 1$) corresponding to the free TLL action $\mathcal{S}[\phi_s]$ ($\mathcal{S}[\theta_c]$) in (3), respectively. The dual model specified by (4) is equivalent to the initial one (1) at low energy. It relates to a Point Scatterer with internal degree of freedom in TLL. Fortunately, this in general rather complicated model may be solved easily through fermionization. This simplification arises from the marginal behavior of the Mott-Hubbard insulator [7]: the charge of the transport carriers does not change on passing from the low energies to the higher ones despite the nature of the carriers does.

Fermionization - From the commutation relations and hermiticity, the Pauli matrices can be written as $\sigma_\alpha = (-1)^{\alpha+1}\frac{i}{2}\sum_{\beta,\gamma}\epsilon^{\alpha,\beta,\gamma}\xi_\beta\xi_\gamma$ with Majorana fermions $\xi_{1,2,3}$ and antisymmetrical tensor ϵ : $\epsilon^{123} = 1$. Since the interaction in (4) is point-like localized and its evolution involves only the appropriate time-dependent correlators, we can fermionize it making use of $\psi_c(0) = \sqrt{\frac{D'}{2\pi}}\xi_1 exp\{i\frac{\theta_c(0)}{\sqrt{2}}\}$, $\psi_s(0) = \sqrt{\frac{E_F}{2\pi}}\xi_3 exp\{i\frac{\phi_s(x_s)}{\sqrt{2}}\}$. Here the charge (spin) fermionic field $\psi_{c(s)}(0)$ is taken at $x = 0$. These fields have linear dispersions taken after their related bosonic fields with momentum cut-offs (equal to the energy ones) D' and E_F, respectively. Substitution of these fields into (4) produces a free-electron Hamiltonian where the interaction reduces to tunneling between the $\psi_{c,s}$ fermions and the Majorana one ξ_2 ($\equiv \xi$, below). Application of a voltage V between the left and right reservoirs forces us to use the real-time representation. Since each instanton tunneling transfers charge $1e$ between the reservoirs [7] the voltage may be accounted for by a shift $\theta_c/\sqrt{2} \to \theta_c/\sqrt{2} + Vt$ of the cos-argument in the real-time form of the action corresponding to (4). Assuming that it remains small enough, $V < T_L < M$, we will neglect its effect on the other parameters. The real-time Lagrangian associated with the fermionized Hamiltonian (4) reads:

$$\mathcal{L}_F = i\xi\partial_t\xi(t) + i\sum_{a=c,s}\int dx\psi_a^+(\partial_t + \partial_x)\psi_a - [(\sqrt{\Gamma_2}\psi_c^+(0,t)e^{-iVt} + \sqrt{\Gamma_1}\psi_s^+(0,t))\xi(t) + h.c.] \quad (5)$$

where the rate of impurity scattering is $\Gamma_1 = \frac{2E_F}{\pi}(\cos(\varphi)V_{imp})^2$ and the rate of the instanton tunneling is $\Gamma_2 = const\sqrt{T_L M \sin^3 \varpi}e^{-2s_0/T_L}$. The current flowing through the channel is [1,7] $J = -\frac{\partial \mathcal{L}_F}{\partial (Vt)} = -i\sqrt{\Gamma_2}[\psi_c^+(0,t)\xi(t)e^{-iVt} - h.c.]$. Its calculation with the non-equilibrium Lagrangian (5) results in:

$$J = \frac{2\Gamma_2(\Gamma_1 + \Gamma_2)}{\pi} \int d\omega \frac{f(\frac{\omega-V}{T}) - f(\frac{\omega+V}{T})}{\omega^2 + 4(\Gamma_1 + \Gamma_2)^2} \tag{6}$$

which is the current passing through a resonant level of half-width $2(\Gamma_1 + \Gamma_2)$ and suppressed by the factor $\Gamma_2/(\Gamma_1 + \Gamma_2)$. The typical features of this current can be illustrated with its zero temperature behavior vs. voltage and the linear bias conductance G vs. temperature, respectively:

$$J = \frac{2\Gamma_2}{\pi} \arctan\left(\frac{V}{2(\Gamma_1 + \Gamma_2)}\right), \quad G = \frac{\Gamma_2}{\pi^2 T}\psi'\left(\frac{1}{2} + \frac{\Gamma_1 + \Gamma_2}{\pi T}\right) \tag{7}$$

where $\psi'(x)$ is the derivative of the di-gamma function, $\psi'(1/2) = \pi^2/2$, and the high temperature asymptotics of G is $\Gamma_2/(2T)$. The zero-temperature conductance $G \to \frac{\Gamma_2}{\pi(\Gamma_1 + \Gamma_2)}$ manifest in (6) is the function of $\Gamma_1/\Gamma_2 = (\cos\varphi V_{imp}e^{so/T_L})^2 E_F/\sqrt{T_L M \sin^3 \varpi}$. Comparison with the initial transmittance of the impurity scatterer shows that $\sqrt{\Gamma_1/\Gamma_2}$ may be conceived as a renormalization of the initial amplitude V_{imp} by the instanton exponent and a power factor of the TLL for $g = 1/2$.

In conclusion, the low energy tunneling through a 1D Mott-Hubbard insulator weakly pinned by an impurity of low scattering rate Γ_1 has been described by an effective model of a point scatterer with pseudo spin imbedded in TLL. The result shows a zero energy resonance in the conductance smeared by the impurity over a large width equal to $\Gamma_1 + \Gamma_2$ where $\Gamma_2 \approx \sqrt{T_L M}e^{-2M/T_L}$ is an exponentially low rate of tunneling of the condensate phase. The linear bias conductance at zero temperature $G = \frac{\Gamma_2}{\pi(\Gamma_1 + \Gamma_2)}$ reveals an exponential enhancement of the amplitude of the impurity potential $\sqrt{\Gamma_1/\Gamma_2}$ by the condensate tunneling. This suppression of the resonant conductance does not affect the saturation of the current at $J = \Gamma_2$ above the crossover as far as $\Gamma_1 + \Gamma_2 \ll T_L$.

REFERENCES

1. V. V. Ponomarenko and N. Nagaosa, Phys. Rev. Lett. **81**, 2304 (1998).
2. F.D.M. Haldane, Phys. Rev. Lett.**47**, 1840 (1981).
3. S. Tarucha, T. Honda, and T. Saku, Solid State Commun. **94**, 413 (1995).
4. A. Yacobi et al, Phys. Rev. Lett. **77**, 4612 (1996).
5. H. Fukuyama, S. Tarucha (private communications).
6. J. Solyom, Adv. Phys. 31, 293 (1979).
7. V. V. Ponomarenko and N. Nagaosa, cond-mat/9806136.
8. U. Weiss, Solid State Commun. **100**, 281 (1996).
9. A. Schmid, Phys. Rev. Lett. **79**, 1714 (1984).

Quantum coherence and decoherence in a superconducting single-electron transistor

Y. Nakamura and J. S. Tsai

NEC Fundamental Research Laboratories,
34 Miyukigaoka, Tsukuba, Ibaraki, 305-8501, Japan*

In a small-Josephson-junction circuit, we can construct a two-level system that consists of two charge-number states in a macroscopic electrode. We have investigated coherence and decoherence in the two-level system with a spectroscopic technique based on a dc-measurement of a photon-assisted Cooper-pair-tunneling current through a superconducting single-electron transistor. We confirmed experimentally the existence of energy-level splitting due to the coherent superposition of the two charge states.

1. INTRODUCTION

Recently, quantum coherence and decoherence have attracted more interest in conjunction with the increasing amount of research into quantum computation [1]. In particular, the coherence in a two-level system, which works as a quantum bit (qubit), that is, the elementary unit of quantum computation, is one of the most basic requirements in this context. In general, microscopic systems such as atomic states and nuclear-spin polarization are known to be coherent two-level systems that have a long decoherence time. Such microscopic systems are well isolated from their environment which has large degrees of freedom and causes decoherence of the system. On the other hand, a coherent two-level system is rarely obtained in a macroscopic system within which large numbers of freedom are interacting strongly. For example, in standard electronic devices which have worked extremely well as classical bits, the coherence of the electronic state is rapidly destroyed mainly due to the existence of the Fermi sea which is a continuum of energy levels. Low-energy excitations among those levels are frequently caused by electron-electron scattering, electron-phonon scattering, and so on. Even in single-electron devices in which we can control the tunneling of a single electron by virtue of the single-electron charging energy in nano-scale

* This work has been supported by the Core Research for Evolutional Science and Technology (CREST) project of the Japan Science and Technology Corporation (JST).

electrodes [2], we still have the same decoherence due to the Fermi sea as long as the devices are made of normal-metal electrodes with a quasi-continuous density of states. The charging effect introduces an energy gap for the inter-electrode excitations, i.e., for the tunneling, but not for the intra-electrode excitations.

To eliminate the low-energy intra-electrode excitations, we can utilize superconductivity. In a superconductor, all the electrons in the electrode condense into a single ground state and the lowest excitation is above the superconducting gap energy Δ. In addition, Cooper-pair tunneling through a Josephson junction coherently couples different charge-number states. By combining superconductivity and the charging effect, we can obtain a two-level system in a macroscopic electric circuit as we will explain in the following.

2. SUPERCONDUCTING SINGLE-ELECTRON BOX

As shown in the enclosed area of Fig.1a, a superconducting single-electron box (S-SEB) circuit consists of superconducting electrodes, a small Josephson junction and a gate capacitor. Because of the charging effect of the sub-micron (but still

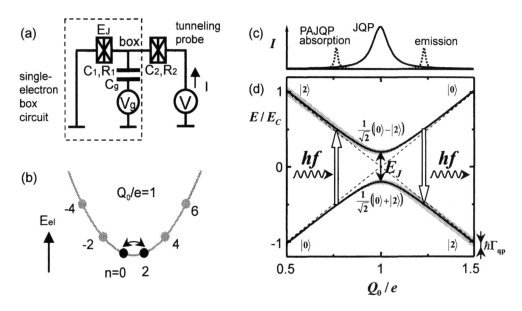

Figure 1. (a) A circuit diagram of an S-SET. (b) Electrostatic energy of each charge state in an S-SEB for $Q_0/e=1$. (c) Schematic Q_0-dependences of the JQP and PAJQP current. (d) An energy-band diagram illustrating the PAJQP process. The solid curves show the energy levels of the two (quasi-) eigenstates of the S-SEB. The energy gap between them is $\Delta E(Q_0)$. The shadowed areas around the curve indicate the broadening of the levels due to the quasiparticle-tunneling decay. The vertical arrows represent the photon-assisted interband transitions.

containing more than 10^9 atoms and electrons) box electrode, the electrostatic energy of each charge state |n⟩ depends on the number n of excess electrons in the box as $E_{el}=E_C(n-Q_0/e)^2$ (Fig.1b), where E_C is the single-electron charging energy of the box electrode and Q0 is the gate-induced charge on the box ($Q_0\equiv C_gV_g$). For the moment, we take into account only the even-n states, since quasiparticle (single-electron) excitation is suppressed by the large Δ. For instance, for $0\leq Q_0/e<2$, the two lowest-energy states |0⟩ and |2⟩ constitute an effective two-level system as long as the conditions $\Delta>E_C\gg E_J,k_BT$ are fulfilled. Here E_J is the Josephson energy which gives the coupling energy between two charge states that differ by 2e. As a function of Q_0, the eigenenergies of the two-level system behave as shown by the solid curves in Fig.1d where there is energy-level splitting at $Q_0/e=1$ due to the coherent superposition of the two charge states. The energy gap between the two eigenstates is given by $\Delta E(Q_0)=(E_J^2+\delta E(Q_0)^2)^{1/2}$, where $\delta E(Q_0)\equiv 4E_C(1-Q_0/e)$ is the difference in the electrostatic energy of the two charge states.

The coherent superposition of the two charge states in the ground state of an S-SEB has been confirmed by the measurement of the average charge number in the S-SEB [3]. However, the excited state and the energy-level splitting have not been explored so far.

3. SPECTROSCOPY OF THE ENERGY-LEVEL SPLITTING

For a spectroscopic measurement of the energy-level splitting, we used a photon-assisted Cooper-pair tunneling current through a voltage-biased superconducting single-electron transistor (S-SET). As shown in Fig.1a, current measurement through an S-SET can be considered a tunnel-probe measurement of the S-SEB state. Here we need to redefine $Q_0\equiv C_gV_g+C_2V$, including the effect of the probe voltage V. Under a certain bias voltage, we can selectively detect the state |2⟩ as two sequential quasiparticle tunnelings which reset the state to |0⟩. Thus, only when there exists a transition from |0⟩ to |2⟩, are the processes repeated and observed as a dc current. The current is called a Josephson-quasiparticle (JQP) current [4]. It shows a resonant peak at $Q_0/e=1$ where the two charge states are degenerate and coherent Cooper-pair tunneling between them occurs efficiently (Figs.1c and 1d). Here we focus on the case where the level broadening (see below) is much smaller than the splitting E_J. Then, the JQP peak has a width of an order of E_J in the energy scale. Within this range of the electrostatic energy difference, the two charge states are significantly superposed.

As shown in Fig.1d, microwave irradiation opens a new channel for the transport. When the microwave photon energy hf coincides the energy gap $\Delta E(Q_0)$, photon-assisted interband transitions become possible, which results in side peaks in the I-Q_0 curve (Fig.1c) corresponding to the photon-absorption and photon-emission processes [5,6]. Therefore, by plotting the photon-assisted JQP (PAJQP) peak position as a function of the microwave frequency, we can trace the energy-dispersion curve. A similar spectroscopic measurement has also been reported in Ref.[7]. In addition, the PAJQP peak has, in principle, a width

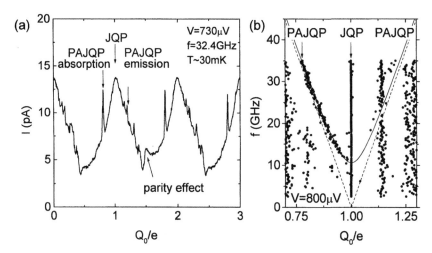

Figure 2. (a) An I-Q_0 curve at V=730µV under 32.4-GHz microwave irradiation. (b) Peak positions (black dots) in the I-Q_0 curves at V=800µV as a function of microwave frequency f. For each frequency, fluctuating offset in Q_0 is compensated for by adjusting the JQP peak to $Q_0/e=1$. The solid curve represents hf=$\Delta E(Q_0)$ with E_J=44µeV. The dashed lines show hf=$|\delta E(Q_0)|$.

corresponding to the energy-level broadening. In the present scheme, the narrowest broadening is limited by the *measurement*, that is, the quasiparticle-tunneling decay from the state $|2\rangle$ with a rate Γ_{qp}.

4. RESULTS AND DISCUSSION

The measured sample was an Al-based S-SET with the parameters: Δ=200µeV; the junction capacitance and the gate capacitance, C_1=490aF, C_2=20.8aF, and C_g=2.7aF; $E_C \equiv e^2/2C_\Sigma$=155µeV where $C_\Sigma \equiv C_1+C_2+C_g$; the tunnel-junction resistance R_1~20kΩ and R_2~30MΩ; $k_B T$~3µeV (T~30mK) and $\hbar\Gamma_{qp}$~0.1µeV.

Figure 2a shows an I-Q_0 curve at V=730µV under 32.4-GHz microwave irradiation. In addition to the broad JQP peak, we observed narrow PAJQP peaks in agreement with the schematics in Fig.1c. A possible reason for the suppression of the PAJQP peak on the photo-emission side may be the difference in the quasiparticle tunneling rate on either side. We also observed a 2e-periodic current step (the periodicity is not shown) at Q_0/e=1.5 caused by a single residual quasiparticle in the box electrode in the odd-n state [8].

The position of the PAJQP peak shifted as a function of the microwave frequency f. The black dots in Fig.2b show the position of peaks in the I-Q_0 curve for each frequency. The PAJQP peak position approximately followed the solid curve hf=$\Delta E(Q_0)$, indicating energy-level splitting between the two quasi-eigenstates. The $E_J=44\mu eV$ used in the fitting was a reasonable value given the junction resistance R_1 estimated independently from the size of the parity-dependent current step. In Fig.2, we also observed several small peaks on the emission side of the JQP peak. The positions of those peaks did not depend on f, for example, as shown in Fig.2b at $Q_0/e\sim0.065$. We may attribute these peaks to the enhanced spontaneous emission due to coupling with some resonant modes in the environment.

The decoherence time in an S-SEB limited by the spontaneous emission from the excited state due to the coupling with an electromagnetic environment could be estimated by means of Fermi's golden rule [9]. In the case of an ohmic environment with impedance R_{env} on the probe electrode, the decoherence time τ_d reads $(\hbar/\pi)(\Delta E(Q_0)/E_J{}^2)(R_Q/\kappa^2 R_{env})$, where $\kappa=C_2/(C_1+C_2+C_g)$ and $R_Q\equiv h/4e^2\sim6.45k\Omega$. For our experimental parameters, and assuming R_{env} of the vacuum impedance $\sim377\Omega$, τ_d was estimated as $\sim100nsec$, $\sim10^3$ times larger than the period of the coherent oscillation ($h/E_J\sim100psec$). In the experiment, however, the observed PAJQP peak width ($\sim7\mu eV$ in the energy scale) was much wider than the corresponding width ($\sim0.01\mu eV$) and than the broadening due to the quasiparticle tunneling for the measurement ($\sim0.1\mu eV$). Though the reason for the extra broadening of the PAJQP peak is not clear, the width is still narrower than E_J, thus reflecting the coherence in the two-level system.

5. CONCLUSION

We have investigated the coherence and decoherence in a two-level system consisting of two charge states in a superconducting single-electron box. Our spectroscopic measurement using a photon-assisted JQP current showed the existence of energy-level splitting due to the coherent superposition of the two charge states in the macroscopic electrode. This means that not only the ground state, but also the excited state of the two-level system, would preserve the coherence for a certain length of time that exceeds the characteristic time of the coherent evolution.

There have been several proposals concerning the implementation of such kind of a two-level system as a qubit into a quantum computer [10-12]. Such a macroscopic qubit might be advantageous in terms of facilitating circuit integration and fabrication. Further studies on the decoherence sources and the operation scheme of the qubits are needed, though, before such an implementation will become practical.

REFERENCES

1. See for example, D. P. DiVincenzo, in *Mesoscopic Electron Transport*, edited by L. L. Sohn, L. P. Kouwenhoven, and G. Schön (Kluwer, Dordrecht, 1997), p.657, and references therein.

2. See for review, *Single Charge Tunneling*, edited by H. Grabert and M. H. Devoret (Plenum, New York, 1992).
3. V. Bouchiat, D. Vion, D. Esteve, and M. H. Devoret, in *Quantum Devices and Circuits*, edited by K. Ismail, S. Bandyopadhyay, and J. P. Leburton, (Imperial College, London, 1997), p. 204.
4. T. A. Fulton, P. L. Gammel, D. J. Bishop, L. N. Dunkleberger, and G. J. Dolan, Phys. Rev. Lett. **63**, 1307 (1989).
5. Y. Nakamura, C. D. Chen, and J. S. Tsai, Czech. J. Phys. **46**, 2301 (1996).
6. Y. Nakamura, C. D. Chen, and J. S. Tsai, Phys. Rev. Lett. **79**, 2328 (1997).
7. D. J. Flees, S. Han, and J. E. Lukens, Phys. Rev. Lett. **78**, 4817 (1997).
8. G. Schön and A. D. Zaikin, Europhys. Lett. **26**, 695 (1994); G. Schön, J. Siewert and, A. D. Zaikin, Physica B **203**, 340 (1994).
9. See for example, A. Maasen van den Brink, A. A. Odintsov, P. A. Bobbert, and G. Schön, Z. Phys. B **85**, 459 (1991).
10. A. Shnirman, G. Schön, and Z. Hermon, Phys. Rev. Lett. **79**, 2371 (1997).
11. D. V. Averin, Solid State Commun., **105**, 659 (1998).
12. Y. Makhlin, G. Schön, and A. Shnirman, cond-mat/9808067.

Quantum Coherence and Decoherence - ISQM - Tokyo '98
Y.A. Ono and K. Fujikawa (Editors)
© 1999 Elsevier Science B.V. All rights reserved.

Superconducting Order Parameters Detected by Single-Electron Transistors

Shingo Katsumoto*, Hideki Sato and Yasuhiro Iye*

Institute for Solid State Physics, University of Tokyo
7-22-1 Roppongi Tokyo 106, Japan

We have studied superconductivity in mesoscopic samples with ring-shaped and simply connected geometry. We have found GL theory is applicable to these systems. Qualitative explanation is given both from giant vortex picture and from confined Abrikosov lattice model (London limit).

1. INTRODUCTION

Superconductivity strongly depends on system geometry such as thin films or fine particles. Averaged properties of superconducting fine particles have long been studied by nuclear magnetic resonance[1]. Recently superconductivity of individual nano-fabricated samples was investigated by resistance measurement[2] and by magnetization measurement[3]. Here, we would like to present the results by tunneling measurement, by which we can get direct information of the amplitude of order parameter (or superconducting energy gap) in zero-resistance states keeping the isolation of samples.

We adopt superconducting single-electron transistor (SSET) to detect the variation in the superconductivity gap in the Coulomb island isolated from the environment by Coulomb blockade.

This study also reveals high duality between single-electron charging effect and fluxoid quantization in mesoscopic superconducting rings or that between single-electron states in quantum dots and fluxoid states in simply connected mesoscopic superconductors. These dualities are originated from that between electric field and magnetic field.

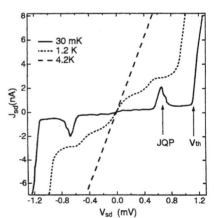

Fig.1 Typical I-V characteristics of an SSET.

2. EXPERIMENT

The principle of measurement of superconducting order parameter is very simple as follows. A single-electron transistor consists of a Coulomb island, two ultra-small tunnel (Josephson) junctions and a gate electrode. This structure

---

*Also at CREST, Japan Science and Technology Corporation.

can be regarded as a measurement system of the superconducting gap(Δ) in the Coulomb island. A typical current-voltage(I-V) characteristics is shown in Fig.1. Δ can, for example, be measured in the following way. We experimentally found that the resistance above the quasi-particle threshold (V_{th}) does not change significantly while the excess current (the current offset) is almost proportional to Δ when the external magnetic field is small and $\delta\Delta(H)/\Delta$ is also small. Variation in source-drain current (J_{sd}) for a fixed bias voltage is thus proportional to $\delta\Delta$. More direct method is to measure $V_{\text{th}} = (4\Delta + 3E_C)/e$ where $E_C \equiv e^2/2C_{\text{tot}}$ is the charging energy of single electron though the voltage-bias method has some advantage in the sensitivity.

Single-Electron transistors were fabricated by electron-beam lithography and oblique angle deposition of 15nm thick Al. In the present case, "samples" are the Coulomb islands and SSET's with the islands of various dimensions and one with a ring-shaped island were prepared. Source-drain biases are applied by a home-made voltage source. Temperature for the measurement was 1.3K, at which the I-V characteristics was rounded compared with that at 30mK (Fig.1). However the response of source-drain current to the magnetic field showed no significant difference for these two temperatures. The critical temperature of the superconductivity (T_C) was 1.9K. The coherence length (ξ) and the penetration depth (λ) at 1.3K are estimated to 50nm and 900nm respectively. Hence these Al thin films are type-II superconductors.

3. RESULTS AND DISCUSSION

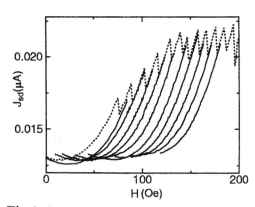

Fig.2 J_{sd} as a function of external magnetic field of an SSET with ring-shaped Coulomb island.

Figure 2 shows the results for a ring-shaped Coulomb island, whose dimensions are $1.3\mu\text{m}\times1.6\mu\text{m}$. Comparatively periodic jumps against the external magnetic field H appeared in J_{sd}. The averaged period is about one flux-quantum per the area of the ring, which manifests that fluxoids are quantized even in mesoscopic islands in which the number of Cooper pairs is fixed by Coulomb blockade[4]. Furthermore just like famous Little-Parks experiment, we can reveal each parabola corresponds to a state with fixed number of fluxoids as demonstrated in Fig.2. This result assures the spatial single-valuedness of order parameters in mesoscopic islands. At first sight this result contradicts with Heisenberg's uncertainty relation between the phase and the number of Cooper pairs[5,6] though the paradox can be easily removed by noticing that the fluctuating phase is that between the island and the electrodes[7].

The above result certifies applicability of Ginzburg-Landau (GL) equations to mesoscopic superconductors with size comparable to ξ. Thus the free energy of a superconducting mesoscopic ring as a function of external magnetic field can be represented as a set of parabolas while the charging energy of a Coulomb island is a set of parabolas against external electric field (gate voltage).

The duality originates from that between electric and magnetic fields. In the case of normal electrons, the magnetic effect is smaller by the fine structure constant $4\pi\hbar c\epsilon_0/e^2$ ("persistent current" in mesoscopic ring). In the superconducting case, it is strengthen by the number of Cooper pairs because of the coherence, and the spectacular magnetic effects appear.

We then move to the case of simply connected islands. This case resembles to that of quantum dots rather than the single-electron charging effect in metallic systems. When the temperature is close to T_C, the order parameter ψ obeys the linearized GL equation[8]

$$\frac{1}{4m}\left(i\hbar\nabla + 2e\boldsymbol{A}\right)^2\psi = \alpha\psi, \tag{1}$$

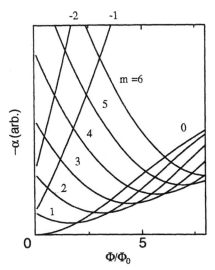

Fig.3 Calculated value of α in eq.(1) as a function of external magnetic flux.

where α is a kind of eigenvalue and a small variation in α is proportional to that in Δ with a negative coefficient. (1) can be solved for simple cylindrical geometry (infinite along z-axis) by assuming the rotational symmetry, i.e., existence of a giant vortex at the center of the cylinder. Calculated value of $-\alpha$ is plotted against external magnetic flux in Fig.3 for the number of flux quanta m. As m increases the bottom of parabola-like branch goes up corresponding to expansion of the giant vortex. Thus m corresponds to the quantum number of single-electron level in a quantum dot.

On the other hand, when $\xi \ll R$ (radius of the cylinder) $< \lambda$ (London limit), separate vortices with single flux quantum exist in the disk lowering the symmetry[9]. This case corresponds to a quantum dot with strongly repulsive electrons.

This picture has a big problem in the explanation of the present experiment that there is no modulation in the order parameter at the edge of the sample which was detected in the experiment. Thus we are forced to adopt a hybrid picture in which the configuration of Abrikosov lattice is not so far from that calculated in London limit but the order parameter at the edge reflects the total free energy.

Figure 4 shows the results for an SSET with dimensions $1.2\mu\text{m}\times0.7\mu\text{m}$. At first sight the behavior is qualitatively similar to that shown in Fig.2 besides the irregularity in the oscillation. We would like to stress that the oscillation pattern is highly reproducible. Again the hysteresis was utilized to reveal each branch, which is well approximated by a parabola.

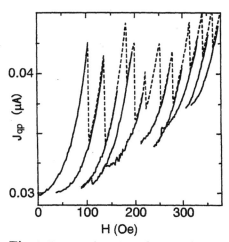

Fig.4 J_{sd} as a function of external magnetic field of an SSET with simply-connected Coulomb island.

228

In Fig.5 we compare the simply connected island and the ring-shaped island for the set of minima of parabolas. Distinct increase in J_{sd} at the minima is observed for the simply connected island, which suggests that the giant vortex picture better describes the present system.

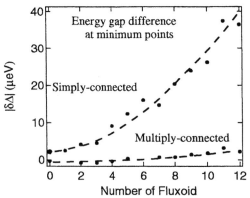

Fig.5 J_{sd} at minimum points of parabolas fitted to the data in Fig.4 as a function of external magnetic flux.

We would like to comment on the origin of the hysteresis. There should be "Bean potential"[10], which forms barriers for both incoming and outgoing of the vortices. However this alone cannot explain the fact that the oscillation pattern is sample-dependent and we should consider some pining potential probably due to disorder besides Bean potential.

In summary, we have studied superconductivity in mesoscopic samples with ring-shaped and simply connected geometry. We have found GL theory is applicable to these systems. Qualitative explanation is given both from giant vortex picture and from confined Abrikosov lattice model(London limit) though there still remains long way to quantitative understanding.

Acknowledgement

We thank M. Kimura for the collaboration in an early stage of this work and H. Akera for valuable discussions. This work was partly supported by a Grant-in-Aid for Scientific Research on the Priority Area of "Single Electron Devices and Their High Density Integration" from the Ministry of Education, Science, Sports and Culture of Japan.

REFERENCES

1. S. Kobayashi, Phase Transitions **24−26**, 463 (1990).
2. V. V. Moshchalkov, L. Gielen, C. Stunk, R. Jonckheere, X. Qiu, C. van Hasendonck and Y. Brunseraede, Nature **373**, 319 (1995).
3. A. K. Geim, I. V. Grigorieva, S. V. Dubonos, J. G. S. Lok, J. C. Maan, A. E. Filippov and A. E. Peeters, Nature **390**, 259 (1997).
4. H. Sato, M. Kimura and S. Katsumoto, Jpn. J. Appl. Phys. **36**, 3978 (1997).
5. M. Matters, W. J. Elion and J. E. Mooij, Phys. Rev. Lett. **75**, 721 (1995).
6. S. Katsumoto and M. Kimura, J. Phys. Soc. Jpn. **65**, 3704 (1996).
7. P. W. Anderson, in "Lectures on the Many Body Problem", ed. E. R. Caianello (Academic, 1964), Vol.2, p.127.
8. D. Saint-James and P. G. de Gennes, Phys. Lett. **7**, 306 (1963).
9. A. I. Buzdin, Phys. Lett. **A196**, 267 (1994).
10. C. P. Bean and J. D. Livingston, Phys. Rev. Lett. **12**, 14 (1964).

Quantum Coherence and Decoherence - ISQM - Tokyo '98
Y.A. Ono and K. Fujikawa (Editors)
© 1999 Elsevier Science B.V. All rights reserved.

Local detection of superconducting energy gap by small tunnel junctions

A. Kanda, M.C. Geisler, K. Ishibashi, Y. Aoyagi and T. Sugano

Frontier Research Program,
The Institute of Physical and Chemical Research (RIKEN)
2-1 Hirosawa, Wako, Saitama 351-0198, Japan

The energy gaps of small superconductors were measured by using small tunnel junctions. The energy gap sensitively reflects the supercurrent underneath the tunnel junction. Transition between flux states was observed and the fluxoid number was determined. The sign of the shielded flux shows that the magnetic response of disks is qualitatively different from that of rings. This can be explained in terms of the Bean-Livingston barrier.

1. INTRODUCTION

Direct measurement of superconducting energy gap through tunnel effect is the most powerful experimental method for the study of superconducting states. In this study, we used highly resistive small tunnel junctions made with e-beam lithography and shadow-evaporation technique [1] to detect the change of superconducting energy gap in small superconductors. Using small tunnel junctions has several advantages: (1) When the junction size is of the order of or smaller than the superconducting coherence length ξ, the *local* energy gap near the tunnel junction can be measured. (2) When the junction resistance is much larger than the quantum resistance, $R_Q = h/(2e)^2 \approx 6.5\text{k}\Omega$, coherent effects in tunnel junctions such as the proximity effect and the Josephson tunneling can be neglected. Furthermore, the superconductor can be isolated from the environment when all leads are attached to it through such tunnel junctions.

In the current-voltage characteristics of superconductor (S)-normal metal (N) or S-S tunnel junctions, there appears a gap voltage below which quasiparticle current is suppressed by the energy gap Δ. Since the voltage for a small current is close to the gap voltage, it is sensitive to the change in Δ. We studied the change in Δ at low temperatures ($\approx 50\text{mK}$) by measuring the voltage at a fixed small current (10-100 pA).

2. FLUXOID QUANTIZATION IN SMALL SUPERCONDUCTING RINGS

It is well-known that in superconductor, Bohr-Sommerfeld quantum condition leads to the fluxoid quantization:

$$\iint_S \mathbf{B}' \cdot d\mathbf{S} + \oint_C \Lambda \mathbf{J}_S \cdot d\mathbf{l} = n\Phi_0 \quad (n : \text{integer}). \tag{1}$$

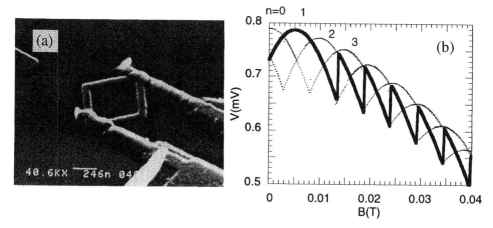

Figure 1. (a) A SEM picture of a small superconducting ring in a S-S-S double tunnel junction structure. The tunneling resistance is 100kΩ. (b) Magnetic-field dependence of the voltage at $I = 100$ pA. Bold line is for increasing B and the dots are the data when the sweep direction is changed at various fields.

Here, B' is the magnetic field, Λ the London coefficient, $\mathbf{J}_S$ the supercurrent density, $\Phi_0 = h/2e$ the flux quantum, and S is the surface defined by the contour C along the ring. The response of superconducting rings to magnetic fields depends on the linewidth of the ring d, the penetration depth λ and the coherence length ξ. In a thin ring where $d \ll \lambda$ and $d < \xi$, the supercurrent is almost uniform and the induced magnetic field is negligible. In this case, the direction and strength of the supercurrent directly correspond to the difference $n\Phi_0 - \iint B dS$. Figure 1(a) is a SEM picture of such a small ring. An aluminum ring with an area $A = 0.4(\mu m)^2$ and a linewidth of 90nm is connected to two aluminum leads through small tunnel junctions. Here, ξ and λ are $0.15 \mu m$ and $0.63 \mu m$, respectively. Figure 1(b) shows B-dependence of the voltage at I=100pA in this sample. When B increases (bold line), the voltage repeats a cycle of a moderate change (in most cases a decrease) followed by a jump.[2] The voltage changes mainly due to the decrease of Δ by the supercurrent and its jump comes from the variation of the fluxoid number n. When the sweep direction is changed at various fields, the hidden parts shown by dots appear, forming many parabolas. Interval of the tops of adjacent parabolas is close to Φ_0/A, suggesting that the fluxoid number n changes by one. Thus, we can assign n for each curve as shown in the figure. This method enables us to determine the fluxoid number n by the voltage measurement.

3. VORTEX PENETRATION INTO SUPERCONDUCTING DISKS

In superconductors whose size is larger than ξ, the energy gap can vary spatially. By using more than one tunnel junction, one can detect the spatial change in Δ. In Fig. 2, a diamond-shaped thin superconducting (aluminum) disk with two superconducting wires is connected to three normal-metal (copper) leads through small tunnel junctions.

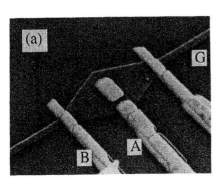
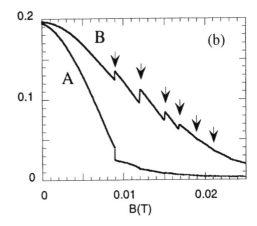

Figure 2. (a) A SEM picture of a superconducting disk. Three normal-metal leads are attached to it through tunnel junctions with $R_t \approx 100$kΩ. (b) Voltage variation at $I = 20$ pA while the magnetic field increases. Current flows (A) from lead A to G and (B) from B to G.

By supplying small constant current both from A to the ground G and from B to G, we detect the change of Δ at two points simultaneously. Figure 2(b) shows the result. With increasing B, both voltages show moderate decrease and several jumps (indicated by arrows), corresponding to the increase of supercurrent near the tunnel junction and single vortex penetration into the disk, respectively. A clear difference between A and B contains information about the vortex arrangement and the supercurrent flow. Although its full interpretation needs numerical simulation, one can understand some features qualitatively. For example, (1) at low magnetic fields, decrease of the voltage in A is faster than that in B. It comes from the antisymmetric shape of the disk. (2) At the first jump, the voltage in A decreases while that in B increases. It is because the vortex is close to the junction of lead A.

4. SUPERCURRENT IN RINGS AND DISKS

To get information on the magnetic response of superconductors, it is useful to define a shielded flux

$$\Phi_S \equiv B \cdot A - n\Phi_0 \approx \oint_{C'} \Lambda J_S \cdot dl, \qquad (2)$$

where B is the applied magnetic field, A the area enclosed by the perimeter C' of the superconductor, and n the fluxoid (vortex) number. When Φ_S is positive (negative), outermost supercurrent flow is in the direction of cancelling (supporting) B. Figure 3 shows Φ_S of a ring (a) and a disk (b). In a ring, $\Phi_S > 0$ for increasing B and $\Phi_S < 0$ for decreasing B, while Φ_S is always positive in a disk. This difference is explained as follows. To penetrate into superconductor, vortex should overcome the Bean-Livingston barrier[3] with the help of the supercurrent. The directions of the current seen by a vortex when

232

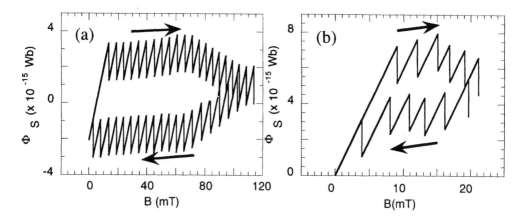

Figure 3. The shielded flux Φ_S of a superconducting ring (a) and a disk (b). Arrows indicate the direction of field sweep.

it enters and exits a superconducting rings are opposite. On the other hand, a vortex can go out of disks in the current direction same as for vortex entry because there is no potential barrier against exit when the vortex pinning is weak enough.

5. CONCLUSION

Magnetic response of small superconductors is studied by the energy gap measurement with small tunnel junctions. Because the sample size is smaller than the penetration depth, the overall response to magnetic field can be detected. Both in small rings and disks, the fluxoid (vortex) number can be determined. Multiprobe measurement in a disk shows spatial variation in the energy gap. The behavior of the shielded flux Φ_S of rings is qualitatively different from that of disks. This is explained in terms of the Bean-Livingston barrier.

This work has been supported in part by CREST project of Japan Science and Technology Corporation (JST) and by a Grant-in-Aid for Encouragement of Young Scientists from the Ministry of Education, Science, Sports and Culture, Japan.

REFERENCES

1. G.J. Dolan, Appl. Phys. Lett. 31 (1977) 337.
2. A. Kanda, K. Ishibashi, Y. Aoyagi and T. Sugano, Physica B227 (1996) 235; Czech. J. Phys. 46 (1996) S4 2297.
3. C.P. Bean and J.D. Livingston, Phys. Rev. Lett. 12 (1964) 14.

Quantum Coherence and Decoherence - ISQM - Tokyo '98
Y.A. Ono and K. Fujikawa (Editors)
© 1999 Elsevier Science B.V. All rights reserved.

Vortex matter in high-T_c superconductor with columnar defects: a new field-driven coupling transition

T. Onogi[a]*, R. Sugano[a], K. Hirata[b], and M. Tachiki[b]

[a]Advanced Research Laboratory, Hitachi, Ltd., Hatoyama, Saitama 350-0395, Japan

[b]National Research Institute for Metals, 1-2-1 Sengen, Tsukuba, Ibaraki 305-0047, Japan

By using Monte Carlo simulation technique, we numerically study the vortex-matter phase diagram of high-T_c cuprate superconductor $Bi_2Sr_2CaCu_2O_8$ in the presence of dense and random columnar defects. We find that a field-driven discontinuous transition occurs between decoupled and well-coupled pancake vortices, at nearly one third the matching field B_Φ. This fractional matching transition is relevant to recent unusual experimental results obtained from Josephson plasma resonance and DC magnetization studies.

1. INTRODUCTION

The cuprate superconductor $Bi_2Sr_2CaCu_2O_8$ (BSCCO) has a quasi-two-dimensional character based on the layered structure of CuO_2 superconducting planes. Accordingly, the quantized flux lines (vortex lines) penetrating under an applied magnetic field are fragile against thermal fluctuations [1], in contrast to the case of conventional metal superconductors. The Abrikosov vortex-line lattice thermally melts into a gaslike phase of 2D "pancake" vortices (or decoupled vortex liquid), rather than into a liquid of rigid vortex lines, because the vortex line tension along the c-axis is extremely weak. It is expected that a straight introduction of columnar defects (CDs) along the c-axis tend to align the pancake vortices straightly to increase the pinning force, resulting in large critical current J_c. Nelson and Vinokur [2] proposed theoretically that a localized vortex phase called Bose glass appears at low temperature, and predicted that Mott insulator with a sharp peak effect in J_c may appear at an *integer* filling of the matching field B_Φ where the vortex density exactly coincides with the defect density. In this article, we examine a disordered vortex matter in spatially random and densely distributed CDs introduced by heavy-ion irradiation to pristine BSCCO crystals, and report a new field-driven vortex phase transition at a *fractionally* 1/3-filled magnetic field of B_Φ [3].

2. MODEL AND SIMULATION METHOD

The vortex system in BSCCO under a magnetic field (B) is described by the Lawrence-Doniach model of Josephson-coupled superconducting layers, using the vortex-variable

---

*The present work was supported by Joint Research Promotion System on Computational Science and Technology (Science and Technology Agency, Japan).

234

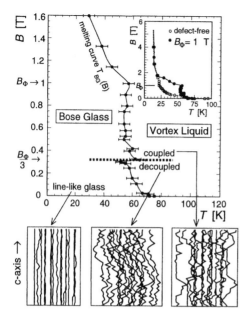

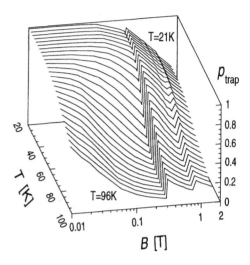

Figure 1. B-T phase diagram of $Bi_2Sr_2CaCu_2O_8$ with columnar defects.

Figure 2. Trapping rate p_{trap} of pancake vortices within columnar defects.

representation. The pancake vortices within an individual layer interact via the repulsive vortex-vortex interaction $\epsilon_0 K_0(r/\lambda_{ab})$ with the modified Bessel function $K_0(x)$ and magnetic penetration depth λ_{ab}. The vortices are also coupled along the c-axis, via the Josephson string between adjacent layers. The finite temperature Monte Carlo simulation based on the Metropolis algorithm has been performed for the system of 16 vortex lines on a $256 \times 222 \times 40$ cubic grid. Here we make both local and global movements of individual vortex lines and also exchange the connection of two given lines at each layer. The CD ($\parallel$ c-axis) is modeled [2] by a cylindrical potential well with radius c_0 equal to the coherence length ξ_{ab} and depth $U_0 = (\epsilon_0/2)\ln[1 + (c_0/\sqrt{2}\xi_{ab})^2]$. For BSCCO, we estimated the energy scales as $\epsilon_0/k_B = 1000$ K and $U_0/k_B = 343$ K, with $\lambda_{ab} = 2000$ Åand $\xi_{ab}=10$ Å.

3. COMPUTATIONAL RESULTS AND DISCUSSIONS

Figure 1 shows a simulated B-T phase diagram, where the density of CDs was fixed at $B_\Phi = 1$ Tesla (T) and ten random sets of CD configuration were used for the sample average. The Bose-glass (BG) melting curve $T_{BG}(B)$ (solid line) was computed from the onset of thermal diffusion of pancake vortices. As shown in Fig. 1, for $B < B_\Phi$, the BG melting curve is upper than the one of the defect-free case, implying that the pinning by CDs is very effective below the matching field B_Φ. The horizontal boundary (bold dotted line) near $B = B_\Phi/3$ denotes a new field-driven transition line, where we find the reentrant behavior of the BG melting curve. As shown in Fig. 2, the average trapping rate

p_{trap} of vortices to CD shows a large distinctive jump at $B = B_\Phi/3$. The jump width of p_{trap} takes a maximum at $T_{\text{BG}}(B \simeq B_\Phi/3)$, and tends to vanish for lower temperature, as is different from the behavior expected from the Mott insulator analogy. Viewed from the c-axis correlation of vortex state, another feature of the present transition is as follows; When the magnetic field is increased towards B_Φ at a fixed temperature, the vortex liquid phase ($T > T_{\text{BG}}$) exhibits a coupling transition of decoupled vortex liquid (or pancake gas) at $B \simeq B_\Phi/3$, whereas the BG phase ($T < T_{\text{BG}}$) exhibits a decoupling transition of line-like glass at $B \simeq B_\Phi/3$. On the other hand, Fig. 3 shows a snapshot of inplane vortex configuration around the field-driven transition. Based on the Delaunay triangulation analysis, we can see development of short-range hexagonal order (coordination number $z = 6$), while the long-range positional order remain completely absent.

These numerical findings are relevant to recent experimental data on heavy-ion irradiated BSCCO (B_Φ =0.3, 1 T). First, Josephson plasma resonance (JPR) measurements [4,5], which can directly probe the interlayer coherence along the c-axis, revealed that a double peak structure appears above the BG melting curve, and they suggested that the decoupled vortex liquid rapidly changes into the well-coupled liquid well below B_Φ. The JPR frequency ω_p is related to the interlayer coherence through $\omega_p^2(B, T) \propto \langle \cos \phi_{z,z+1} \rangle$, where $\phi_{z,z+1}$ represents the gauge-invariant phase difference between two adjacent superconducting layers. We calculated the field dependence of interlayer Josephson fluctuation energy $E_J \propto (1 - \langle \cos \phi_{z,z+1} \rangle)$ by monitoring the c-axis wandering length of vortex lines, and the result is shown in Fig. 4. If a discontinuous upward jump exists in $\langle \cos \phi_{z,z+1} \rangle$, the resonance at $\omega_p(B) = \omega_f$ should occur twice in the field-sweep measurement with the

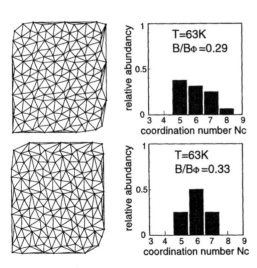

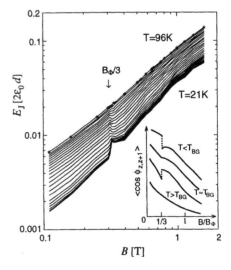

Figure 3. Delaunay triangulation analysis of inplane vortex configuration.

Figure 4. Interlayer Josephson fluctuation energy E_J vs. B ($\Delta T = 3$K).

fixed microwave frequency ω_f, thus leading to a double-peak structure. Our simulation gives a direct numerical evidence for the existence of such a recoupling transition of vortex liquid. In addition, the double peak has been experimentally observed at a narrow field regime around $B_\Phi/3$. From the significant growth in the trapping rate (Fig. 2), we can expect enhancement of vortex pinning force and J_c for the BG phase. Very recently, irreversible magnetization measurements [6] found the existence of a new peak ("third peak") of $J_c(B)$ just near $B_\Phi/3$ within the BG phase, in contrast to the second peak of the pristine BSCCO. Moreover, they also observed a reentrant behavior of the melting line at $B \simeq B_\Phi/3$, in excellent agreement with our numerical prediction. The non-monotonic field dependence of J_c can not be explained from the conventional collective-pinning theory which predicts that $J_c(B) \propto 1/B$ for $B < B_\Phi$. The above-mentioned sudden change in the c-axis vortex correlation should give a crucial clue to understand the peak effect.

In order to examine the mechanism of the field-driven coupling transition, we carried out additional simulation of the single layer model. As a result, we found that the transition is peculiar to the 2D CuO_2 superconducting layer, implying a subsidiary role of the interlayer Josephson coupling. Important questions, why it occurs at the fractional filling $(B/B_\Phi \simeq 1/3)$ and why it is prominent at rather high temperatures, remain challenging issue for strongly interacting vortices in disordered media. When the inplane repulsion between vortices is strong and long-ranged at high fields where the average vortex spacing is smaller than λ_{ab}, one-to-one occupation of all available CDs (just at $B/B_\Phi = 1$) is energetically unfavorable, because some neighboring CDs happens to be close in random defect configuration. Thus the effective trapping could occur below the matching field, in contrast to the Mott insulator valid for weakly interacting vortices. In addition, thermal fluctuations assist each pancake vortex to search for unoccupied CDs to gain the pinning energy, while local hexagonal order of vortex positions is enhanced to lower the energy cost of inplane repulsion, as is shown in Fig. 3. Therefore we consider that the interplay of 2D intervortex interaction and entropy plays crucial role in the transition.

To conclude, we proposed an existence of fractional matching transition accompanied by drastic change in the interlayer vortex coupling in the vortex state of $Bi_2Sr_2CaCu_2O_8$ containing columnar defects.

REFERENCES

1. G. W. Crabtree and D. R. Nelson, Physics Today 50, No. 4 (1997) 38.
2. D. R. Nelson and V. M. Vinokur, Phys. Rev. Lett. 68 (1992) 2398.
3. R. Sugano, T. Onogi, K. Hirata, and M. Tachiki, Phys. Rev. Lett. 80 (1998) 2925; Phys. Rev. B 60 (1999) 9734.
4. M. Kosugi, Y. Matsuda, M. B. Gaifullin, L. N. Bulaevskii, N. Chikumoto, M. Konczykowski, J. Shimoyama, K. Kishio, K. Hirata, and K. Kumagai, Phys. Rev. Lett. 79 (1997) 3763.
5. M. Sato, T. Shibauchi, S. Ooi, T. Tamegai, and M. Konczykowski, Phys. Rev. Lett. 79 (1997) 3759.
6. N. Chikumoto, M. Kosugi, Y. Matsuda, M. Konczykowski, and K. Kishio, Phys. Rev. B 57 (1998) 14507; K. Hirata, T. Mochiku, and N. Nishida, Advances in Superconductivity X (Springer-Verlag, Berlin, 1998) p. 553.

Quantum Coherence and Decoherence - ISQM - Tokyo '98
Y.A. Ono and K. Fujikawa (Editors)
© 1999 Elsevier Science B.V. All rights reserved.

Charging Effects on Superconducting Proximity Correction in Normal-Metal Wire Conductance.

Hayato Nakano and Hideaki Takayanagi

NTT Basic Research Laboratories,
3-1, Morinosato-Wakamiya, Atsugi-shi, Kanagawa 243-0198, Japan

The influence of the charging effect on the proximity correction in the conductance of a mesoscopic superconductor(S)/normal metal(N) coupled system is theoretically investigated. The proximity correction in the local conductivity is caused by Andreev reflection at the S/N interface. If the S/N interface is very small and low transparent, the Andreev reflection is strongly suppressed by the Coulomb blockade at a low temperature ($k_B T < E_C$, $E_C = e^2/2C$) in an exponential manner $\exp[-4E_C/(k_B T)]$, where C is the capacitance of the S/N junction. Nevertheless, the proximity correction in the conductance is only suppressed with power law of $k_B T/(4E_C)$ because the charged state is an intermediate state in the process of the proximity correction in the conductivity.

1. INTRODUCTION

In a mesoscopic superconductor(S)/normal metal(N) coupled system, the normal conductance is sensitive to the macroscopic phase difference between superconducting electrodes. The phenomenon arises from the interference of phase coherent quasiparticles. [1] In this decade, it has been well established that the most important contribution for the the phase-sensitive conductance component in a diffusive normal metal comes from the superconducting proximity correction in the local conductivity $\delta\sigma(r)$. [2]

On the other hand, the charging effect in a small tunnel junction is another very interesting topic in mesoscopic physics.[3] If two metal electrodes make contact via a very small-area interface with a very small capacitance, the charging effect evokes the so-called Coulomb Blockade; that is, an exponential suppression of electron transfers through the interface. A perspective description of these charging effects is given by the introducing the quantity "phase" (φ), which is canonical to the number of the charges (n)that cause a charging energy. They satisfy the relation $[\varphi, n] = i/2$,

In this paper, we discuss the the influences of the charging effect on the proximity correction in the conductance of a mesoscopic S/N coupled system.

2. LINEAR RESPONSE EXPRESSION of the PROXIMITY CORRECTION in CONDUCTANCE

We consider the proximity correction in the conductance of a normal metal wire like that in Fig. 2(a). In the quasiclassical approximation, for a finite volume and nearly spatially uniform system, the conductance G is given by $G = \frac{1}{L^2} \int_{\text{vol.}} \sigma(r) dr$. Here, $\sigma(r)$ is the local component of the conductivity $\sigma(r, r')$. In a diffusive normal metal/superconductor coupled system, the proximity correction of this local conductivity is the most important factor in the correction of the conductance.

From linear response theory,[5] the lowest order contributions in the local conductivity can be illustrated as in Fig. 1. The first term (a) is the conductivity without proximity

238

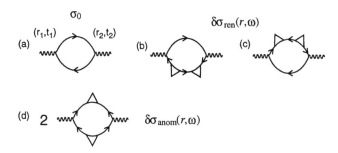

Figure 1:
Perturbation expansion by Andreev reflections of the proximity correction in the local conductivity $\sigma(r)$. $\triangle$ means Andreev reflection. (a) without perturbation, (b)and(c) conductivity suppression, and (d) conductivity enhancement. (b)+(c)+(d)= 0 at the absolute zero temperature and zero bias voltage.

corrections. The second (b) and the third (c) terms are corrections that diminish the conductivity. They exactly correspond to the decrease in conductivity by the exclusion of the quasiparticle density of states in the normal region by proximity effects. The last term gives conductivity enhancement by the proximity effect. It should be noted that the second, the third, and the last terms exactly cancel out each other at the absolute zero temperature and at the zero bias voltage limits. In early investigations of proximity correction in conductance by using Kubo approach, [5] S/N systems are considered at such limits. Therefore, the importance of these terms was not emphasized. At a finite temperature or at a finite bias voltage, however, the last term exceeds the the second and the third and gives a large conductance enhancement. Hereafter, we concentrate our attention on this term.

3. CHARGING EFFECTS on PROXIMITY CORRECTIONS in CONDUCTIVITY

The lowest order contribution to the local conductivity is given from linear response theory as

$$\delta\sigma(\omega, r) = \frac{(v_F \tau_{el})^3}{V_N} \frac{K_{anom}(\omega, r) - K_{anom}(0, r)}{\omega}, \tag{1}$$

where τ_{el} is the elastic scattering time, and V_N is the volume of the normal-metal wire. Following the imaginary time treatment by Bruder et al., [6] we can get the proximity correction in the kernel with the temperature Green's functions [Fig.2(b)],

$$K_{anom}(i\omega_\nu, r) = \frac{V_N e^2 v_F^2}{h} \sum_{\omega_l, \varepsilon_n \geq 0}{}' (k_B T)^2 \mathcal{F}_N(r, \omega_l + \frac{1}{2}\omega_\nu + \frac{1}{2}\varepsilon_n) \mathcal{F}_N(r, \omega_l - \frac{1}{2}\omega_\nu + \frac{1}{2}\varepsilon_n) h(i\varepsilon_n), \tag{2}$$

where $\sum'$ means to take sum with ω_l's and ε_n's that satisfy the conditions $|\omega_l - \omega_\nu/2| > \varepsilon_n/2$ and $|\omega_l + \omega_\nu/2| > \varepsilon_n/2$. D is the diffusion constant in the normal region, and

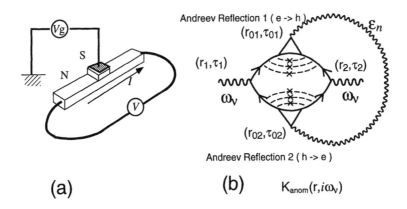

(a)   (b) $K_{anom}(r, i\omega_\nu)$

Figure 2:

(a) The model for the analysis.

(b) The kernel for the conductivity enhancement (including the charging effect).

$$\mathcal{F}_N(r, \omega_m) = \lim_{r_1, r_2 \to r} \mathcal{F}_N(r_2, r_1, \omega_m) = \frac{G_T}{A_I V_N} \int_{A_I} dr_0 \sum_q \frac{e^{iq \cdot (r_0 - r)}}{Dq^2 + |2\omega_m|}. \tag{3}$$

Because of the retro-property of Andreev reflection and the long-rangeness of a Cooperon (p-p ladder), Andreev reflection at the S/N interface affects the local conductivity at a point in the normal region far from the interface (within the phase coherence length L_ϕ) Here, $h(i\varepsilon_n)$ is the Fourier transform of the phase-phase correlation function such that

$$h(i\omega_n) = \int_0^\beta d\tau \langle e^{i\varphi(\tau)} e^{-i\varphi(0)} \rangle_{av} e^{i\omega_n \tau}, \tag{4}$$

where $\langle \cdots \rangle_{av}$ means to take the thermal average. $\beta = 1/(k_B T)$, and G_T is the conductance (unit of e^2/h) of the interface when the both electrodes are normal states. A_I is the area of the interface.

4. RESULT and DISCUSSION

In a strict sense, the phase-phase correlation function $h(i\varepsilon_n)$ in Eq. (2) should be estimated by the effective action for the phase variable. [3, 6] In the high-impedance limit of the environment, we can easily estimate the average with the Hamiltonian of the charging energy such that $h(\tau) = \langle \exp\{-4E_C[1 + (q - q_g)]\tau\} \rangle_q$. Here $\langle \cdots \rangle_q$ means the average over the discrete charge states, $q = 0, \pm 1, \pm 2 \cdots$, and q_g is the average charge number induced by the Gate voltage $q_g = C_g V_g$. For simplicity, moreover, we consider the case where the gate voltage is not applied ($q_g = 0$), and E_C is very large so that we have to take into account only the $q = 0$ state and the $q = \pm 1$ states. In this case, the phase correlation function is given by

$$h(i\varepsilon_n) = \frac{4E_C}{(4E_C)^2 + \varepsilon_n^2}. \tag{5}$$

This gives an overestimation of the charging effect. In other words, the true influence of the charging effect should be smaller than the one by Eq. (5).

Now we can easily find in Eq. (2) that the terms with $\varepsilon_n \sim 0$ contribute even in the presence of the charging effect. Physically, the meaning of this is as follows. An Andreev reflection needs a two-particle tunneling through the S/N interface and the tunneling evokes a charged state in a Coulomb blockade situation. However, in the process in Fig. 2(b), the charged state can be included as a quantum intermediate state because each two-particle tunneling is in the opposite direction.

We can roughly estimate the charging effects on the DC conductivity ($\omega \to 0$) near the Thouless temperature ($k_B T \sim D/L^2$) and $k_B T < E_C$ like

$$\frac{\delta\sigma(0,r)}{\delta\sigma(0,r)|_{E_C=0}} \sim \sum_n \frac{4E_C\beta(1 - e^{-4\beta E_C})}{(2n\pi)^2 + (4E_C\beta)^2} \exp[-2\sqrt{2n\pi}\frac{r}{L_T}], \tag{6}$$

where $L_T = \sqrt{D/k_B T}$ is the thermal diffusion length.

Equation (6) shows that, in contrast to the Coulomb Blockade, which gives an exponential suppression $\exp[-4E_C/(k_B T)]$, the charging causes a decrease in the proximity correction in the normal metal conductance with a power law of $k_B T/(4E_C)$. The charging effect does not strongly suppress the proximity correction in the local conductivity and the resulting correction in the conductance.

REFERENCES

[1] B. Z. Spivak and D. E. Khmelnitskĭi, Pis'ma Zh. Eksp. Theo. Fiz. 35, 334 (1982) [JETP Lett. 35,412 (1982)], H. Nakano and H. Takayanagi, Solid State Commun., 80, 997 (1991).

[2] A. I. Larkin and Yu. N. Ovchinnikov, in Nonequilibruim Superconductivity, ed. by D. N. Langenberg and A. I. Larkin, (North-Holland, 1986) 493, J. Rammer and H. Smith, Rev. Mod. Phys. 58, 323 (1986). A. V. Zaitsev, Phys. Lett., A 194, 315 (1994), Yu. V. Nazarov and T. H. Stoof, Phys. Rev. Lett. 76, 823 (1996), A. Volkov and H. Takayanagi, Phys. Rev. B 56, 11184 (1997), A. Volkov, N. Allsopp, and C. J. Lambert, J. Phys., Condens. Matter 8, L45 (1996).

[3] Single Charge Tunneling, NATO ASI Series, Les Houches, eds. H. Grabert and M. Devoret (Plenum, New York, 1992), G. Schön and A. D. Zaikin, Phys. Rep. 198, 237 (1990), F. Guinea and G. Schön, Physica B 152, 165 (1988), H. Higurashi, S. Iwabuchi, and Y. Nagaoka, Phys. Rev. B 51, 2387 (1995).

[4] R. Kubo, J. Phys. Soc. Jpn, 12, 570 (1957), C. L. Kane, R. A. Serota, and P. A. Lee, Phys. Rev. B 37, 6701 (1988).

[5] Y. Takane and H. Ebisawa, J. Phys. Soc. Jpn., 61, 3466 (1992), F. W. J. Hekking and Yu. V. Nazarov, Phys. Rev. Lett., 71, 1625 (1993), Y. V. Nazarov, Phys. Rev. Lett., 73, 1420 (1994), A. F. Volkov, A. V. Zaitsev, and T. M. Klapwijk, Physica C 210, 21 (1992).

[6] C. Bruder, R. Fazio, and G. Schön, Phys. Rev. B 50, 12766 (1994), C. Bruder, R. Fazio, and G. Schön , Physica B 203, 240 (1994), C. Bruder, R. Fazio, A. van Otterlo, and G. Schön, Physica B 203, 247 (1994).

Quantum Coherence and Decoherence - ISQM - Tokyo '98
Y.A. Ono and K. Fujikawa (Editors)
© 1999 Elsevier Science B.V. All rights reserved.

Andreev Reflection at High Magnetic Fields

T.D. Moore[a] and D.A. Williams[b]

[a]Microelectronics Research Centre, University of Cambridge, Cavendish Laboratory, Madingley Road, Cambridge CB3 0HE U.K.

[b]Hitachi Cambridge Laboratory, Cavendish Laboratory, Madingley Road, Cambridge CB3 0HE U.K.

Measurements of superconductor – semiconductor hybrid devices have shown a change in the character of electron transport as an applied magnetic field is varied in strength. At zero field, the transport across the junction is single-particle in nature, whereas at higher fields correlated transport (Andreev reflection) is observed, contrary to expectation.

1. INTRODUCTION

Charge transport across hybrid structures composed of mixtures of superconducting regions and other materials (such as metals, insulators and semiconductors) has been of considerable recent interest. One critical mechanism in the transport of charge across the interface between a superconductor and a normal metal or semiconductor is the process of Andreev reflection at the interface.[1] In this, two electrons from the conduction band of the non-superconducting region pair and enter the superconductor, or a Cooper pair breaks to enter the normal region. Clearly, this requires the matching of energy, wavevector, spin and phase of the two electrons in the normal region to those of the pair. This was previously thought to imply that Andreev reflection was a relatively unlikely process, and one which would be very sensitive to environmental factors such as the presence of disorder and applied magnetic fields. However, recent experimental and theoretical studies have shown that Andreev reflection can be a more robust effect, the probability of which can actually be increased by such environmental disturbances.[2,3]

One area that has received increasing theoretical attention recently is the effect of a magnetic field on Andreev reflection.[4-6] The low-field response is now well understood, with a wealth of experimental evidence, mostly based on the clean Nb:InAs structure. However, these contacts tend to have a low critical magnetic field, with the result that the regime above ~50mT has been little studied experimentally.

We present results from the first study of the high magnetic field behaviour of electron transport through superconducting contacts and a two-dimensional electron gas (2-DEG). Alloyed contacts with a high critical magnetic field were used with a GaAs:AlGaAs heterostructure 2-DEG channel. An unexpected increase in the relative probability of Andreev reflection up to a field of 1.5T was observed, with excess normalised conductance in the range 60mT-3T.

2. EXPERIMENTAL RESULTS

A connection between a high-mobility two-dimensional electron gas and high critical field superconductor has been achieved with a GaAs/AlGaAs heterostructure and a sintered alloy superconductor. A careful choice of starting material and rapid annealing conditions creates a contact containing multiply-connected regions of superconducting alloy (predominantly $AuSn_4$) in a normal metal alloy matrix, with a critical field typically above 4T.[7] The specimen structure is shown in Figure 1. The refractory silicon nitride and chromium layers are to prevent mass flow of the tin alloys during sintering, and the aluminium oxide prevents wetting of the channel region. After annealing, the contacts are alloyed downwards into the 2-DEG.

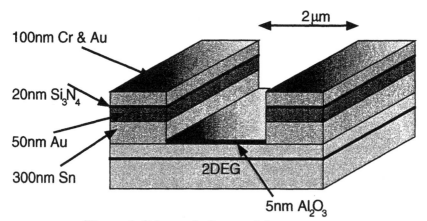

Figure 1: Schematic figure of the contact structure.

Figure 2 a) and b) show differential resistance traces of such a device at 50mK as a function of applied bias, for no applied field and an applied magnetic field perpendicular to the 2-DEG of 1T, respectively. When there is no field applied, transport at zero bias is by single particle tunnelling only, and a characteristic decrease in resistance is observed with increasing bias, as the quasiparticle distribution in the superconductor progressively overlaps the Fermi distribution in the normal material.

For a finite applied field, however, the characteristics are seen to be quite different, with an *increase* in resistance with increasing bias. This is characteristic of transport which is predominantly by Andreev reflection at zero bias. The magnitude of the dip in resistance around zero bias is seen to increase with field up to ~0.9T, then decrease, disappearing totally by~4T.

As the applied field increases, so the absolute resistance increases. This is because the measurements are 4-terminal from the superconductors, and so 2-terminal through the junction. Control devices with exactly the same geometry, but non-superconducting contacts show very similar general magnetoresistance, but the devices with superconducting contacts show relatively decreased resistance in the field region from 60mT to 4T, when the contacts are below their critical temperature. This is again indicative of transport by Andreev reflection.

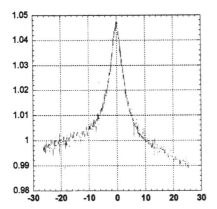 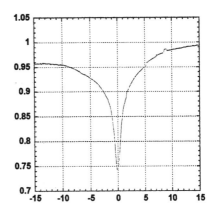

Figure 2: Normalised differential resistance traces at 0T and 1T applied perpendicular magnetic field, as a function of bias (mV).

3. DISCUSSION

The response of these devices was unexpected, but can be explained as follows: The device can be understood as containing five distinct regions; the two superconducting contacts with high H_{C2}, the high-mobility 2-DEG channel, and two disordered interface regions. The disordered regions tend to increase the probability of Andreev reflection at the interface, but also contribute to the resistance by weak localisation.

In the field regime from 0-60mT, this weak localisation is broken and the normalised resistance will decrease. As the field is increased, the overall resistance increases, as the number of transport channels decreases, and charge injection and removal takes place preferentially at the corners of the structure. As this happens, however, the momentum space for the available transport states is reducing, and changing in nature from two-dimensional to quasi-one-dimensional. This increases the probability of Andreev reflection, as the wavevector matching conditions become more likely to be satisfied, thus there is a greater likelihood of two incoming electrons matching the conditions of $E_1-E_{pair} = -E_2-E_{pair}$, $k_{1\uparrow}=-k_{2\downarrow}$, where the subscripts refer to the electrons 1 and 2 in the normal region and the pair in the superconducting region, and to the spin.

In conclusion, we have observed an increase in the probability of Andreev reflection in a superconductor – semiconductor hybrid junction as a perpendicular magnetic field is applied. This effect has been observed in many devices with several different contact alloy compositions, and is believed to be due to the progressive reduction in momentum-space freedom for interface transport as the field is increased. However, detailed analysis of the transport mechanisms is still needed. In particular, the temperature dependence of the transport at fixed applied field is currently being investigated.

The authors would like to thank H. Ahmed for support in this project.

REFERENCES

[1] For a review, see: C.J. Lambert and R. Raimondi cond-mat/9708056
[2] B.J. van Wees et al. Phys. Rev Lett. **69,** 510 (1992)
[3] I.K. Marmorkos et al. Phys. Rev. B **48,** 2811 (1993)
[4] P.W. Brouwer and C.W.J. Beenakker, Phys. Rev. B **52,** R3868 (1995)
[5] M.P. Fisher Phys.Rev. B **49,** 14550 (1994)
[6] Y. Takagaki Phys. Rev. B **57,** 4009 (1998)
[7] A.M. Marsh and D.A. Williams J. Vac. Sci. Technol. A **14,** 2577 (1996)

Quantum Coherence and Decoherence - ISQM - Tokyo '98
Y.A. Ono and K. Fujikawa (Editors)
© 1999 Elsevier Science B.V. All rights reserved.

Towards an experiment of MQC: determination of the decoherence time from energy level quantization in rf SQUIDs

C. Cosmelli[a,b], P. Carelli[a,c], M.G. Castellano[a,d], F. Chiarello[a,b], R. Leoni[a,d], G. Torrioli[a,d]

[a] Istituto Nazionale di Fisica Nucleare, Italy,
[b] Dip. di Fisica, Universita' di Roma, 00185 Roma, Italy,
[c] Dip. di Energetica, Università dell'Aquila, Monteluco di Roio, 67040 L'Aquila, Italy,
[d] Istituto di Elettronica dello Stato Solido, CNR, 00156 Roma, Italy

In view of realizing an experiment of Macroscopic Quantum Coherence with SQUIDs we have measured the effective resistance of a SQUID cooled at a temperature of 35mK. From a fit of the data with a simplified solution of the master equation describing the system dynamics we found an effective resistance R≈4 MΩ. This value, in the same system cooled at 4 mK, should lead to a decoherence time of approximately 1.4 μs allowing to perform MQC measurements with a tunneling frequency of the order of few MHz.

1. INTRODUCTION

The success in the description of microscopic world given by Quantum Mechanics is undoubtedly one of the most complete in the history of physics. Since the very beginning, however, there was a debate on the interpretation of Quantum Mechanics versus Macrorealism predictions given by Classical Mechanics or realistic theories. The milestone in this debate is the paper by Einstein, Podolski and Rosen [1] published in 1935. In 1980 Anthony Leggett [2,3] discussed the same problem for a macroscopic system, proposing an experiment to test Bell inequalities in a quantum system performing tunnelling oscillations between two macroscopically distinct states [a Macroscopic Quantum Coherence experiment]. The goal is to detect a non-classical feature or to falsify the predictions of Quantum Mechanics for a macroscopic system. The proposed system is an rf SQUID, i.e. a superconducting ring interrupted by a Josephson junction, biased by a half flux quantum. The corresponding variable associated to the SQUID dynamics is the magnetic flux ϕ linked to the SQUID, related to the collective motion of a macroscopic number of particles This system can be described by a double well potential with two degenerate energy levels in the left and right well. The flux then would perform tunnelling oscillation between the two equivalent flux eigenstates of the SQUID. All the proposed tests are then performed by measuring the probability to find the SQUID flux in one of the two wells at different times.

In the last decade there was a big efforts in realizing and studying system that in principle could be used to make the tests proposed by Leggett [4-8]. These experiments were successfully in observing the tunnelling to a continuum or the resonant tunnelling between distinct quantum levels. They were not able however to test the superposition of macroscopically distinct states, showing a non-classical feature of the macroscopic system observed. The problem arise because to realize the experiment is necessary both an ad hoc experimental set-up [3,9,10], very low dissipation and very low temperatures. Now, while the realization of the experimental set-up and the cooling to low temperatures can be done quite easier, a low dissipation macroscopic quantum system has never been realized. This goal however is essential to perform the experiment. A high degree of dissipation infact is a strong "link" with the external world, that act as an "observer" of the quantum system we want to study. Such "observer" will cause a very fast decoherence making impossible any test on its free evolution. The realization of a low dissipative system is therefore the main problem in any experiment involving tests on the quantum behavior of a macroscopic system such a Macroscopic Quantum Coherence [MQC] experiment or tests on quantum computing by means of Josephson devices [11].

2. DISSIPATION AND DECOHERENCE TIME FOR AN RF SQUID

The theoretical analysis of the dissipation for an rf SQUID in the quantum regime has been done by many authors [12-15]. In our case we can simplify the problem by analyzing the rf SQUID biased by a half flux quantum, i.e. in the condition were the MQC tests must be done. For this system we can write the conditional probability of finding the flux at a time t in one of the two wells, once the system has been prepared in the same well at a time t=0, as [13]: $P(t) \cong \cos(\omega t - \varphi) \exp(-t/\tau)/\cos(\varphi)$, where the tunnelling frequency ω, the phase φ, and the decoherence time τ are: $\varphi=\arctan(1/\omega\tau)$; $\omega=[(T^*/T)^{2-2\alpha}-1]^{1/2}$; and $\tau=h/2\pi^2\alpha kT$; having defined the dissipation factor $\alpha=\delta\phi^2/hR$, and the crossover temperature $T^*=h\omega_0/2\pi^2 k\alpha$, $\delta\phi$ is the difference between the flux in the two wells, ω_0 the tunnelling frequency at T=0, k the Boltzmann constant and h the Planck constant. All these relations are valid in the limit of low dissipation ($\alpha\ll1$), and low temperature ($T\ll T^*$). With the proper substitutions the decoherence time can be written as:

$$\tau = T^* / \omega T \cong 1.4 \ \mu s \cdot [R / 1M\Omega] \cdot [1mK / T] \tag{1}$$

Now, the operating temperature for such a system being typically the base temperature of a dilution refrigerator, i.e. few millikelvin, we can see that the decoherence time is determined essentially from the value of the overall dissipation R. In the following paragraph we will show how we made a measurement of the intrinsic dissipation of an rf SQUID realized for an MQC experiment in Rome [10].

3. THE RF SQUID AND THE EXPERIMENTAL SET UP

The rf-SQUID can be described by a single dynamical variable, the total magnetic flux Φ linked to the ring, which is subjected to a potential $U=(\Phi-\Phi_x)^2/2L-(i_c\Phi_0/2\pi)\cos(2\pi\Phi/\Phi_0)$, where Φ_x is the applied flux, L is the SQUID inductance, i_c is the junction critical current and $\Phi_0=h/2e=2.07 \ 10^{-15}$ Wb is the flux quantum [16]. The system equation is the same of a particle of mass C (the junction capacitance) with friction coefficient 1/R, subjected to the same potential. If the parameter $\beta_L=2\pi Li_c/\Phi_0$ is greater than one, this potential is a succession of wells, corresponding to metastable flux states of the SQUID, superposed to a quadratic term. If β_L is large, more than one metastable state is available to the system, once the energy is given [Fig.1]. In this paper we to study the process of escape from a metastable well. By sweeping the externally applied flux, the potential barrier on one side of the well is decreased, until, at a critical value of $\Phi_{xc,}$ the particle can overcome the barrier by thermal fluctuations or quantum tunneling, and it rolls down along the potential until it gets trapped in a nearby well.

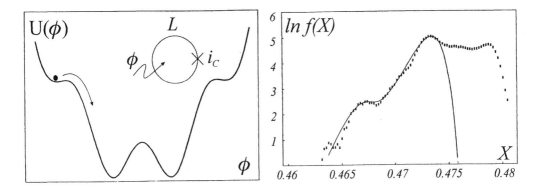

Figure 1. Scheme of the rf SQUID potential for an external bias $\phi_x = \phi_0/2$.

Figure 2. Experimental data (•) and best fit of the switching probability vs. the external flux at a temperature of 35 mK.

The experimental set-up is the following. The main chip is directly inserted into the mixing chamber of a dilution refrigerator to enhance the thermal contact between the cold mixture and the SQUID body. In the chip, realized at the Istituto di Elettronica dello Stato Solido in Roma, we have the rf SQUID, a gradiometric double loop device having an inductance L=235pH, capacity C=0.45pF and a critical current $i_c \sim 11\mu A$, giving a factor $\beta_L \sim 8$. The flux from the rf SQUID is read by a nearby dc SQUID having a geometry identical to the rf SQUID. The coupling between rf and dc SQUIDs is direct, without any flux transformer. The devices are all fabricated using Nb/AlOx/Nb trilayer technology and the quality of the Josephson junctions has been tested in a separate experiment [17]. The signal coming from the dc SQUID is sent through a couple of two 2Ω resistors to a second dc SQUID placed near the 1K pot cold point of the refrigerator at about 50 cm of distance from the main chip.

This second chip is then connected to the room temperature electronics by standard shielded twisted pairs of copper wires. To measure the statistical distribution of the switching values Φ_{xc} we used the technique first proposed by Fulton and Ducklemberger[18]: the ϕ_x sweep cycle is repeated $\sim 10^4$ times: when the rf-SQUID undergoes a transition, the sudden flux jump in the characteristic is read by the dc-SQUID and triggers the data acquisition system, so that the corresponding value of the sweeping current is recorded and converted in a flux value. With all the recorded values we can make the histogram representing the switching probability $f(\phi_x)$, and with some algebra we can derive the escape rate out of the well and any other quantities related to $f(\phi_x)$.

4. RESULTS AND CONCLUSIONS

On this system we made escape measurements at a temperature of 35 mK obtaining the data shown in Fig.2 where we show the switching probability $f(X=\phi_x/\phi_0)$ and the related best fit. We can see clearly peaks due to energy level quantization in the potential well of the SQUID. The data can be analyzed by solving the complete master equation describing the system dynamics [20]. This procedure however can be simplified if we analyze just two nearby peaks at a time; in this case we can find the equivalent temperature and the resistance associated to the overall dissipation of the rf SQUID. From the fit [Fig.2] we can see that the system dissipation can be described by an effective resistance $R\approx 4M\Omega$. From this value we can calculate the decoherence time that we would have in a MQC experiment performed at the planned temperature of 4 mK and in the (pessimistic) assumption that the resistance do not increase by lowering the temperature from 35mK to 4 mK. The decoherence time calculated from Eqs.(1) results $\tau\approx 1.4\mu s$. This time should be enough to perform an MQC experiment having a frequency of oscillation between the two wells of few MHz.

REFERENCES

1. Einstein, B. Podolski and N. Rosen, Phys. Rev. 47, 777 (1935).
2. Leggett, Suppl.Prog. Theor. Phys. 69, 80(1980).
3. Caldeira and A.J. Leggett, Phys. Rev. Lett., 46, 211 (1981)
4. Voss and R. A. Webb, Phys. Rev. Lett. 47, 265 (1981).
5. Martinis, M.H. Devoret and J. Clarke, Phys. Rev. Lett. 55, 1543 (1985).
6. D.B.Schwartz, B. Sen, C.N.Archie, and J.E. Lukens, Phys. Rev. Lett. 55 , 1547(1985).
7. Bol and R. De Bruyn Ouboter, Physica B160, 56 (1989)
8. R.Rouse, S. Han and J.E. Lukens, Phys. Rev. Lett. 75, 1624 (1995).
9. Tesche, Physica B, 165&166, 925 (1990).
10. C.Cosmelli, Proc. Int. Conf. on Macroscopic Quantum Coherence, Boston(1997), in press.
11. Bocko, A.M. Herr and M.J. Feldmann, IEEE Tr. On Appl. Sup. 7, 3638 (1997).
12. Caldeira and A.J. Leggett, Ann. Phys., 149, 374 (1983).
13. Weiss, H. Grabert and S. Linkwitz, Jap. Journ. of Appl. Phys. 26, 1391 (1987).
14. Grabert and H. Weiss, Phys. Rev. Lett. 53, 1787 (1984).
15. Garg, Phys. Rev. B, 32, 4746 (1985).
16. Barone and G. Paterno', Physics and Applications of the Josephson Effect, (J. Wiley & Sons Pub., New York, 1982).
17. M.G.Castellano et al., J. Appl. Phys. 80, 2922 (1996).
18. Fulton and L.N. Dunkleberger, Phys. Rev. B9, 4760 (1974).
19. Kramers, Physica 7, 284 (1940).
20. Larkin and Y.N. Ovchinnikov, Zh. Eksp. Teor. Fiz. 91, 318 (1986).

Quantum Coherence and Decoherence - ISQM - Tokyo '98
Y.A. Ono and K. Fujikawa (Editors)
© 1999 Elsevier Science B.V. All rights reserved.

Property of atomic structures fabricated on the hydrogen-terminated Si(100) surface

T. Hashizume[a], S. Heike[a], T. Hitosugi[b], S. Watanabe[c], Y. Wada[a], M. Ichimura[a], T. Onogi[a], T. Hasegawa[d] and K. Kitazawa[b]

[a]Advanced Research Laboratory, Hitachi, Ltd., Hatoyama, Saitama 350-0395 Japan

[b]Department of Superconductivity, The University of Tokyo, Tokyo 113-8656 Japan

[c]Department of Materials Science, The University of Tokyo, Tokyo 113-8656 Japan

[d]Materials and Structures Laboratory, Tokyo Institute of Technology, Nagatsuda, Yokohama 226-8503 Japan

Scanning tunneling microscopy (STM) has been used to fabricate and characterize atomic-scale structures on a hydrogen-terminated Si(100)–2×1–H surface. Hydrogen atoms are extracted by STM current and atomic-scale dangling-bond (DB) wires are fabricated. The DB wires composed of both unpaired and paired DBs show a finite density of states at the Fermi energy. Thermally deposited Ga atoms preferentially adsorb on DBs. Atomic-scale Ga wires on the silicon surface are fabricated by thermally depositing Ga atoms on the DB wires.

1. Introduction

The idea of switching devices on an atomic scale [1] has become more realistic by scanning tunneling microscopy/spectroscopy (STM/STS) [2] and atom manipulation. [3] A hydrogen-terminated Si(100)–2×1–H surface [4] is one of the promising substrates for atomic-scale device fabrication. Hydrogen atoms on the surface were extracted by the STM tunneling current and atomic-scale dangling-bond (DB) patterning was demonstrated. [5] Adsorption of Ga on the clean Si(100)–2×1 surface has been studied by STM. [6,7] We report on the fabrication and characterization of atomic-scale structures on hydrogen-terminated Si(100)–2×1–H surfaces using STM/STS. A row of fabricated DBs forms a one-dimensional atomic structure that we call a DB wire. We analyze our results based on the recent first-principles calculations. [8] Selective adsorption of thermally deposited Ga atoms is used to demonstrate a method of fabricating atomic-scale Ga wires on the silicon surface.

2. Experimental

All the experiments were carried out in ultra-high-vacuum STM systems. Rectangular Si(100) samples of 2×15 mm^2 were cut from As-doped n-type 7 to 18-mΩ·cm silicon wafers.

The STM images were taken at a sample bias voltage V_s ranging from -3.0 to $+3.0$ V and at a constant tunneling current I_t ranging from 10 to 100 pA. All the experiments were carried out at room temperature. The detailed experimental methods are described elsewhere. [9,10]

3. Results and Discussion

A typical STM image of the hydrogen-terminated Si(100)–2×1–H monohydride phase is shown in Fig. 1(a). We observe cocoon-shaped hydrogen-terminated Si dimers similar to those observed on a clean Si(100)–2×1 surface. Missing hydrogen defects (DBs) are observed as white protrusions, since the DB states near the Fermi energy are recovered. [4] Most of the protrusions we observe are unpaired DBs, which are imaged off-center on a dimer row (Fig. 1(b)). [9] A paired DB would be imaged at the center of the dimer row as is shown in Fig. 1(c). Their ball stick models are shown in Fig. 1(d). The dark spots in Fig. 1(a) are missing Si-dimer defects.

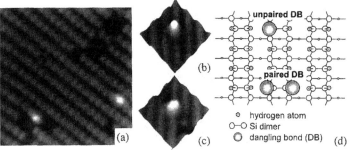

Figure 1. (a) A typical filled-state gray-scale STM image of the Si(100)–2×1–H surface (7 nm × 7 nm, $V_s = -2.0$ V, $I_t = 20$ pA). An unpaired DB (b) and a paired DB (c) can be distinguished by their symmetry against the Si dimer row. (d) Schematics of an unpaired and a paired DB.

A DB wire fabricated perpendicular to the dimer-row direction is shown in Fig. 2(a) and the observed tunneling spectra are shown in Fig. 2(b). The DB wire is made of both unpaired and paired DBs. The $I - V$ curve in Fig. 2(b) shows that the DB wire has finite density of states at the Fermi energy. Characteristic peaks at -1.6, -0.9, and $+0.8$ eV are seen in the normalized conductance. A DB wire made of both unpaired and paired DBs fabricated parallel to the dimer-row direction also showed a finite density of states at the Fermi energy, although a DB wire made of only paired DBs showed a band gap of 0.5 eV. [10]

The first-principles calculations [8] predicted a mid-gap band resulting from the DBs for the case of the DB wire made of unpaired DBs. It also showed the Peierls instability with an energy of 14 meV, which is small compared with the thermal energy at room temperature. [8] Thus, we expect a finite density of states at the Fermi energy for the DB wires made of unpaired DBs. Two characteristic peaks at -0.9 and $+0.8$ eV in Fig. 2(b) can be assigned to the bonding and antibonding states, resulting from the π interaction of two DBs on a Si dimer. We here propose that the electronic states of the DB wire shown

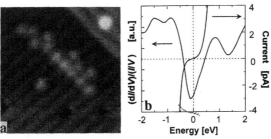

Figure 2. (a) Filled-state STM image of a DB wire perpendicular to the dimer-row direction (7.5 nm × 7.5 nm, $V_s = -2.0$ V, $I_t = 100$ pA) and (b) an $I - V$ and a normalized conductance curve recorded on the DB wire. The tunneling junction is stabilized at $V_s = -2.0$ V and $I_t = 100$ pA.

in Fig. 2 are a mixture of the electronic states of unpaired and paired DBs. [10] However, the calculations do not fully treat the finite-length, disorder or electron correlation effects.

Chemical reactions on the semiconductor surfaces are frequently dominated by the nature of the DBs. We expect that some of the metal atoms show selective adsorption on the DBs, which can be used to fabricate atomic-scale metal structures on the silicon surface. When the hydrogen-terminated surface was covered with a small amount of thermally-deposited Ga, the Ga atoms indeed showed preferential adsorption on the DBs. [9] The first-principles calculation showed that the barrier height of Ga atoms for surface migration is approximately 0.3 eV on the hydrogen-terminated area and 1.2 eV on the DB. [8] Several metal atoms such as Au and Ni showed similar adsorption features as Ga atoms, although Ti atoms reacted with hydrogen-terminated Si dimers and showed random adsorption.

We have fabricated atomic-scale DB wire by the tunneling current of the STM (Fig. 3(a)) and have thermally deposited Ga atoms in order to demonstrate a method of fabricating atomic-scale metal-atom structures on the silicon surface (Fig. 3(b)). The width of the Ga wire, however, is not constant, indicating that part of the wire is two or more Ga-atoms wide. The electronic structures of several kinds of atomic-scale Ga wires were examined by the first-principles calculations by Watanabe *et al.* [8,11] Depending on the number of Ga atoms per one DB in the DB wire, semiconducting and metallic features were predicted. More interestingly, part of the mid-gap energy-band structures showed very small dispersion (flat bands) and ferromagnetic ground states were predicted for the specific metastable structures of the metallic Ga wires. [11,12] The flat-band ferromagnetism was also predicted theoretically by Yajima *et al.* [13] and Arita *et al.* [14] for atomic-scale wires made of As. We believe that new physics in these atomic-scale structures on the silicon surface will give us new ideas for atomic-scale electronic devices. Connecting the atomic structures to the bulk electrodes is one of the important issues for evaluating the atomic-scale devices. We used thermal deposition of Au *in situ* through a micron-order metal mask and formed metal wires and bonding pads on the Si(100)–2x1–H surface. A detailed method of intermediate connection between the bulk electrodes and the atomic structures is under investigation.

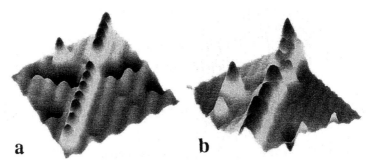

Figure 3. Three-dimensional views of (a) an atomic-scale DB wire and (b) a Ga wire, demonstrating a method of fabricating atomic-scale metal wires on the silicon surface (6.5 nm× 6.5 nm, $V_s = -2.0$ V, $I_t = 20$ pA).

4. Summary

We have applied STM/STS and atom manipulation to fabricate and characterize atomic-scale structures on a hydrogen-terminated Si(100)–2×1–H surface. We fabricated atomic-scale DB wires by extracting hydrogen atoms from the surface. Tunneling spectra of the DB wires made of both unpaired and paired DBs showed a metallic density of states. The Ga atoms on the Si(100)–2×1–H surface showed preferential adsorption on the DBs. We demonstrated a method of fabricating atomic-scale Ga patterns on the silicon surface using DB patterns.

REFERENCES

1. R. Feynman, reprinted in J. Microelectromechanical Systems, **1** (1992) 60.
2. G. Binnig, H. Rohrer, Ch. Gerber and E. Weibel, Phys. Rev. Lett. **49** (1982) 57.
3. D. M. Eigler and E. K. Schweizer, Nature **344** (1990) 524.
4. J. J. Boland, Adv. Phys. **42** (1993) 129.
5. J. W. Lyding, T. C. Shen, J. S. Hubacek, J. R. Tucker and G. C. Abeln, Appl. Phys. Lett. **64** (1994) 2010.
6. J. Nogami, Sang-il Park and C. F. Quate, Appl. Phys. Lett. **53** (1988) 2086.
7. A. A. Baski, J. Nogami and C. F. Quate, J. Vac. Sci. Technol. **A8** (1990) 245.
8. S. Watanabe, Y. A. Ono, T. Hashizume, Y. Wada, Phys. Rev. **B54** (1996) R17308.
9. T. Hashizume *et al.*, Jpn. J. Appl. Phys. **35** (1996) L1085.
10. T. Hitosugi *et al.*, Jpn. J. Appl. Phys. **36** (1997) L361.
11. S. Watanabe, M. Ichimura, T. Onogi, Y. A. Ono, T. Hashizume, and Y. Wada, Jpn. J. Appl. Phys. **36** (1997) L929.
12. M. Ichimura, K. Kusakabe, S. Watanabe and T. Onogi, Phys. Rev. **B58** (1998) 9595.
13. A. Yajima *et al.*, Phys. Rev. **B** in press.
14. R. Arita, K. Kuroki, H. Aoki, A. Yajima, M. Tsukada, S. Watanabe, M. Ichimura, T. Onogi and T. Hashizume, Phys. Rev. **B57** (1998) R6854.

Quantum Coherence and Decoherence - ISQM - Tokyo '98
Y.A. Ono and K. Fujikawa (Editors)
© 1999 Elsevier Science B.V. All rights reserved.

Coulomb Blockade Oscillations in Single Atom Nanotips

F. Yamaguchi[a], D. H. Huang[b], Y. Zhang[a] , K. J. Cho[c] and Y. Yamamoto[a,d]

[a]ERATO Quantum Fluctuation Project,
Edward L. Ginzton Laboratory, Stanford University, Stanford, CA 94305, USA

[b]ERATO Quantum Fluctuation Project,
NTT Musashino R&D Center 3-9-11 Midoricho, Musashino,
Tokyo 180-0012, Japan

[c]Department of Mechanical Engineering,
Stanford University, Stanford, CA 94305, USA

[d]NTT Basic Research Laboratories,
3-1, Morinosato-Wakamiya, Atsugi-shi, Kanagawa 243-0198, Japan

The single charging effect of 1 eV was observed in a system which consists of a single atom nanotip made of tungsten and a sample surface by STM experiment. In the apex of such a single atom nanotip which ends with a single atom followed by a pyramid-like structure, localized electronic states exist. Once a localized state is charged by electron tunneling from or to it, the electronic charge is not screened by delocalized electrons as in a bulk metal. This causes steps in the current between a single atom nanotip (STM tip) and a sample surface as a function of an applied bias voltage or oscillations in the dI/dV.

1. INTRODUCTION

The Coulomb blockade effect can be observed in a system where there is a central island that is electronically isolated from adjacent electrodes by tunnel barriers.[1] Because of the small capacitances of these tunnel junctions, the energy to charge the island with a single electron can be large. If this energy is not supplied by a bias voltage, the tunnel current is blocked. Once the bias voltage exceeds the charging energy, one electron can tunnel through the island at a time, producing a step in the I-V characteristics or a peak in the dI/dV. Since the peak width is determined by either the thermal energy or the finite lifetime of the island, large charging energies are needed to operate Coulomb blockade devices at room temperature. To obtain large charging energies, we need small capacitances. One way to get small capacitances is to make the islands smaller. A natural question is how small we can make these islands.

In this paper, we will present theoretical and experimental studies to show that there exist a possible way to make a Coulomb blockade device with a single charging energy of 1eV.

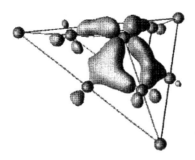

Figure 1. Localized state $5d_{xy}$ calculated by the ab initio calculation at the apex of a single atom nanotip which ends with a single tungsten atom and is followed by three tungsten atoms.

2. SINGLE ATOMS AS CENTRAL ISLANDS

2.1. Single charging energy

Ionization energies are known to a high degree of accuracy for removal of the first, second, third, fourth and fifth electrons for most elements and listed in many books.[2] If we hypothesize the ionization energy is the sum of a single particle energy and an interaction term, the change in the electro-static energy when one electron is removed, the difference in adjacent ionization energies corresponds to the single charging energy of an atom. The single charging energy of a single tungsten atom depends on the electron orbits from which an electron is removed. The charging energies when electrons are removed from the 5d level and the 6s levels can be estimated as 12eV and 10eV, respectively. This suggests that if the central island is made out of a single atom, the single charging energy can be of the order of 10eV.

2.2. Mechanical stabilities

One way to hold a single atom stably between two electrodes is to make a pyramid-like structure as in a tungsten single atom nanotip on top of a <111> base tip.[3] Such a tip has a sharp apex which ends with a single tungsten atom, followed by three tungsten atoms in the next layer and six atoms in the third layer as observed by FIM analysis.[3] It shows a high stability in time over days under an applied bias field up to 1V/Å. The stability originates from the fact that the topmost single atom is chemically bonded to the pyramid-like structure at the apex of the tip. Because of this chemical bonding, the confinement of electrons to the topmost atom is weakened compared with the case of an isolated atom, the single charging energy is smaller than the one for an isolated atom, which is on the order of 10eV, as is seen below.

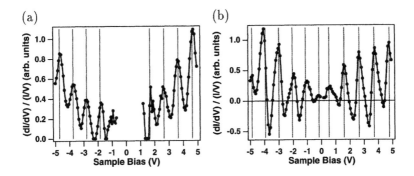

Figure 2. Normalized differential conductance vs. applied bias voltage calculated from measured currents between a single atom nanotip (STM tip) and (a) a silicon reconstructed surface Si(100)2×1 and (b) a gold surface (111).

3. LOCALIZED STATES AT THE APEX OF A SINGLE ATOM NANOTIP

It has been confirmed experimentally that there are localized electronic states at the apex of a single atom nanotip by field emission microscopy[3]. The feature of the localized states were calculated by the ab-initio calculation and the tight binding calculation. The theories show that there are two main localized modes, which consist of 5d (the highest occupied level of a tungsten atom) electrons and have ten–fold approximate degeneracies. The localized states maintain their original electronic orbitals such as the 5d states. An example of the localized states which maintain the 5d orbitals is shown in Fig.1 One of the modes shows stronger localization to the apex atom for a larger positive sample bias voltage, while the other mode becomes strongly localized for a negative sample bias voltage.

Given the experimental and theoretical confirmations of the localized states at the apex of a single atom nanotip, one would expect that when an electron tunnels from or into the localized states, the electronic charge cannot immediately be screened by delocalized electrons and a charging effect remains at the apex for a while after the tunneling.

4. OBSERVATION OF A SINGLE CHARGING EFFECT IN A SINGLE ATOM NANOTIP

The experiment was performed using an ultra-high vacuum STM with an operating pressure of 2×10^{-8} Pa at 77K and 300K. The single atom nanotip was made from a single crystal tungsten wire with (111) orientation.

Figure 2 shows oscillations in the dI/dV due to the single charging effect which remains at the tip apex. Currents between the single atom nanotip and (a) a silicon reconstructed surface and (b) a gold surface were measured. In both figures (a) and (b), peak separations, which correspond to the single charging energy of the localized states at the apex of of the tip, are about 1V. The same period of the oscillating differential conductance $\sim$ 1V was repeatedly obtained at each point and at different points on a sample surface as far as we used the same STM tip. This fact concludes that the oscillation differential conductance attributes the bare STM tip that are chemically stable. Atoms or molecules adsorbed on the STM tip or a sample surface would not give the same period of the oscillation every time or at every point we measure the differential conductance by applying a large bias voltage $\sim$ 1V/Å. The same periods of the oscillation were observed both on a silicon surface and on a gold surface.

The observed oscillations in the differential conductance are the first experimental proof of the single charging effects in a single atomic junction (a single atom nanotip and a sample surface).

5. CONCLUSION

In conclusion, electrons are localized at the apex of a single atom nanotip at certain energies. The localized states are not immediately screened once they get charged after electrons tunnel. Thus the localized states at the apex of such a tip acts as a central island that is electronically isolated from adjacent electrodes by tunnel barriers in a sense that the single charging effect remains at the apex, giving rise to oscillations in the dI/dV.

REFERENCES

1. M. H. and H. Grabert, in *Single Charge Tunneling: Coulomb Blockade Phenomena in Nanostructures,* Vol. 294 of *NATO Advanced Study Institute, Series B: Physics,* edited by H. Grabert and M. H. Devoret (Plenum Press, New York, 1992), chap. 1.
2. J. Emsley, *The Elements* (Clarendon Press, Oxford, 1998), p 219.
3. Vu Thien Binh, N. Garcia and S. T. Purcell, in *Advances in imaging and electron physics* (Academic Press, San Diego, 1996), Vol. 95, p 63.

Quantum Coherence and Decoherence - ISQM - Tokyo '98
Y.A. Ono and K. Fujikawa (Editors)
© 1999 Elsevier Science B.V. All rights reserved.

Electronic Structure of Pseudo-one-dimensional Dangling-bond Structures Fabricated on a Hydrogen-terminated Si(100)2x1 surface

T. Hitosugi[a], S. Heike[b], H. Kajiyama[b], Y. Wada[b], T. Onogi[b], T. Hashizume[b],
S. Watanabe[c], T. Hasegawa[d], K. Kitazawa[a], Z.-Q. Li[e], K. Ohno[e], Y. Kawazoe[e]

[a]Department of Superconductivity, University of Tokyo, Japan
[b]Advanced Research Laboratory, Hitachi, Ltd., Japan
[c]Department of Materials Science, University of Tokyo, Japan
[d]Materials and Structures Laboratory, Tokyo Institute of Technology, Japan
[e]Institute for Materials Research, Tohoku University, Japan

Various atomic-scale dangling-bond (DB) structures on a hydrogen-terminated Si(100)-2 × 1-H surface, made of paired and/or unpaired DBs, are fabricated and studied by scanning tunneling microscopy/spectroscopy (STM/STS). We compared our experimental results with first-principles calculations and concluded that the structures, made of only paired DBs and of only unpaired DBs along a dimer row, show semiconductive electronic states with an energy gap. For the DB structure made of only unpaired DBs, we observed charge redistribution in the DB structure due to Peierls distortion. We propose that in fabricating a structure that has finite density of states at the Fermi energy on this surface, the unpaired and paired DBs should be mixed properly to suppress Peierls distortion.

1. INTRODUCTION

Electrons and atoms in pseudo-one-dimensional structures have been attracting a lot of attention, since many new physical properties are expected. Recently, not only fabricating atomic structures but also characterizing the physical properties of those fabricated structures have been key issues. Our attempts to measure STS and investigate electronic properties are also along these lines. The hydrogen-terminated Si(100)-2×1 surface [1] is a promising substrate for the fabrication of atomic-scale structures. The structures are made by extracting hydrogen atoms and fabricating DB structures using STM atom manipulation techniques [2]. Hashizume et al. reported on the formation of atomic-scale Ga structures on this surface using DB structures [3].

In this paper, we present electronic structures of atomic-scale DB structures fabricated on a hydrogen-terminated Si(100)-2×1 surface studied by STM/STS. DB structures are fabricated and characterized at room temperature and at low temperatures (96 K to 107 K). We used the STM manipulation technique to extract hydrogen atoms, and routinely obtained DB structures of one to two dimers (0.8 to 1.6 nm) in widths. Experimental procedures and methods of hydrogen desorption are described elsewhere [4]. The DB structures on this surface are made up of two types of DBs; paired and unpaired DBs [4]. A row of fabricated DBs forms a pseudo-one-dimensional atomic structure, and confines electrons. We categorized the DB structures into several types and used STS to measure the electronic structures [5]. The experimental results were analyzed based on recent first-principles calculations by Watanabe et al [6]. We think that this structure serves as an instructive example of an atomic-scale structure on a semiconducting surface, as it is the first step in fabricating useful atomic structures, such as electrically conducting atomic wire.

2. RESULTS AND DISCUSSIONS

Three different types of DB structures were fabricated: DB structures made of only paired DBs; DB structures made of only unpaired DBs; and a mixture of both paired and unpaired DBs. First, we discuss the electronic structures of artificial DB structures which are made of paired DBs only. Figure 1(a) shows part of a long DB structure, made of only paired DBs, fabricated at room temperature. The dimers in this DB structure are buckled (Fig. 1(b)) and the structure is observed as a zigzag line similar to the asymmetrical dimers seen in $p(2 \times 2)$ or $p(4 \times 2)$ reconstruction. The tunneling spectra in Fig. 1(c), obtained from another DB structure made of only paired DBs, have an energy gap of 0.5 eV. First-principles calculations predict that DB structures made of DB pairs have an energy gap of 0.5 eV, which is in good agreement with experiments. There are two explanations for this energy gap. One of the explanation is the splitting between π bonding and π^* antibonding states caused by the interaction between DBs of Si atoms on a dimer. The other is that it is due to the pair of surface states that result from charge transfer between DBs on the lower and upper atoms of the buckled dimer [7].

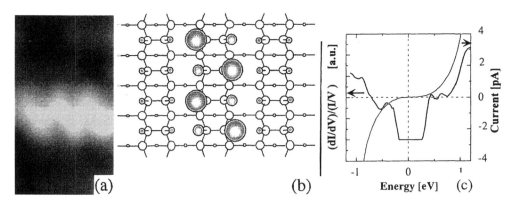

Fig. 1. (a) Filled-state STM image of DB structure made of only paired DBs (2.6 nm $\times$ 3.8 nm, sample bias voltage is -2.0 V, tunneling current is 100 pA). (b) Schematics of a buckled DB structure. (c) STS results.

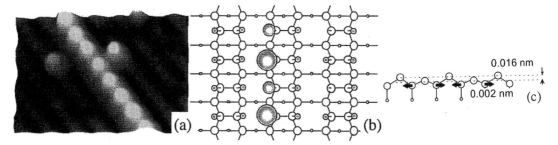

Fig. 2. (a) Filled-state STM image of a DB structure fabricated parallel to the dimer row. (96 K, 7 nm×4 nm, sample bias voltage is -2.0 V, tunneling current is 20 pA). (b) Schematics of a DB structure shown in Fig.2(a). (c) Cross-sectional view of DB structure.

A filled-state STM image of the DB structure, fabricated at a low temperature (96 K), made of unpaired DBs is shown in Fig. 2(a). The structure is 15 dimer long, but has only 8 peaks. The distance between the peaks is approximately 0.77 nm, which means that DB is seen alternately, as shown in Fig. 2(b). There was a difference in the lateral positions of the peaks between the empty (sample bias voltage of +2.0 V) and filled-state (sample bias voltage of -2.0 V) images. In contrast with the filled-state image, the dark (lower in height) DBs seen in the filled-state image are brighter in the empty-state image, and the bright DBs seen in the filled-state image are darker in the empty-state image. Considering these STM results, we conclude that a change in the electronic structure, accompanying charge redistribution, has occurred in this DB structure.

A surface-state band in the bulk band gap resulting from the DBs is predicted by theoretical calculation [6]. This surface-state band is half-filled when Peierls distortion [8] is not taken into account. When the Peierls distortion is considered, the system is stabilized by 14 meV. From this distortion, energies of the DBs and charge distribution may vary, so that certain DBs may be seen at some tip-sample voltages but not at others. The direction of this distortion is not along the DB structure but rather perpendicular to it. As a result, a height difference of 0.016 nm arises between the neighboring Si atoms in the DB structure (Fig. 2(c)). Since the period of this distortion is twice as large as the atomic distance between neighboring Si atoms, it is reasonable to assign the charge redistribution to the Peierls transition. We also observed this distortion at room temperature.

The above discussion confirms that the periodic structures either made of only unpaired or paired DBs are always stabilized and an energy gap will be generated. Random arrangement of unpaired and paired DBs may result in the suppression of stabilization. Figure. 3(a) shows the DB structure made of both unpaired and paired DBs, which is not periodic. Figure 3(b) shows conductance curves (dI/dV) recorded on a hydrogen-terminated Si dimer (narrow line) and on the DB structure (broad line). This DB structure has finite density of states at the Fermi energy, whereas no density of states is observed for the hydrogen-terminated dimer. Similar results are obtained for the tunneling spectra at different positions of the DB structure in Fig. 3(a).

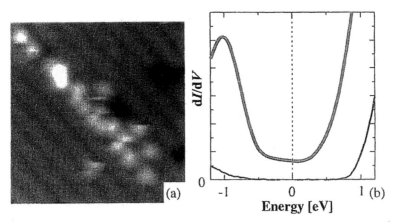

Fig. 3. (a) Filled-state STM image of a DB structure. (9 nm×9 nm, sample bias voltage is -2.0 V, tunneling current is 100 pA) (b) Conductance curve obtained from DB structure shown in Fig. 3(a) (broad line) and hydrogen-terminated Si(100)2×1 surface (narrow line).

We speculate that if the unpaired and paired DBs are suitably mixed, the Peierls distortion and Si dimer buckling is suppressed and the DB structures will show finite density of states at the Fermi energy. The best arrangement that gives the maximum density of states is not clear. However, this experiment clearly indicates that randomness is required to avoid stabilization.

3. CONCLUSIONS

On the hydrogen-terminated Si(100)-2×1 surface, DB structures composed of only paired DBs show a semiconductive band gap. Further, the charge redistribution was seen for the DB structures made of only unpaired DBs. We compared the results with the first-principles calculations and concluded that the origin of this redistribution is Peierls distortion. When fabricating atomic-scale wire having finite density of states at the Fermi energy, controlling the Peierls transition is essential.

[1] J. J. Boland, Phys. Rev. Lett. 67 (1991) 1539.; J. J. Boland: Adv. Phys. 42 (1993) 129.
[2] J. W. Lyding, T. -C. Shen, J. S. Hubacek, J. R. Tucker and G. C. Abeln, Appl. Phys. Lett. 64 (1994) 2010.
[3] T. Hashizume, S. Heike, M. I. Lutwyche, S. Watanabe, K. Nakajima, T. Nishi and Y. Wada, Jpn. J. Appl. Phys. 35 (1996) L1085.
[4] T. Hitosugi, T. Hashizume, S. Heike, S. Watanabe, Y. Wada, T. Hasegawa, K. Kitazawa, Jpn. J. Appl. Phys. 36 (1997) L361.
[5] T. Hitosugi, T. Hashizume, S. Heike, Y. Wada, S. Watanabe, T. Hasegawa, K. Kitazawa, Appl. Phys. A66 (1998) S695.
[6] S. Watanabe, Y. A. Ono, T. Hashizume, Y. Wada, J. Yamaguchi and M. Tsukada, Phys. Rev. B52 (1995) 10768; S. Watanabe, Y. A. Ono, T. Hashizume, Y. Wada, Phys. Rev. B54 (1996) 17308; S. Watanabe , Y. A. Ono, T. Hashizume, and Y. Wada, Surf. Sci. 386 (1997) 340.
[7] J. Pollman, P. Kruger, A. Mazur, J. vac. Sci, technol. B5 (1987) 945.
[8] P. E. Peierls, Quantum Theory of Solids, (Oxford University Press, London, 1955).

Quantum Coherence and Decoherence - ISQM - Tokyo '98
Y.A. Ono and K. Fujikawa (Editors)
© 1999 Elsevier Science B.V. All rights reserved.

Nano-scale lithography using scanning probe microscopy

Masayoshi Ishibashi, Seiji Heike, Nami Sugita, Hiroshi Kajiyama,
Yasuo Wada, and Tomihiro Hashizume

Advanced Research Laboratory, Hitachi, Ltd., Hatoyama, Saitama 350-0395, Japan

An AFM lithography system that uses a constant current-controlled exposure system is developed. Using this system, several kinds of patterns are fabricated on a negative-type resist and the characteristics required for a lithography are evaluated. We found that the resolution depends on the resist thickness and that a minimum linewidth of 27 nm is obtained for a 15-nm-thick resist. The proximity effect is smaller than that of the electron beam lithography. The cross-sectional shape of the developed resist pattern depends on the amount of the exposure dose. The characteristics of the AFM exposure system based on the proposed exposure mechanism is explained by evaluating the electric-field mapping inside the resist.

1. INTRODUCTION

Nanofabrication technique is indispensable for fabricating quantum-effect electronic devices, such as single electron transistors (SETs). Scanning probe lithography (SPL), which is based on scanning tunneling microscopy (STM), atomic force microscopy (AFM), or other scanning-probe methods, is one of the most powerful nanofabrication techniques. After the first demonstration of an STM-based electron exposure of a Langmuir-Blodgett (LB) film [1], several kinds of SPL were reported [2-5].

In this paper, we report on the superior characteristics of AFM-based lithography using a current-controlled exposure system. In AFM lithography, the tip position is controlled by an AFM cantilever and the exposure of the resist is performed by a field-emission current between a conductive substrate and a metal-coated tip [3]. Our system incorporates a second feedback system so that we can control the bias voltage between the metal-coated tip and the conductive substrate. We fabricated some patterns on a negative-type resist, RD2100N (Hitachi Chemical Co.), using this method and we evaluated the characteristics for making devices [4].

2. EXPERIMENTAL

A contact-mode AFM system was used to control the tip position. A tip with a cantilever (0.1 to 1 N/m) was coated with 20 to 40 nm of Ti by a dc sputter deposition. The experimental set up was similar to earlier work [3], except that we used an additional feedback system to adjust the negative-bias voltage and applied it to the metal-coated tip to control the emission current constant.

A sample substrate, an 8-Ωcm to 12-Ωcm p-type Si (100) wafer, was spin-coated with a negative resist, RD2100N (Hitachi Chemical Co.), and was pre-baked on a hot plate at 75°C

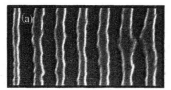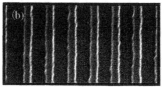

500nm

Figure 1. SEM top images of the typical line-and-space resist patterns using (a) the regular constant-bias feedback system and (b) the present constant-current feedback system. The resists are all 40-nm-thick.

500nm

Figure 2. Cross-sectional SEM micrographs of the developed resist patterns for the line dose of (a)105 nC/cm, (b) 28 nC/cm, and (c) 13 nC/cm. The bias voltage of exposure was (a) 85 V, (b) 81 V, and(c) 77 V, respectively.

for 7 minutes. A negative tip-to-substrate bias of 10 to 100 V was used to obtain a constant current of 10 to 100 pA so that the resist with an electron flux from the tip was exposed. A line dose of 0.3 to 110 nC/cm at an AFM scanning speed of 10 to 100 mm/s was used for resist patterning. After pattern exposure, the samples were developed NMD-W (Tokyo Oka Kogyo). The developed resist pattern was observed with a scanning electron microscope (SEM).

3. RESULTS AND DISCUSSION

Figure 1(a) shows an SEM top image of the typical line-and-space resist pattern using a regular constant-bias feedback system and figure 1(b) shows an SEM image of the typical line and space resist pattern using our constant-current feedback system. The constant-current feedback system significantly reduces variation in the width of the line patterns compared with the constant-bias feedback system. These results show that by using the constant-current feedback system, the exposure dose was kept constant even though the resist-thickness was not uniform or the shape of the tip was changed [5].

Cross-sectional SEM micrographs of the developed line patterns are shown in Fig. 2(a), 2(b), and 2(c). The resist profiles are steep even though the top of the resist is deformed. An AFM image of the resist surface after pattern exposure showed that the resist surface was deformed even before the development. The depth of the indentation was measured to be approximately 10 nm, which is 10% of the resist thickness. These results indicate that the indentation made by the tip into the surface of the resist is due to the strong Coulomb force supplied from the bias voltage between the tip and the substrate. Since we are using contact-mode AFM, this attractive force cannot be canceled out by the feedback of the AFM system. However, our current-controlled exposure system allows us to control the variation in the attractive force which is caused by the local variation in the resist thickness or other causes, and which results in a constant current exposure.

It can also be seen from Fig. 2 that for a large exposure, the cross-sectional shape of the resist is trapezoidal and that for a small exposure dose, the cross-sectional shape of the resist is an upside-down trapezoidal. The ability to control the resist shape by changing the amount of exposure dose is another important feature of this system.

We evaluated the electric field in the resist region in order to discuss the exposure mechanism and the cross-sectional shape of the resist. Figure 3 shows the calculated electric-flux lines with the contour lines of the electric field inside the resist. Based on our analysis of the resist shape and mapping of the electric field, the mechanism of the latent-image formation can be explained as follows. Electrons, which are field-emitted from the tip into the resist, travel mainly along the electric-flux lines while gaining energy from the electric field. If an electron obtains enough energy to cause a chemical reaction with a molecule of the resist, it reacts with the resist at some probability. Therefore, the latent image is formed along the electric-flux lines at a large exposure dose, and the trapezoidal shape of the developed resist is based on the shape of the electric-flux lines. However, the number of electrons which gain enough energy to cause a chemical reaction is limited at a small exposure dose, and the number of chemical reactions per unit volume of the resist may not exceed the critical value. This is more significant near the substrate, especially where the distance from the tip is larger, because the electric-flux lines diverge there. Thus, the upside-down trapezoidal latent image is obtained.

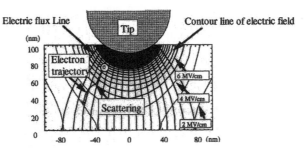

Figure 3. Calculated electric-flux lines and contour lines of the electric field inside the resist. We assumed a spherical tip and used the typical tip radius of 50 nm, which was measured with an SEM. The relative dielectric constant of the resist was assumed to be 3, which is a typical value of various kinds of resist. The tip-substrate bias voltage was -82 V, the resist thickness was 100 nm, and a 10-nm-deep indentation of the tip into the resist surface was assumed.

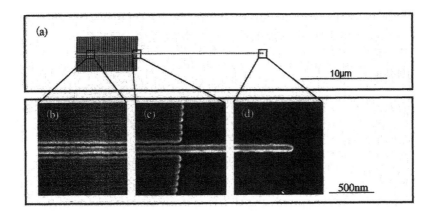

Figure 4. (a) Evaluation patterns, (b), (c) and (d) are SEM top views of the developed patterns.

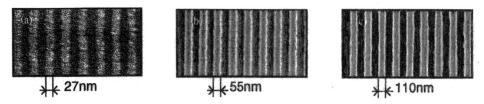

Figure 5. SEM top views of the patterns showing the minimum linewidth for resist thicknesses of (a) 15 nm, (b) 40 nm, and (c)100 nm.

The resist patterns containing an area of low and high density were fabricated with a uniform dose to investigate the proximity effect in SPL. The proximity effect is one of the major factors limiting the resolution of electron-beam (EB) lithography. Figure 4(a), 4(b) and 4(c) show the SEM micrographs of the developed resist patterns. If the proximity effect occurs, the linewidth at the isolated region (Fig. 4(d)) becomes narrower than that at the dense region (Fig. 4(a)). However, the linewidth is constant from the isolated region to the dense region. This indicates that the proximity effect in SPL is negligible compared with that in EB lithography. The proximity effect is small because the scattered electrons have little energy (momentum) and they also accelerate along the electric-flux lines.

The SEM images of the minimum line width we obtained are shown in Fig. 5. For the resist thickness of 15, the minimum linewidth was 27 nm, for the resist thickness of 40, the minimum linewidth was 55 nm, and for the resist thickness of 100, the minimum linewidth was 110 nm. These results indicate that the resolution depends on the resist-thickness.

4. CONCLUSION

We fabricated line-and-space patterns on a commercial negative-type resist, RD2100N (Hitachi Chemical Co.), and we evaluated the lithography's characteristics. The cross section of the pattern was an upside-down trapezoid at a low-exposure dose and was a trapezoid at a high-exposure dose. For the resist thicknesses of 15, 40, and 100 nm, we obtained minimum resist linewidths of 27, 55, and 110 nm. The advantage of using AFM lithography compared to EB lithography in terms of the proximity effect was discussed. We evaluated electric-field mapping inside the resist and we explained the characteristics of the AFM exposure system.

REFERENCES

1. M. A. McCord and R. F. W. Pease, J. Vac. Sci. Technol. B4, 86 (1986).
2. E. A. Doboisz and C. R. K. Marrian, Appl. Phys. Lett. 58, 2526 (1991).
3. A. Majumdar, P. I. Oden, J. P. Carrejo, L. A. Nagahara, J. J. Graham, and J. Alexander, Appl. Phys. Lett. 61, 2293 (1992).
4. M. Ishibashi, S. Heike, H. Kajiyama, Y. Wada, and T. Hashizume, Appl. Phys. Lett. 72, 1581 (1998).
5. H. T. Soh, K. Wilder, A. Atalar, and C. F. Quate, Digest of Technical Papers, 1997 Symposium on VLSI Technology 129 (1997); K. Wilder, H. T. Soh, A. Atalar, and C. F. Quate, J. Vac. Sci. Technol. B15, 1811 (1997).

Quantum Coherence and Decoherence - ISQM - Tokyo '98
Y.A. Ono and K. Fujikawa (Editors)
© 1999 Elsevier Science B.V. All rights reserved.

Chaos and quantum interference effect in semiconductor ballistic micro-structures

Shiro Kawabata*

Physical Science Division, Electrotechnical Laboratory,
1-1-4 Umezono, Tsukuba, Ibaraki 305-8568, Japan

We study the quantum-interference effect in the ballistic Aharonov-Bohm (AB) billiard. The wave-number averaged conductance and the correlation function of the non-averaged conductance are calculated by use of semiclassical theory. Chaotic and regular AB billiards have turned out to lead to qualitatively different semiclassical formulas for the conductance with their behavior determined only by knowledge regarding the underlying classical scattering.

1. INTRODUCTION

Recently a growing number of works accumulated around the subject of an interplay between quantum chaos and ballistic quantum transport in mesoscopic systems [1,2]. Particularly, an important role of quantum chaos is addressed in quantum interference in Aharonov-Bohm (AB) billiards. In this paper, we shall focus our attention to the latest issue of our research on the $h/2e$ Altshuler-Aronov-Spivak (AAS) oscillation [3,4] and h/e AB oscillation [5]. We shall present the analysis of semiclassical theory of AAS and AB oscillation in an open chaotic system, e.g., the Sinai billiard. Comparative study on other regular billiards will also be presented.

2. $h/2e$ AAS OSCILLATION

In this section, we shall consider an open single AB billiard in uniform normal magnetic field B penetrating *only* through the hollow. The ballistic weak localization (BWL) correction [6] is most easily discussed in terms of the reflection coefficient $R = \sum_{n,m=1}^{N_M} |r_{n,m}|^2$, where N_M and $r_{n,m}$ are the mode number and the reflection amplitude, respectively. Therefore, our starting point is the reflection amplitude [7]

$$r_{n,m} = \delta_{n,m} - i\hbar\sqrt{v_n v_m} \int dy \int dy' \psi_n^*(y') \psi_m(y) G(y', y, E_F), \tag{1}$$

*E-mail: shiro@etl.go.jp

where $v_m(v_n)$ and $\psi_n(\psi_m)$ are the longitudinal velocity and transverse wave function for the mode $m(n)$ at a pair of lead wires attached to the dot. G is the retarded Green's function connecting points (x, y) and (x', y') on the left and right leads, respectively.

In order to carry out the semi-classical approximation, we replace G by its semiclassical Feynman path-integral expression [8],

$$G^{sc}(y', y, E) = \frac{2\pi}{(2\pi i\hbar)^{3/2}} \sum_{s(y,y')} \sqrt{D_s} \exp\left[\frac{i}{\hbar} S_s(y', y, E) - i\frac{\pi}{2}\mu_s\right], \qquad (2)$$

where S_s is the action integral along classical path s, $D_s = (v_F \cos\theta')^{-1}|(\partial\theta/\partial y')_y|$, θ (θ') is the incoming (outgoing) angle, and μ_s is the Maslov index.

Substituting eq.(2) into eq.(1) and carrying out the double integrals by the saddle-point approximation, we obtain

$$r_{n,m} = -\frac{\sqrt{2\pi i\hbar}}{2W} \sum_{s(\bar{n},\bar{m})} \text{sgn}(\bar{n})\text{sgn}(\bar{m})\sqrt{\tilde{D}_s} \exp\left[\frac{i}{\hbar}\tilde{S}_s(\bar{n}, \bar{m}; E) - i\frac{\pi}{2}\tilde{\mu}_s\right], \qquad (3)$$

where W is the width of leads and $\bar{m} = \pm m$. The summation is over trajectories between the cross sections at x and x' with angles $\sin\theta = \bar{m}\pi/kW$ and $\sin\theta' = \bar{n}\pi/kW$. In eq.(3), $\tilde{S}_s(\bar{n}, \bar{m}; E) = S_s(y'_0, y_0; E) + \hbar\pi(\bar{m}y_0 - \bar{n}y'_0)/W$, $\tilde{D}_s = (m_e v_F \cos\theta')^{-1}|(\partial y/\partial\theta')_\theta|$ and $\tilde{\mu}_s = \mu_s + H\left(-(\partial\theta/\partial y)_{y'}\right) + H\left(-(\partial\theta'/\partial y')_\theta\right)$, respectively, where H is the Heaviside step function.

Taking the diagonal approximation, there is a natural procedure for finding the average of $\delta R_D = \sum_{n=1}^{N_M} \delta R_{n,n}$ over wave-number k, which is denoted by $\overline{\delta R_D}$. Therefore the contribution to the BWL correction term $\delta\mathcal{R}_D(\phi)$ is just given by the pair of time reversal paths. With use of the extended semiclassical theory [9], we can take account of the offdiagonal part and the influence of the small-angle diffraction as $\delta\mathcal{R}(\phi) = \delta\mathcal{R}_D(\phi)/2$, for the case that the width of the lead wires are equal. Then we obtain the full quantum correction of the conductance for chaotic AB billiards as

$$\delta g(\phi) = -\frac{e^2}{\pi\hbar}\frac{1}{4}\frac{\cosh\eta - 1}{\sinh\eta}\left\{1 + 2\sum_{n=1}^{\infty} \exp(-\eta n)\cos\left(4\pi n\frac{\phi}{\phi_0}\right)\right\}, \qquad (4)$$

where $\eta = \sqrt{2T_0\gamma/\alpha}$. [3] System-dependent constants α, T_0 and γ correspond to the variance of the winding number distribution [10], the dwelling time for the shortest classical orbit and the escape rate, i.e., $N(T) \sim \exp(-\gamma T)$, [6,11] respectively. In eq. (4), the period of the conductance oscillation is $h/2e$, analogous to the AAS oscillation in diffusive systems. [12] In chaotic case, the oscillation amplitude decays exponentially with increasing the rank of higher harmonics n, so that the main contribution to the conductance oscillation comes from $n = 1$ component which oscillates with the period of $h/2e$.

On the other hand, for regular AB billiards, we obtain

$$\delta g(\phi) = -|\delta g(0)|\frac{1 + 2\sum_{n=1}^{\infty} F\left(\beta - \frac{1}{2}, \beta + \frac{1}{2}; -\frac{n^2}{2\alpha}\right)\cos\left(4\pi n\frac{\phi}{\phi_0}\right)}{1 + 2\sum_{n=1}^{\infty} F\left(\beta - \frac{1}{2}, \beta + \frac{1}{2}; -\frac{n^2}{2\alpha}\right)}, \qquad (5)$$

where F and z are the hypergeometric function of confluent type and the exponent of dwelling time distribution $N(T) \sim T^{-z}$ [6,11], respectively. In eq.(5) the oscillation amplitude decays algebraically for large n, therefore the higher-harmonics components give noticeable contribution to magneto-conductance oscillation. These discoveries indicate that *the $h/2e$ AAS oscillation occurs in both ballistic and diffusive systems forming AB geometry and the behavior of higher harmonics components reflects a difference between chaotic and non-chaotic classical dynamics.*

In real experiments [13], magnetic field would be applied to *all* region (both the hollow and annulus) in the billiard. In this situation, we will observe $h/2e$ oscillation together with the negative magneto-resistance and dampening of the oscillation amplitude with increasing magnetic field as in the case of diffusive AB rings. [14]

3. h/e AB OSCILLATION

In previous section, we have investigated the $h/2e$ AAS oscillation for *energy−averaged* magneto-conductance. The result of quantum-mechanical calculations [4] indicated that the period of the energy averaged conductance,

$$< g(\phi) >_E = \frac{1}{\Delta E} \int_{E_F - \Delta E/2}^{E_F + \Delta E/2} g(E, \phi) dE, \tag{6}$$

changed from $h/2e$ to h/e, when the range of energy average ΔE is decreased. In this section, we shall calculate the correlation function $C(\Delta\phi)$ of the *non − averaged* conductance by using the semiclassical theory and show that $C(\Delta\phi)$ is qualitatively different between chaotic and regular AB billiards. [5]

The fluctuations of the conductance $g = (e^2/\pi\hbar)T(k) = (e^2/\pi\hbar)\sum_{n,m} |t_{n,m}|^2$ are defined by their deviation from the classical value; $\delta g \equiv g - g_{cl}$. In this equation $g_{cl} = (e^2/\pi\hbar)T_{cl}$, where T_{cl} is the classical total transmitted intensity. In order to characterize the h/e AB oscillation, we define the correlation function of the oscillation in magnetic field ϕ by the average over ϕ, $C(\Delta\phi) \equiv \langle \delta g(\phi)\delta g(\phi + \Delta\phi)\rangle_\phi$. With use of the ergodic hypothesis, ϕ averaging can be replaced by the k averaging, i.e., $C(\Delta\phi) = \langle \delta g(k, \phi)\delta g(k, \phi + \Delta\phi)\rangle_k$. The semiclassical correlation function of conductance is given by

$$C(\Delta\phi) = \left(\frac{e^2}{\pi\hbar}\right)^2 \frac{1}{8} \left(\frac{\cosh\eta - 1}{\sinh\eta}\right)^2 \cos\left(2\pi\frac{\Delta\phi}{\phi_0}\right) \left\{1 + 2\sum_{n=1}^{\infty} e^{-\eta n} \cos\left(2\pi n\frac{\Delta\phi}{\phi_0}\right)\right\}^2, \tag{7}$$

where $\eta = \sqrt{2T_0\gamma/\alpha}$. In deriving eq. (7) we have used the exponential dwelling time distribution and the Gaussian winding number distribution.

On the other hand, for the regular cases, we use $N(T) \sim T^{-\beta}$. [6] Assuming as well the Gaussian winding number distribution , we get

$$C(\Delta\phi) = C(0) \cos\left(2\pi\frac{\Delta\phi}{\phi_0}\right) \left\{\frac{1 + 2\sum_{n=1}^{\infty} F\left(\beta - \frac{1}{2}, \beta + \frac{1}{2}; -\frac{n^2}{2\alpha}\right) \cos\left(2\pi n\frac{\Delta\phi}{\phi_0}\right)}{1 + 2\sum_{n=1}^{\infty} F\left(\beta - \frac{1}{2}, \beta + \frac{1}{2}; -\frac{n^2}{2\alpha}\right)}\right\}^2. \tag{8}$$

Therefore, these results indicate that the difference of $C(\Delta\phi)$ of these ballistic AB billiards can be attributed to the difference between chaotic and regular classical scattering dynamics.

4. SUMMARY

The statistical aspect of classical open AB billiards is characterized by the dwelling time distributions $N(T)$ which reflects the integrability of the system. For chaotic billiards, $N(T)$ obeys the exponential distribution. On the other hand, for regular AB billiards, $N(T)$ obeys the power-law distribution. The difference of the distributions affects the AAS and AB oscillation in the ballistic regimes.

In conclusion, we indicated a way to observe a quantum signature of chaos through the novel quantum interference phenomena in ballistic AB billiards.

I would like to acknowledge Y. Takane, Y. Ochiai, H.A. Weidenmüller, R.P. Taylor, F. Nihey and K. Nakamura for valuable discussions and comments.

REFERENCES

1. For a variety of recent topics in this field, see K. Nakamura (ed.) *Chaos and Quantum Transport in Mesoscopic Cosmos*, Special issue of Chaos, Solitons and Fractals **8**, No.7/8 (1997).
2. H.U. Baranger, R.A. Jalabert and A.D. Stone, Chaos **3** (1993) 665; N. Argaman, Phys. Rev. B **53** (1996) 7035.
3. S. Kawabata and K. Nakamura, J. Phys. Soc. Jpn. **65** (1996) 3708.
4. S. Kawabata and K. Nakamura, Chaos, Solitons and Fractals **8** (1997) 1085.
5. S. Kawabata, Phys. Rev. B **58** (1998) 6704.
6. H.U. Baranger, R.A. Jalabert and A.D. Stone, Phys. Rev. Lett. **70** (1993) 3876.
7. D.S. Fisher and P.A. Lee, Phys. Rev. B **23** (1981) 6851.
8. M.C. Gutzwiller:*Chaos in Classical and Quantum Mechanics*, Springer-Verlag, New York, 1991.
9. Y. Takane and K. Nakamura, J. Phys. Soc. Jpn. **66** (1997) 2977.
10. M.V. Berry and J.P. Keating, J. Phys. A **27** (1994) 6167.
11. W.A. Lin, J.B. Delos, and R.V. Jensen, Chaos **3** (1993) 655.
12. B.L. Altshuler, A.G. Aronov, and B.Z. Spivak, JETP Lett. **33** (1981) 94.
13. R.P. Taylor, R. Newbury, A.S. Sachrajda, Y. Feng, P.T. Coleridge, C. Dettmann, N. Zhu, H. Guo, A. Delage, P.J. Kelly, and Z. Wasilewski, Phys. Rev. Lett. **78** (1997) 1952; R.P. Taylor, A.P. Micolich, R. Newbury, and T.M. Fromhold, Phys. Rev. B **56** (1997) R12733.
14. S. Kawabata and K. Nakamura: Phys. Rev. B **57** (1998) 6282; Solid State Electronics **42** (1998) 1131.

Quantum Coherence and Decoherence - ISQM - Tokyo '98
Y.A. Ono and K. Fujikawa (Editors)
© 1999 Elsevier Science B.V. All rights reserved.

Thermodynamics and transport properties of a dissipative particle in a tight-binding model

T. Kato and M. Imada

Institute of Solid State Physics, University of Tokyo,
7-22-1 Roppongi, Minato-ku, Tokyo 106-8666, Japan

Thermodynamics and transport properties of a dissipative carrier are studied based on the Caldeira-Leggett phenomenological theory. A weak coupling theory is constructed to study the crossover behavior between the low-temperature and the high-temperature region. We found that the coherent part of the optical mobility disappears for $0 < s < 2$, where s is an exponent of a spectral function of the environment. Detailed calculation is performed for ohmic damping ($s = 1$). In this case, the specific heat shows unusual T-linear behavior, and the optical mobility takes a non-Drude form even at zero temperature.

1. INTRODUCTION

The dissipative dynamics of a quantum particle coupled to a heat bath has been studied for several decades in many areas of physics [1]. Properties of a dissipative quantum particle has been studied in detail based on a phenomenological model

$$H = H_S(p,q) + \sum_j \left[\frac{p_j^2}{2m} + \frac{1}{2}m_j\omega_j^2 \left(x_j - \frac{C_j}{m_j\omega_j^2}q \right)^2 \right].$$ (1)

Here, H_S is a system Hamiltonian, and the system is coupled to harmonic oscillators through the second term. The influence of the environment is determined by the spectral density

$$J(\omega) = \frac{\pi}{2}\sum_j \frac{C_j^2}{m_j\omega_j}\delta(\omega - \omega_j).$$ (2)

This model has been used to study quantum tunneling effects by Caldeira and Leggett [2], Brownian motion [3] and the decoherence problem in two-state systems [4].

Here, we discuss thermodynamics and transport properties of a dissipative particle for the case that H_S is truncated to a one-dimensional tight-binding Hamiltonian,

$$H_S = -\Delta \sum_l \left(c_{l+1}^\dagger c_l + c_l^\dagger c_{l+1} \right),$$ (3)

where Δ is a hopping amplitude, and c_l ($c_l^\dagger$) is an annihilation (creation) operator at the site l. We focus mainly on the weak-damping region, and calculate the specific heat and optical mobility to study how the dissipation influences density of state and transport properties of the particle.

2. FORMULATION

The partition function is formulated from (1) and (3) as

$$Z = \sum_{m=0}^{\infty} \frac{\Delta^{2m}}{2m!} \sum_{\{\xi_l\}'} \prod_{n=1}^{2m} \int_0^\beta d\tau_n \exp\left[\sum_{k<l} \xi_k \xi_l \phi(\tau_l - \tau_k)\right]. \tag{4}$$

Here, $\xi_l = \pm 1$, and the prime in $\{\xi_l\}'$ denotes the summation in accordance with the constraint $\sum_l \xi_l = 0$, and $\phi(\tau)$ is defined by

$$\phi(\tau) = \frac{a^2}{\pi} \int_0^\infty d\omega \frac{J(\omega)}{\omega^2} \left(\coth(\beta\omega/2) - \frac{\cosh\left[\omega(\beta/2 - |\tau|)\right]}{\sinh(\beta\omega/2)}\right), \tag{5}$$

where a is the lattice constant. The expression (4) may be interpreted as the statistical model of the classical interacting particles with the potential $\phi(\tau)$. Hence, the weak coupling theory may be constructed by applying the Debye-Hückel theory to calculate the partition function [5]. Thus, we may calculate the specific heat.

The specific heat is directly calculated from the partition function. We formulate the optical mobility $\mu(\omega)$ by the Kubo formula. As well as for the partition function, we may write the exact expression for $\mu(\omega)$, and may interpret it as the partition function of classical interacting particles including two fixed charges [5]. Thus, we may perform analytical calculation for $\mu(\omega)$ in the weak coupling region by the Debye-Hückel theory.

3. CONTINUUM LIMIT

In the continuum limit $(a \to 0)$, the system can be described by the continuum classical equation

$$\langle \ddot{q}(t)\rangle + \int_{-\infty}^t dt' \gamma(t - t')\langle \dot{q}(t')\rangle = F(t)/M, \tag{6}$$

where $\langle q(t)\rangle$ is the expectation value of the particle position. The band mass is denoted by M, and $F(t)$ is an external force. The kernel $\gamma(t)$ is defined by

$$\gamma(t) = \int_{-\infty}^\infty \frac{d\omega}{2\pi} e^{-i\omega t} \hat{\gamma}(z \to -i\omega + \delta), \tag{7}$$

$$\hat{\gamma}(z) = \frac{2z}{\pi M} \int_0^\infty d\omega' \frac{J(\omega')}{\omega'(\omega'^2 + z^2)}. \tag{8}$$

From these equations, the optical mobility $\mu(\omega)$ is obtained for an arbitrary form of $J(\omega)$.

Here, we consider the case that the spectral function is taken as $J(\omega) = A\omega^s e^{-\omega/\omega_c}$, where ω_c is a cut-off frequency. In the case $2 < s$, we obtain $\mu(\omega) = D\delta(\omega)$, where $\delta(\omega)$ is a delta function, and D is the weight of the coherent part of the mobility. Thus, for $s > 2$, the coherence of the particle dynamics is retained. On the other hand, for the case $0 < s < 2$, the delta function in $\mu(\omega)$ disappears, and $\mu(\omega)$ takes nonzero values for $\omega > 0$. The dissipation strongly influences the coherence of the particle, and the particle moves incoherently for $0 < s < 2$.

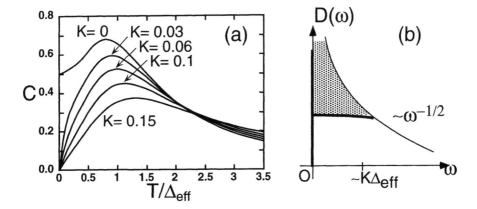

Figure 1. (a) the specific heat C and (b) the density of states $D(\omega)$ of a dissipative particle for the ohmic damping case. Thick curve in (b) is for the dissipative case while thin curve illustrates the case without the dissipation.

4. OHMIC DAMPING

Here, we discuss in detail the case $s = 1$ called the ohmic damping. We take $J(\omega) = 2\pi K \omega e^{-\omega/\omega_c}/a^2$, where K is a dimensionless damping strength. The ohmic damping has been studied in a tight-binding model by several authors [6]. The weak-damping region, however, has not been studied yet.

The specific heat C for small K is shown in Fig. 1 (a). Here, the temperature is normalized by an effective hopping amplitude $\Delta_{\text{eff}} = \Delta(\Delta/\omega_c)^{K/(1-K)}$. Even for the weak damping, the specific heat is suppressed at low temperatures, and goes to zero in the limit $T \to 0$. We can calculate the density of states $D(\omega)$ from the partition function. The change in $D(\omega)$ caused by the dissipation is shown schematically in Fig. 1 (b). For the dissipationless case, $D(\omega)$ behaves as $\omega^{-1/2}$, while for the ohmic damping case, $D(\omega)$ has a delta function peak at $\omega = 0$. Roughly speaking, because of the dissipation, some parts of low energy states up to $\omega \sim K\Delta_{\text{eff}}$, shown in the shaded area in the figure, amounts to the weight of the delta function. This indicates that the dispersion of the particle is modified by dissipation to become flat.

The optical mobility $\mu(\omega)$ for high-temperature region ($\Delta_{\text{eff}} \ll T$) is calculated for arbitrary K. The result for $K = 0.5$ is shown in Fig. 2 (a). On the high-frequency side ($\omega > KT$), the optical mobility is temperature-independent, and behaves as ω^{2K-2}. On the low-frequency side ($\omega < KT$), the optical mobility depends on the temperature, and behaves as T^{2K-2}.

The optical mobility at zero temperature for the weak damping region is shown in Fig 2 (b). The result on the low-frequency side corresponds to the Drude form

$$\mu(\omega) = \frac{a^2}{2\pi K}\frac{\gamma^2}{\omega^2 + \gamma^2},\tag{9}$$

272

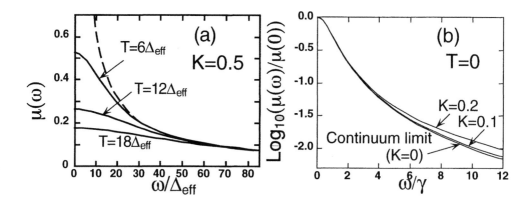

Figure 2. The optical mobility $\mu(\omega)$ (a) at high temperatures and (b) at zero temperature. The broken line in (a) shows the asymptotic behavior for $\omega \to \infty$.

where $\gamma = 4\pi K \Delta_{\text{eff}}$, while $\mu(\omega)$ deviates from the Drude form in the limit $\omega \to \infty$, and behaves as ω^{2K-2}, which is consistent with the result for the high-temperature region. The limit $K \to 0$ corresponds to the continuum limit.

The results of this model with Ohmic damping is directly related to the I-V characteristics of a Josephson junction with Ohmic shunt [7]. This physical realization will allows us experimental test of the predictions.

5. SUMMARY

We have studied the thermodynamics and transport properties of the dissipative particle in the tight-binding model based on the Caldeira-Leggett phenomenological model. We have constructed the weak coupling theory for this model, and calculated the specific heat and the optical mobility for the ohmic damping case. The results determine how the dissipation influences the coherence of the quantum particle.

REFERENCES

1. U. Weiss, Quantum Dissipative Systems, World Scientific, Singapore, 1993.
2. A. O. Caldeira and A. J. Leggett, Ann. Phys. (N. Y.) 149 (1983) 374; 153 (1984) 445(E).
3. H. Grabert, P. Schramm and G. -L. Ingold, Phys. Rep. 168 (1988) 115.
4. A. J. Leggett, S. Chakravarty, A. T. Dorsey, M. P. A. Fisher, A. Garg and W. Zwerger, Rev. Mod. Phys. 59 (1987) 1; 67 (1995) 725(E).
5. T. Kato and M. Imada, J. Phys. Soc. Jpn. 67 (1998) 2828.
6. M. Sassetti, M. Milch and U. Weiss: Phys. Rev. B 46 (1992) 4615.
7. G. Schön and Z. D. Zaikin: Phys. Rep. 198 (1990) 237.

Quantum Coherence and Decoherence - ISQM - Tokyo '98
Y.A. Ono and K. Fujikawa (Editors)
© 1999 Elsevier Science B.V. All rights reserved.

Spectral flow and gauge invariance of the effective field theory of charge density wave with topological defects *

M. Hayashi[a,b,†]

[a]Department of Applied Physics/DIMES, Delft University of Technology,

Lorentzweg 1, 2628 CJ Delft, The Netherlands

[b]Graduate School of Information Sciences (GSIS)

Tohoku University, Aramaki, Aoba-ku, Sendai 980-8579, Japan

The conventional Ginzburg-Landau type treatment of the condensate of charge density wave (CDW) can result an effective action which violates gauge invariance in the presence of topological defects (dislocations). We show that this originates from the *spectral flow* in the core of the dislocation, which causes a charge transfer between the condensed and uncondensed (normal) component of the electrons. By explicitly incorporating the contributions of the spectral flow, we can derive the gauge invariant effective action of the dislocations.

1. INTRODUCTION

The topological defect (dislocation) mediated sliding conduction in the three dimensionally ordered charge density waves (CDW's) [1] is one of the most intriguing phenomenon in which we can expect quantum tunneling of a macroscopic object [2]. In this case, the dislocations are expected to nucleate near the electric contacts to add or remove wave fronts from the CDW condensate, thus allowing the sliding motion of the condensate. This phenomenon has been studied by several authors and a phenomenological treatment, which enables us to understand some of the important natures of the phenomenon, has been developed [3-5]. Although these theories capture the global feature of the defect dynamics, microscopic basis of the effective theory has not been investigated precisely. The preceding treatments are based on the effective action which is obtained from some kind of Ginzburg-Landau type perturbative treatment, where the effect of the symmetry breaking (or order parameter) is treated within an adiabatic perturbation scheme. In the imaginary time formulation, the action has a well known form at low temperatures where

---

*This work is partially supported by the research program of the "Stichting voor Fundamentele Onderzoek der Materie (FOM)" which is financially supported by the "Nederlandse Organisatie voor Wetenschappelijk Onderzoek (NWO)".

†Present address: Graduate School of Information Sciences, Tohoku University, Aramaki, Aoba-ku, Sendai 980-8579, Japan.

the amplitude of the order parameter is constant outside the cores of dislocations:

$$S = \frac{N_\perp f_c}{4\pi} \int_0^{\beta\hbar} d\tau \int_V d\mathbf{r}\Big\{\hbar v_F \left(c_0^{-2}(\partial_\tau\theta)^2 + (\partial_x\theta)^2 + \gamma_y^2(\partial_y\theta)^2 + \gamma_z^2(\partial_z\theta)^2\right)$$
$$+ i\,4e\,(\varphi\partial_x\theta + A_x\partial_\tau\theta)\Big\} \tag{1}$$

where $N_\perp$, f_c, c_0, v_F and e are the areal density of the conducting chains, fraction of the electrons in the condensate ($f_c = 1$ at $T = 0$), the phason velocity, the Fermi velocity in the chain direction and the electronic charge ($e < 0$), respectively; $\theta(\mathbf{r},\tau)$, $\varphi(\mathbf{r},\tau)$ and $A_x(\mathbf{r},\tau)$ are the phase of condensate, the scalar and vector potential, respectively; γ_y and γ_z are the anisotropy parameters. Here the conducting chain is assumed to be parallel to the x-axis.

It has been believed that the effective action of the dislocations is obtained by introducing topological singularities in θ. This, however, violates the gauge invariance of the system, which can be seen in the following way. From Eq. (1), the electric current and charge density of the system can be estimated as,

$$\mathbf{j} = -\mathbf{e}_x \frac{eN_\perp f_c}{\pi}\partial_\tau\theta, \qquad \rho = \frac{eN_\perp f_c}{\pi}\partial_x\theta, \tag{2}$$

where $\mathbf{e}_x$ is the unit vector in x-direction. It is clear that the continuity relation of the charge (i.e., $\partial_\tau\rho + \nabla\cdot\mathbf{j} = 0$) yields the relation, $\partial_\tau\partial_x\theta - \partial_x\partial_\tau\theta = 0$. Although this cross derivative vanishes for the smooth function of $\theta(\mathbf{r},\tau)$, it becomes non-zero when there are dislocations which are moving across the chains. In this case a delta function contribution appear at the cores of the dislocations:

$$\partial_\tau\partial_x\theta - \partial_x\partial_\tau\theta = -2\pi\sum_\nu \int dl\,(\partial_l\mathbf{R}_\nu \times \partial_\tau\mathbf{R}_\nu)\cdot\mathbf{e}_x\,\delta^{(3)}(\mathbf{r} - \mathbf{R}_\nu), \tag{3}$$

where $\mathbf{R}_\nu(l,\tau)$ denotes the position of the ν-the dislocation, l is the parameter along the line, and $\delta^{(3)}(\mathbf{r})$ is the three dimensional delta function. We may consider that this delta function actually has an extension of the size of the dislocation core. This is nothing but the charge conversion between the condensed and uncondensed electrons [6], however, we should also note that the effective action, Eq. (1), is no longer gauge invariant when this kind of process takes place. Therefore, in order to analyze the dislocation mediated sliding of CDW, we need to derive the gauge invariant effective action of the dislocations.

2. SPECTRAL FLOW IN THE CORE OF DISLOACTIONS

First we discuss the microscopic origin of the violation of gauge invariance mentioned above. It is useful to examine the energy levels of the quasiparticles in the presence of the moving dislocations. Here we direct our attention to one conducting chain and neglect the inter-chain electron hopping. The Hamiltonian is, then, given by,

$$\mathcal{H} = \sum_\sigma \int dx \,\left(R_\sigma^\dagger, L_\sigma^\dagger\right)\cdot\begin{pmatrix} -\hbar v_F i\partial_x & \Delta(\mathbf{r},\tau) \\ \Delta^*(\mathbf{r},\tau) & \hbar v_F i\partial_x \end{pmatrix}\cdot\begin{pmatrix} R_\sigma \\ L_\sigma \end{pmatrix}, \tag{4}$$

where R_σ, L_σ and $\Delta(\mathbf{r}, \tau)$ are the right and left moving electron operators, and the order parameter of the condensate, respectively; σ is the spin index. Here the dispersion of the electrons are linearized near two Fermi points and divided into right and left movers.

We assume that the chain has a finite length, $0 < x < l$ and the phase of the condensate is fixed at both ends. We use an approximate order parameter, $\Delta(\mathbf{r}, \tau)$, which has a topological singularity (dislocation) parallel to z-axis at $x = l/2$, $y = Y_d$ and the eigenvalues of the Hamiltonian are numerically estimated for various Y_d. Our result is shown in Fig. 1. In this case, we assumed that the chain is locate at $y = 0$. We can see a characteristic behavior of the spectrum: one energy level is transferred from the upper side to the lower side of the gap when the dislocation crosses the chain. This is so-called *spectral flow* behavior [7]. It should be noted that this process is completely different from the ordinary particle-hole type excitation, since in the present case only one quasiparticle is created (or annihilated). Therefore the charge conservation relation is not satisfied within the condensed and uncondensed component (*i.e.*, the quasiparticle component) of electrons separately. We may recover the gauge invariance of the effective action of dislocations by including the contribution of the spectral flow.

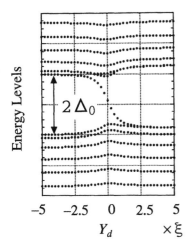

Fig. 1: The energy levels of a chain as a function of the position of the dislcoation, Y_d. Δ and ξ are the energy gap and the coherence length in y-direction, respectively.

3. GAUGE INVARIANT EFFECTIVE ACTION OF DISLOCATION

The gauge invariant effective action is obtained by including the contribution of the spectral flow explicitly. Here we especially pay attention to the contribution of the spectral flow to the charge and current, which we denote by $\delta\rho$ and $\delta\mathbf{j}$, respectively. We assume that these quantities obey diffusion type equation, $\delta\mathbf{j} = -D\nabla\delta\rho$ with D being the diffusion constant. Together with charge conservation relation, $\partial_\tau \delta\rho + \partial_x \delta\mathbf{j} = -(eN_\perp f_c/\pi)\{\partial_\tau \partial_x \theta - \partial_x \partial_\tau \theta\}$, we can determine $\delta\rho$ and $\delta\mathbf{j}$.

Here we confine ourselves to the small quasiparticle mobility limit, *i.e.*, $D \to 0$. This is justified when the characteristic time of the diffusion Dl^2 is much smaller than that of the propagation of phasons l/c_0, which is always satisfied in macroscopic samples ($l \to \infty$). In this case we obtain, $\delta\rho = -(eN_\perp f_c/\pi) \int_0^\tau d\tau' \{\partial_{\tau'} \partial_x \theta(\mathbf{r}, \tau') - \partial_x \partial_{\tau'} \theta(\mathbf{r}, \tau')\}$. Addition of the electro static energy term due to this into Eq. (1) yields the substitution,

$$\varphi \partial_x \theta \to \varphi(\mathbf{r}, \tau) \int_0^\tau d\tau' \partial_x \partial_{\tau'} \theta(\mathbf{r}, \tau'). \tag{5}$$

The resulting effective action is gauge invariant in the presence of the moving dislocations.

4. THE DISLOCATION MEDIATED SLIDING MOTION OF CDW

In the original effective action Eq. (1) a dislocation is just a polarization, whose polarization vector is perpendicular to the chains and the dislocation line, because the charge density is proportional to $\partial_x \theta$. Therefore the nucleation of dislocation does not cause energy gain when the electric field is parallel to the chains.

In the corrected action, Eq. (5), if we simply substitute the dislocation loop solution to $\theta(\mathbf{r}, \tau)$, we obtain $\int d\mathbf{r} d\tau \int_0^\tau d\tau' \partial_x \partial_{\tau'} \theta \propto S_\phi$ where S_ϕ is the area of the dislocation loop. This gives the finite energy gain due to the nucleation of the dislocation loop and it may give a correct estimate for the nucleation rate of the dislocation loop.

5. SUMMARY

We discussed the effective action of the dislocations in CDW. The gauge invariance of the action ,which is violated by the spectral flow in the core of dislocations, is recovered by taking into account the excess charge and current due to the spectral flow explicitly.

REFERENCES

[1] For a review, see G. Grüner, *Density Waves in Solids* (Addison-Wesley, 1995).
[2] S. V. Zaitzev-Zotov, Phys. Rev. Lett. **71**, 605 (1993).
[3] S. Ramakurishuna *et al.*, Phys. Rev. Lett. **68**, 2066 (1992).
[4] Ji-Min Duan, Phys. Rev. **B48**, 4860 (1993)
[5] K. Maki, Phys. Lett. **A 202**, 313 (1995); Ferroelectrics **176**, 353 (1996).
[6] N. P. Ong, G. Verma and K. Maki, Phys. Rev. Lett. **52**, (1984) 663; N. P. Ong and K. Maki, Phys. Rev. **B32**, (1985) 6582.
[7] The spectral flow in dislocations of CDW was first pointed out in M. Stone, Phys. Rev. **B54** (1996) 13222.

Quantum Coherence and Decoherence - ISQM - Tokyo '98
Y.A. Ono and K. Fujikawa (Editors)
© 1999 Elsevier Science B.V. All rights reserved.

Decoherence and Control of Electrons at a Metal Surface

S. Ogawa, H. Nagano and H. Petek

Advanced Research Laboratory, Hitachi, Ltd., Hatoyama, Saitama 350-0395, Japan

The polarization dynamics induced by an optical excitation of the Shockley surface state on Cu(111) are studied by the interferometric time-resolved two-photon photoemission. The polarization decay due to phase breaking collisions of the charge carriers on a ~20 fs time scale is observed and modeled by three-level optical Bloch equations. The control of electron distribution excited either by a pair of phase-locked ~15 fs pulses or by a single chirped pulse also is demonstrated. As a consequence of optical coherence in the two-photon excitation process, the photoemission spectra depend not only on the frequency, as in conventional spectroscopy, but also on the phase of the excitation light.

1. INTRODUCTION

Manipulation of quantum mechanical systems by means of coherent light-matter interactions is the subject of a rapidly growing field of coherent control (CC) [1]. Control of quantum dynamics in atoms [2], molecules [3, 4], and semiconductors [5] has been achieved with phase engineered light pulse sequences and chirped pulse excitation. Interference between one- and two-photon excitation pathways has been used for CC of photocurrents in semiconductors [6] and direction of photoemission from a surface [7]. CC of carrier dynamics in metals and at metallic interfaces also is of great interest in a variety of fields including solid state physics, surface science, and for applications in opto-electronics. However the carrier decoherence in metals has been considered to be too fast to observe/control due to a strong electron-electron (e-e) and electron-phonon (e-p) scattering. Recent development of ultrafast interferometric techniques has opened the way to study the coherent dynamics with sub-femtosecond resolution. Here we report the time scale of optical decoherence in Cu(111) measured by the interferometric time-resolved two-photon photoemission (ITR-2PP) [8, 9], and demonstrate CC of photo-excited distribution at the Cu(111) surface [10].

2. EXPERIMENTAL

Details of the ITR-2PP experiment and data evaluation method are described elsewhere [8, 9]. Briefly, the frequency-doubled output of a Ti:sapphire laser (λ=400 nm (3.1 eV); 15 fs FWHM of the pulse intensity envelope) is split into equal pump and probe pulses in a Mach-Zehnder interferometer. The pump-probe delay scanning by ±150 fs with an accuracy and reproducibility of $< \lambda/25$ corresponding to <50 as (5×10^{-17} s) is achieved by a feedback control system. The pulse pair is focused colliniarly on the sample surface mounted in a UHV chamber (base pressure: 5×10^{-11} Torr). Two-photon photoemission is induced with p-polarized light incident at 30° from the surface normal. The photoelectron energy is measured with a hemispherical energy analyzer (5° angular, and <30 meV energy resolution) along surface normal. Cu(111) single crystal surface is cleaned by cyclic procedures of Ar^+ sputtering (500 V) and annealing at 700 K.

Measurement of the photoemission intensity at a fixed photoelectron energy, while interferometrically scanning the pump-probe delay, gives an interferometric two-pulse correlation (I2PC), such as in Fig. 1. When the population and polarization decay are not instantaneous, the phase average, and envelopes of the ω (laser frequency) and 2ω components in the I2PC scans provide the information on the energy relaxation of the intermediate state (E_1) and phase relaxation times among initial (E_0), intermediate (E_1) and final (E_2) states in the 2PP process. Fits of the phase averaged envelopes give the population decay time T_1 for E_1, ω envelopes give average e-h pair decoherence

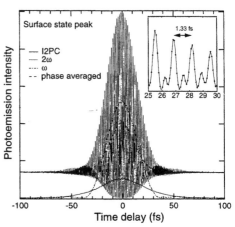

Figure 1. I2PC for two-photon excitation of the occupied surface state in Cu(111).

times $T_2{}^{01}$ and $T_2{}^{12}$ for the linear polarization (designated as $T_2{}^{\omega}$), and 2ω envelopes give the decoherence time $T_2{}^{02}$ between E_0 and E_2 [9].

3. RESULTS AND DISCUSSION

The schematic energy diagram relevant to the two-photon photoemission (2PP) process in Cu(111) is shown in Fig. 2. Cu(111) has an occupied surface state (SS) at -0.4 eV below E_F and an image-potential (IP) state series at the Γ point. The polarization dynamics are studied for excitation from the occupied SS by measuring the I2PC scans at the SS maximum, as shown in Fig. 1. The oscillations due to the interference between the pump induced polarization and the E field of the probe persist for >50 fs. Interference oscillation at the second harmonic (2ω; 1.33/2 fs) of the excitation laser also appears in the wings of the I2PC (Fig. 1 inset). The observed features are

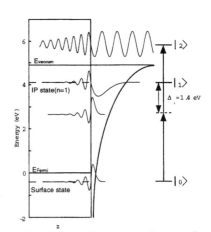

Figure 2. Schematic energy diagram for 2PP from SS on Cu(111).

qualitatively reproduced by a three-level optical Bloch equations model. The amplitude of the 2ω component depends strongly on the non-linear polarization decoherence time between initial and final state, $T_2{}^{02}$. The fit of the exponential decay of the 2ω envelope gives a $T_2{}^{02}$ of

20 fs for the excitation from the SS. The long decoherence time is a consequence of the decoupling of the surface carriers from the bulk, and delocalization of the photoelectron in the vacuum [8, 9].

Since the sample retains the memory of a phase of the excitation laser, the photoemission signal is sensitive to the pump-probe delay (τ_d). In two-pulse excitation ($\tau_d \neq 0$), the relative phase varies across the excitation spectrum with a period of $1/\tau_d$. Thus, some frequency components will be in-phase while others are out-of-phase resulting in a frequency dependent constructive and destructive interference of the excitation fields in the sample. Fig. 3 shows 2PP spectra measured with two-pulse excitation, for four values of t_d between 27.4-28.4 fs corresponding to the phase delay of $21*2\pi-\pi/2$ to $21*2\pi+\pi$ optical cycles. When the phase-locked pulse pair is in phase ($\tau_d=21*2\pi$) the surface state peak intensity is maximum and its width is narrower than the single pulse reference spectrum. By contrast, when the delay is out of phase, the SS peak intensity is minimum and its width is maximum, while for $\tau_d=21*2\pi+\pi/2$ / $-\pi/2$, the intensity of SS is intermediate and the peak is shifted to lower/higher energy. However, the featureless photoemission from the bulk bands with larger binding energy than the SS is not affected by the phase indicating that the observed effect for the SS is mainly from the coherence in the excitation process, rather than from optical interference in the interferometer. This demonstrates that the intensity and energy distribution of 2PP spectra and therefore electron distributions excited in a metal, can be manipulated through the optical phase of the excitation light [10].

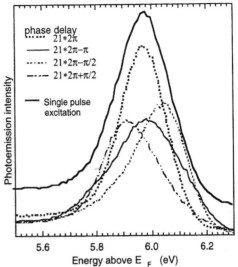

Figure 3. Surface state 2PP spectra excited by pulse pairs with different phase delays.

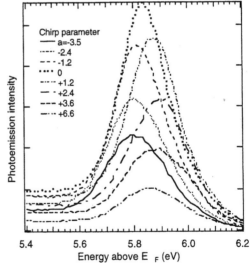

Figure 4. 2PP spectra of the SS excited by a single chirped pulse.

Coherent control of 2PP is also demonstrated by chirped pulse excitation in Fig. 4, which shows a series of 2PP spectra with single pulse excitation for different values of the chirp. Though the spectrum of the chirped pulse does not change, the position of the SS peak in the 2PP spectra red-/blue- shift from the position of unchirped peak by a maximum of approximately +/-50 meV for down/up chirp of +/-2.4 [8]. Further increase in the chirp reduces the SS shift and its intensity. This effect is caused by interference between the polarization induced by the rising part of the pulse and the electric field of the falling part of the pulse, which has an ω dependent phase shift with respect to the induced polarization. This effect only can occur if the excitation proceeds through a real intermediate state with $T_2=5 \sim 10$ fs. The three-level optical Bloch equation model reproduces qualitatively the change of 2PP spectra due to the chirp [10]. Recent observation of the d-band decoherence on 34 fs time scale will make possible extension of these experiments to bulk carriers [11].

4. CONCLUSIONS

The polarization dynamics induced by the optical excitation of the Shockley surface state on Cu(111) are studied by interferometric time-resolved two-photon photoemission. The polarization decay due to phase breaking collisions of the charge carriers on a ~20 fs time scale is modeled by the optical Bloch equations. The control of electron distribution excited either by a pair of ~15 fs pulse or by a single chirped pulse also is demonstrated. As a consequence of optical coherence in the two-photon excitation process, the photoemission spectra depend not only on the frequency, as in conventional spectroscopy, but also on the phase of the excitation light. The optical phase effect in the 2PP spectra clearly demonstrates that the coherent control of electrons in metal is possible, despite of strong Coulomb interaction among electrons and holes, and quasi-elastic electron phonon scattering. Further developments in coherent control of electron dynamics in more complex materials such as at metal/semiconductor interfaces may be useful for ultrafast opto-electronics.

ACKNOWLEDGMENT

The authors thank W. Nessler and A. Heberle for collaboration in the CC experiments, M. Moriya, S. Matsunami, and S. Saito for technical support and NEDO International Joint Research Grant for partial funding of this project.

REFERENCES

1. M. Shapiro and P. Brumer, J. Chem. Soc. Faraday Trans. 93 (1997) 1263; W. S. Warren, H. H. Rabitz and M. Dahleh, Science 259 (1993) 1581; B. Kohler et al., Acc. Chem. Res. 28 (1995) 133; H. Kawashima et al., Annu. Rev. Phys. Chem. 46 (1995) 627.
2. M. M. Wafers, H. Kawashima, and K. A. Nelson, J. Chem. Phys. 102 (1995) 9133.
3. N. F. Scherer et al., J. Chem. Phys. 95 (1991) 1487.
4. B. Kohler et al., Phys. Rev. Lett. 74 (1995) 3360.
5. A. Heberle et al., Phys. Rev. Lett. 75 (1995) 2598.
6. G. Kurizuki, M. Shapiro, and P. Brumer, Phys. Rev. B 39 (1989) 3435; E. Dupont et al., Phys. Rev. Lett. 74 (1995) 3596; R. Atanasov et al., Phys. Rev. Lett. 76 (1996) 1703.
7. N. B. Baranova et al., Opt. Commun. 79 (1990) 116.
8. S. Ogawa, H. Nagano, H. Petek and A. Heberle, Phys. Rev. Lett. 78 (1997) 1339.
9. H. Petek and S. Ogawa, Prog. Surf. Sci. 54 (1997) 239.
10. H. Petek et. al. Phys. Rev. Lett. 79 (1997) 4649.
11. H. Petek, H. Nagano, S. Ogawa, Phys. Rev. Lett. (submitted).

Quantum Coherence and Decoherence - ISQM - Tokyo '98
Y.A. Ono and K. Fujikawa (Editors)
© 1999 Elsevier Science B.V. All rights reserved.

Disorder and the Superfluid Transition

M. H. W. Chan

The Pennsylvania State University, University Park, PA 16802 U.S.A.

The effect of disorder on the superfluid transition was studied by introducing liquid helium into porous media, specifically porous gold and silica aerogel. In spite of the random environment, the transitions were found to be exceedingly sharp. Inside porous gold, the nature of the superfluid transition was found to be identical to that of bulk helium. Inside aerogel, where the silica network constitutes as little as half a percent of the available volume, the nature of the transition, unexpectedly, is completely altered.

1. INTRODUCTION

The influence of quenched disorder on the critical behavior of a pure system is a topic of considerable current interest. Owing to its exceptional purity and the absence of mechanical strains, liquid helium is an ideal system for such studies. This is the case because of quenched disorder can be easily introduced by placing the liquid into porous medium and the resultant effects can be easily identified since the nature of the superfluid transition of pure ^{4}He is very well characterized[1-3].

2. SUPERFLUID TRANSITION IN POROUS GOLD

In Fig. 1, the result of a superfluid density and heat capacity measurement of ^{4}He entrained in a porous gold[4] sample is shown. Porous gold is made by selectively leaching silver out of a gold-silver alloy with 70 at.% silver[5], and as a consequence, the gold has a porosity of 70%. Scanning electron microscope (SEM) pictures show that the substrate consists of interconnected gold strands of a uniform diameter. The remaining open space forms an interconnected pore structure. A methane vapor pressure isotherms at 77K is made on the sample and the surface area and total pore volumes were determined. If a uniform cylindrical model is assumed for the pore space, then a pore diameter of 750Å was found for the sample. The superfluid density was measured by a torsional oscillator technique. The sample in the shape of a coin of 1cm in diameter and 0.05 cm thick, is encapsulated in a thin magnesium shell and then glued onto a hollow beryllium copper torsion rod 1.4 cm long and 0.89mm in diameter, which also served as a filled line. The torsional oscillator has a resonant frequency of 578 Hz with a mechanical Q of 7×10^5 at low temperatures. We were able to resolve a mass loading or superfluid decoupling of 1×10^{-6}g/cm^3. The superfluid density, ρ_s, is proportional to $-\Delta P$, the decrease in the period of the oscillator. An ac technique was used for the heat capacity measurement. A careful configuration of thermal reservoirs provides a temperature stability of 100nK for the calorimeter. A resolution in heat capacity of 1.5 parts in 10^3 is achieved. The porous gold sample used in the heat capacity experiment is cut from the same ingot as that used for the torsional oscillator experiment and was subjected to the same leaching treatment. The porous structure of these two samples, as characterized by SEM and ^{4}He adsorption isotherms are indistinguishable from each other. Fig. 1 shows a well defined superfluid transition at a T_c that is 1.90 ± 0.05mK below that of bulk ^{4}He. The superfluid density is analyzed in the power law form,

$$\rho_s(t) = \rho_{so} t^\varsigma (1 + bt^\Delta) \tag{1}$$

In equation (1), $t=(T_c-T)/T_c$ is the reduced temperature, the correction to scaling exponent Δ is fixed at 0.5[6]. The other parameters are found to be $\rho_{so}=0.14\pm0.01$ gm/cm^3; $\varsigma=0.67\pm0.01$ and $b = 1.6\pm0.5$. The value of the exponent at 0.67 is indistinguishable from that of bulk ^{4}He at 0.6705[2]. The correction to scaling term, it should be noted, contributes 1.6, 4.5 and 16% to the value of $\rho_s(t)$ at $t=10^{-4}$, 10^{-3}, 10^{-2}, respectively.

In Fig. 2, $\rho_s(t)$ vs. t for bulk ^{4}He, ^{4}He in porous gold and Vycor[7,8] are shown in logarithm scales. The fact that ρ_s are parallel to each other is a visual confirmation that all the systems have the same superfluid critical exponent.

The temperature of the heat capacity peak shown in Fig. 1 is found to be roughly 200μK lower than the T_c deduced from ρ_s measurements. The agreement is T_c may, in fact, be

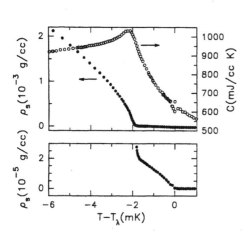

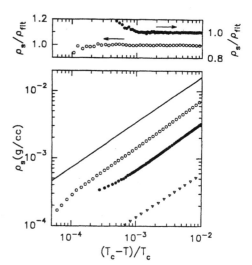

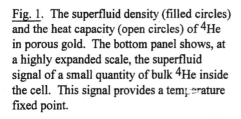

Fig. 1. The superfluid density (filled circles) and the heat capacity (open circles) of ^{4}He in porous gold. The bottom panel shows, at a highly expanded scale, the superfluid signal of a small quantity of bulk ^{4}He inside the cell. This signal provides a temperature fixed point.

Fig. 2. Log-log plots for ρ_s in porous gold (open circles), another porous gold sample of smaller pores (filled circles, ref. 4), Vycor (triangle, ref. 7) and bulk helium (solid line, ref. 2) as a function of reduced temperature. The top panel shows fractional deviations of the ρ_s in porous gold samples from the power law fit.

better[4]. If the ideal heat capacity peak has a shape that resembles the λ peak of bulk ^{4}He, which has a much sharper drop off on the high temperature side, a shift of the apparent peak to lower temperature, in addition to a rounding of the peak, is expected when a finite ac heat is applied to the calorimeter. The coincidence or near coincidence of the heat capacity peak with T_c as determined by superfluid density measurements suggest the transition in porous gold is a genuine phase transition. The fact that the value of the exponent is identical to that of the bulk suggest the transition resides in the same universality class as that of bulk ^{4}He. The heat capacity result shown in Fig. 1, unfortunately, is not of sufficient quality for the deduction of the heat capacity exponent α. However, the magnitude of the singular contribution related to the superfluid transition appears to be in reasonable agreement[4] with that predicted by the hyperuniversality hypothesis[9].

The experimental findings of ^{4}He in porous gold, as outlined above, while truly remarkable, are consistent with theoretical analysis by Harris and others[10] They showed that if the introduction of impurities into a pure system corresponds to imposing random unfrustrated coupling in the system, then the phase transition of the impurity diluted system will remain sharp. The nature of the transition of the diluted system depends on the value of the specific heat exponent α of the pure system. If $\alpha > 0$, then the dilution of impurity will lead to a new class of phase transition with a new set of exponents. This statement was confirmed in a site diluted three dimensional Ising antiferromagnet[11]. If, on the other hand, α is less than zero in the pure system (such systems are said to satisfy the Harris criterion), the diluted system resides in the same universality class as the pure system with no change in the critical exponents. Since $\alpha = -0.01285 \pm 0.00038$ for the superfluid transition[3], the findings in porous gold and in porous vycor glass provide firm confirmation for the Harris argument.

3. SUPERFLUID TRANSITION IN AEROGEL

Silica aerogel is a highly porous solid, consisting of silica strands interconnected at random sites. The porosity can range from 85% to as much as 99.8%[12]. In an aerogel of 99.8% porosity, only 0.2% of the volume is occupied by silica. Small angle neutron and x-ray scattering (SAXS) studies on aerogel show fractal-like correlation in the mass distribution, on length scale up to 100nm for aerogel containing 2% solid silica. In lighter aerogel this extends to longer length scale and conversely to shorter lengths for denser aerogels[12,13].

The first experiments on the critical behavior of the superfluid transition of entrained in aerogel were carried out in samples where the silica network constitutes 6 and 9% of the available volume[8,14-16]. As in the case of porous gold, very sharp transitions were found and the superfluid density vanishes at the same temperature where the heat capacity has a maximum. In contrast to porous gold, the superfluid density exponent is found to be 0.81[8,14,15], significantly larger than the bulk exponent. The most likely explanation for the contrasting critical behavior (in aerogel vs. in porous gold) is the difference in pore structure of these media. It is known theoretically that correlated impurities can alter the critical behavior of the pure parent system[17]. A Monte Carlo simulation study, which mimic aerogel as a percolating cluster with fractal disorder found a superfluid density exponent of 0.72, provided the helium correlation length does not exceed the fractal correlation length[18]. Attributing a nonbulk-like exponent in aerogel to long range correlation in silica, however, is not entirely satisfactory because of the behavior at very small t has not been explained. Simple power law dependence in ρ_s in 6 and 9% aerogel was found to extend to at least $t=10^{-4}$. If we assume Josephson's relation[19] to be valid for ^{4}He in aerogel

$$\xi(t) = \xi_0|t|^{-\varsigma} = \frac{k_B T_c m^2}{\hbar^2 \rho_s(t)} \qquad (2)$$

then the correlation length in 6% aerogel is calculated to be 480nm at $t=10^{-4}$. In equation (2), m, k_B and $\hbar$ are, respectively, the mass of helium atoms, Boltzmann's and Planck's constants. Since a length of 480nm exceeds the upper limit of the fractal-like regime, one would expect ρ_s to exhibit a crossover behavior such that the effective exponent ς smoothly changes towards the bulk value, when t is reduced from 10^{-2} towards 10^{-4}. Such a crossover is not seen.

Even more surprising are the results of the heat capacity[14] and the thermal expansion coefficient[16] (thermodynamically equivalent to heat capacity) measurements. Whereas the data above T_C show singular behavior which can be described with an exponent of $\alpha= - 0.6$, the data below T_C show a linear dependence on T. This implies α' is close to -1 and a violation of the scaling law, $\alpha=\alpha'$.

In order to follow the evolution of the critical behavior in the limit of dilute impurities, measurements were made in aerogel where the volume fraction of silica are 5, 2 and 0.5%[20]. The superfluid transition temperature of these samples was found to be 2.16985 ± 10^{-5}K, $2.17078 \pm 7 \times 10^{-5}$K and $2.1717 \pm 1 \times 10^{-4}$K, respectively. When the superfluid density are analyzed according to Eq. (1), the superfluid density exponent ς are found to be 0.79 ± 0.01, 0.76 ± 0.01, and 0.72 ± 0.015 respectively, for the 5, 2 and 0.5% samples. The value of ς is insensitive to the reduced temperature range of the data included in the analysis, indicating there is no evidence of a crossover to bulk-like behavior even as t approaches 10^{-4}. The log-log plots of ρ_s vs. t for the 5% and 0.5% samples are shown in Fig. 3. Instead of a crossover with respect to t, the exponents in these samples show a trend towards the bulk value as the volume fraction of silica in aerogel is reduced towards zero.

The heat capacity of ^{4}He in aerogel samples with 5%, 2%, and 0.5% silica near the superfluid transition are shown in Fig. 4. For comparison, the heat capacity of bulk ^{4}He, and ^{4}He in porous gold are also shown. The heat capacity is normalized by the coarse-grain-averaged volume of the entire experimental cell, which includes both the solid structure of aerogel and liquid helium contained therein. A small amount of bulk liquid was found in all three heat capacity cells. The presence of the bulk signal at T_λ provides a convenient fixed point in the temperature scale. The "scatters" near T_λ in Fig. 4 are due to imperfect subtraction of the bulk contribution.

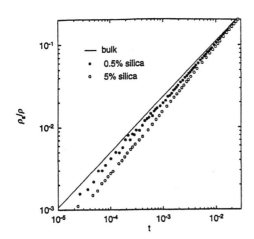

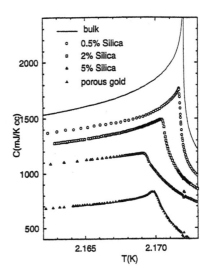

Fig. 3   Log-log plots for the superfluid density as a function of reduced temperature for bulk ⁴He and for ⁴He in 0.5% and 5% aerogel. The 2% data would lie between that of 0.5 % and 5% aerogel.

Fig. 4. The heat capacity near the superfluid transition of ⁴He in aerogel with 0.5%, 2% and 5% silica. The data for bulk ⁴He (ref. 1), and ⁴He in porous gold (fig. 1) are shown for comparison. T_λ for ⁴He in porous gold is at 2.17K because the experiment is performed under pressure. The data for porous gold is shifted down by 200mJ/Kcm³ for clarity.

The heat capacity of ⁴He in 5% aerogel as shown in Fig. 4 is in quantitative agreement with the results of 6% aerogel of references 14 and 16. In particular there is no evidence of any singular feature below T_c, in contrast to the divergent feature found in porous gold of 70% porosity. As the volume fraction of silica in aerogel is reduced below 5%, the heat capacity value increases dramatically and the singular contribution to the heat capacity becomes prominent below as well as above T_c.

When the data are analyzed in the form

$$C(t) = \frac{A'}{\alpha'}|t|^{-\alpha'} + B, \quad \text{for } T < T_c, \quad \text{and } C(t) = \frac{A}{\alpha}|t|^{-\alpha} + B, \quad \text{for } T > T_c, \tag{3}$$

The values of α and α' are determined to be -0.6±0.02 and -0.93±0.04 for 5% aerogel, -0.41±0.01 and -0.50±0.02 for 2% aerogel and -0.14±0.02 and -0.12±0.03 for 0.5% aerogel. While the best values of α differ from α' for the 5% and 2% samples, in 0.5% aerogel the scaling law $\alpha=\alpha'$ is satisfied. Furthermore, the hyperscaling relation[19] $2-\alpha=d\nu=d\zeta$ violated in denser aerogel, is also satisfied in the lightest aerogel.

4. CONCLUSION

Whereas the critical behavior in porous gold is consistent with theoretical expectations; the behavior found in aerogel, where the volume fraction for silica is as little as half-a-percent, is unexpected. The difference appears to be related whether or not the disorder is correlated to long distances. The experiments reviewed in this paper were carried out by Jongsoo Yoon, Norbert Mulders, Dmitri Sergatskov and Jian Ma. I am indebted for their dedication and grateful for their friendship. We thank Lawrence Hrubesh for introducing us to the fine art of aerogel making. The author is delighted to acknowledge useful discussions with D. S. Fisher, Gane Wong, G. Ahlers and J. D. Reppy. The work reviewed here was supported by the National Science Foundation of the United States under grant DMR-9630736.

REFERENCES

1. G. Ahlers, Phys. Rev. A3, 696 (1971).
2. L. S.Goldner, N. Muldes and G. Ahlers, J. Low Temp. Phys. 93, 131 (1993).
3. J. A. Lipa, et al., Phys. Rev. Lett. 76, 944 (1996).
4. J. Yoon and M. H. W. Chan, Phys. Rev. Lett. 78, 4801 (1997).
5. R. Li and K.Sieradzki, Phys. Rev. Lett. 68, 1168 (1992).
6. F. J.Wegner, Phys. Rev. B5, 4529 (1972).
7. A. Tyler, H.A.Cho and J.D. Reppy, J. Low Temp. Phys. 89, 57 (1992).
8. M. H. W. Chan, et al., Phys. Rev. Lett. 61, 1950 (1998).
9. D. Stauffer, M. Ferer and M. Wortis, Phys. Rev. Lett. 29, 345 (1972);
    P. C. Hohenberg, et al., Phys. Rev. B13, 2986 (1976).
10. A. B. Harris, J. Phys. C7, 1671 (1974); J. T. Chayes, et al, Phys. Rev. Lett. 57, 2999 (1986).
11. R. J. Birgeneau, et al., Phys. Rev. B27, 6747 (1983).
12. R. W. Pekala and L. W. Hrubesh, eds. Aerogel 4, Proc. of the Fourth Intern. Symp. On Aerogel, North Holland, Amsterdam [J. of Non-Crys. Solids, Vol. 186 (1995)].
13. M. Chan, N.Mulders and J. D. Reppy, Phys. Today 49, No. 8, 30 (1996).
14. G. K. S. Wong, et al., Phys. Rev. B48, 3858 (1993).
15. N. Mulders, et al., Phys. Rev. Lett. 67, 695 (1991).
16. M. Larson, N. Mulders and G. Ahlers, Phys. Rev. Lett. 68, 3896 (1992).
17. A. Weinreb and B. I. Halperin, Phys. Rev. B27, 413 (1983).
18. K. Moon and S.M. Girvin, Phys. Rev. Lett. 75, 1328 (1995).
19. B. D. Josephson, Phys. Lett. 21, 608 (1966).
20. J. Yoon, et al., Phys. Rev. Lett. 80, 1461 (1998).

Superfluid interferometry: detection of absolute rotations and pinned vorticity

E. Varoquaux[a], O. Avenel[b], P. Hakonen[c], and Yu. Mukharsky[b]

[a] CNRS–Laboratoire de Physique des Solides,
Bât. 510, Université Paris-Sud, 91405 Orsay (France)

[b] CEA–DRECAM, Service de Physique de l'État Condensé,
Centre d'Études de Saclay, 91191 Gif-sur-Yvette Cedex (France)

[c] Low Temperature Laboratory,
Helsinki University of Technology, 02150 Espoo (Finland)

The velocity circulation induced in a superfluid loop by a rotation can be detected in phase-slippage experiments. The rotation of the Earth, $\Omega_\oplus$, is presently measured to better than 1%. This experiment is the superfluid counterpart of interferometric measurements based on the Sagnac effect. The influence of pinned vorticity is discussed. Experiments on the pinning and unpinning of single, nanometric-size vortices at milliKelvin temperatures have shown that unpinning takes place by quantum tunnelling.

1. PHASE SLIPPAGE

Superfluids display quantum mechanical properties over a macroscopic scale. This paper discusses this fundamental aspect in the light of phase slippage experiments aiming at the measurement of small rotation velocities.

Slippage of the quantum-mechanical phase in superfluid ^{4}He takes place *via* vortex motion. If a vortex crosses the entire cross-section of a duct or an orifice, it causes an overall change of the phase along the duct by 2π, *i.e.* a 2π phase slip [1]. Incomplete crossings, in which the vortex gets pinned along the way, lead to partial phase slips. Phase slips are detected with a hydromechanical resonator of the Helmholtz type. When the critical flow velocity v_c is exceeded in a micro-aperture constituting the vent of the resonator, phase slips occur and cause sudden dissipation events which are detected as drops of the resonance amplitude. If the aperture size is of the order of the superfluid coherence length, as can be achieved in superfluid ^{3}He, the phase slippage process takes place by tunnelling and is non-dissipative: it constitutes the superfluid analogue of the Josephson effect [2] and is detected by its dispersive effect on the resonance motion [3].

Let us consider a resonator with *two openings*, a micro-aperture where phase slips take place and a long duct in the shape of a loop. The phase slips induce a quantised change of the state of velocity circulation in the loop: they change the loop quantum number by ± 1. The quantum of circulation in ^{4}He, κ_4, is $2\pi\hbar/m_4 = 10^{-3}$ cm^2 s^{-1}, m_4 being the mass of one ^{4}He atom.

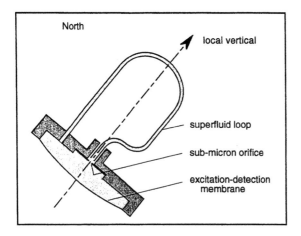

Figure 1. Sketch of the superfluid resonator equipped with a micro-aperture and a rotation sensing loop. The local vertical in Paris makes an angle of ~ 45 deg with the North-South axis of rotation of the Earth.

2. DETECTION of the ROTATION of the EARTH

If the apparatus is subjected to a rotation, the superfluid tends to remain at rest in the inertial frame of reference, much like Foucault's pendulum. In the laboratory frame, a steady flow develops in the loop which changes the apparent critical velocities for phase slippage measured along or against it. The measurement of this 'bias' of the circulation κ_b forms the basis of the detection of rotation by superfluid interferometry, proposed first by Cerdonio and Vitale [4]. The net effect of the rotation on the superfluid loop is to cause a shift of the phase of the superfluid wavefunction $\delta\varphi$ which has the same expression as that involved in a light interferometer and which is known as the Sagnac effect:

$$\frac{\delta\varphi}{2\pi} = \left(\frac{m_4}{2\pi\hbar}\right) 2\vec{\Omega} \cdot \vec{A} = \frac{\kappa_b}{\kappa_4} , \tag{1}$$

in which $\vec{\Omega}$ is the applied rotation vector and $\vec{A}$ the effective area of the loop [5,6].

The experimental set-up is sketched in Fig. 1. The hydrodynamical resonator used to detect phase slippage is fitted with a 4.0 cm^2 loop placed in a vertical plane in the laboratory. By orientating the whole cryostat about the local vertical axis, the loop can cut none of the 'rotation flux' of the Earth when its plane contains the North-South axis, or a sizable fraction of it when its perpendicular points North (or South). The magnitude of the change in the bias circulation κ_b is about one-fifth of a quantum of circulation per cm^2 of loop area for the rotation of the Earth at our latitude (~ 45 deg). The sensitivity of the method can be made very high because the measurement of v_c can be repeated many times: the quantum state of circulation in the loop is changed at each phase slip by $+1$ or -1 according to how the persistent current trapped in the loop adds to the flow driven by the diaphragm; the overall circulation thus suffers no net long term drift due to the detection scheme.

The circulation measured as a function of orientation in the laboratory is shown in Fig. 2 in units of κ_4 for various 'fixed' trapped circulation in the loop, that is the trapped

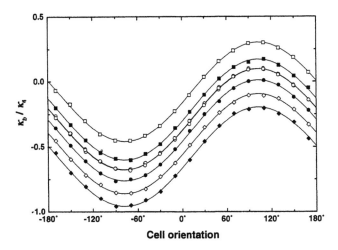

Figure 2. Variation of the phase shift in the loop as a function of the cell orientation about the local vertical axis, for various background contributions of the remanent vorticity.

circulation which is not due to the rotation of the Earth, and hence, which should not change with the cryostat orientation. Here lies a possible cause of experimental artifact: this part is not constant because stray supercurrents also have 'gyroscopic properties' of their own and tend not to follow the cryostat when it is rotated. If orientating the cryostat gives the same effect for different values of the 'fixed' bias current as in Fig. 2, then this effect is likely to be due solely to that of $\Omega_\oplus$ as described by eq. (1). The first successful operation of a superfluid gyrometer was presented at the 21st International Conference of Low Temperature Physics, in Prague in August 1996 [5,6]. Similar experiments were repeated afterwards at Berkeley [7], and are also under way in Trento [8] and in Tübingen.

These results have aroused great interest, in part because they illustrate the unifying point of view which links the law of inertia in fluids to the wave mechanics of photons, for another part, because of the hope that, since they constitute a direct analogue of the detection of magnetic flux by a superconducting interferometer, they ought to achieve a comparably high sensitivity. Such a hope might be deceived because of two factors, among others: *i)* the sampling frequency at which the measurement of v_c can be repeated is low, of tens of Hz at the time being; *ii)* the critical velocity in ^{4}He results from a nucleation process, either thermally activated above *circa* 150 mK, or due to quantum tunnelling below [9] and the resulting fluctuations on v_c are quite significant.

We have performed numerical simulations of the operation of a double-hole hydromechanical resonator with a very large rotation-sensing loop and operated at high frequencies [10], taking explicit account of the finite time of propagation along the loop of the pressure burst associated with the sudden velocity reduction caused by a phase slip. Acoustic waves propagate undamped and interfere with one another, generating pseudo-random noise which limits the resolution. It appears likely that new schemes of detection must be developed if the resonator frequency is to be considerably stepped up.

For these reasons, it would be overoptimistic, in the present state of the art [11], to forecast improvements by more than 2 orders of magnitude over the resolution presently achieved of about 1% of the rotation of the Earth in a one hour time. It must be recalled that laser gyros are presently at 5×10^{-8} of the rotation of the Earth in one hour. The main potential asset of superfluid gyros seems to be that their principle of operation makes them inherently driftless. Long term operation, over periods of several months, can be envisioned without having to resort to dedrifting techniques or other means to minimize the thermal drift as is necessary with laser gyros [12].

Another fundamental process which comes into play in superfluid gyrometers is remanent vorticity. Were it possible to obtain vortex-free superfluid, the circulation which would be found in the loop would stem solely from the rotation of the laboratory frame with respect to the distant stars: then a superfluid gyrometer would measure *absolute* rotations without having to reorient the sensing loop with respect to the rotation axis.

As crossing the λ transition always lead to remanent vorticity, created when the order parameter fluctuations are very large, we tried to reach a vortex-free superfluid state by melting solid at low temperature. These attempts have so far been unsuccessful [13]. In our cell at least, solid can not be made to melt slowly, hence vorticity cannot be prevented to form. The solid-liquid interface gets pinned on wall asperities and unpins suddenly, creating large local flow velocities. The pressure 'noise' due to this sudden unpinning has been directly observed and studied experimentally [13]. However, these findings do not rule out that, in a more suitable geometry, such a slow-melting technique could possibly work and produce a zero vorticity superfluid sample.

We also attempted to 'sweep' vorticity off the gyrometric loop by applying asymmetric flow currents in the hope of gradually pushing the vortices out of the duct into regions where the superfluid remains stagnant. Although this technique appears to clean up part of the remanent vorticity, a zero vorticity state could not be systematically produced in our apparatus.

3. VORTEX PINNING in SUPERFLUID ^{4}He

As it may well prove impossible to obtain vortex-free superfluid, the question of the stability of the remanent vorticity arises. The concept of pinned vorticity is quite prevalent in superfluids, in ^{4}He as here, in ^{3}He where it co-exists with other topological defects of the order parameter, even in neutron superfluids where it is invoked to explain the 'glitches' in the period of pulsars. However, a few hard-won experimental facts only are known, both in ^{4}He [14] and in ^{3}He [15].

We conducted a detailed study of the pinning properties in superfluid ^{4}He of a very small pinned vortex following the rather fortuitous discovery that nanoscopic dust, composed of clusters of hydrogen or air sputtered from the cell walls by an electrical discharge, could be formed *in situ*. Small particles are known to act as pinning sites for vortices [14]. They also greatly altered the pattern of formation of phase slips, causing in particular an increase by two to three orders of magnitude of the rate of 'superflow collapses' [16]. The peak amplitude charts of the resonator usually became quite complicated but, in some instances, *two reproducible* peak amplitude plateaus, instead of the usual single one, that is, two apparent critical velocity levels, could be observed. The existence of two plateaus is accompanied by other features, not as conspicuous but even more robust, which also signal the existence of a process which strongly perturbs vortex nucleation. In particular, the nucleation noise appeared greatly enhanced for one flow direction on the lower plateau compared to that for the other flow direction, which was the same as for both flow directions on the upper plateau.

As discussed in Ref. [17], these features and the existence of the two plateaus can be explained as follows. In the wake of a collapse, when the flow velocity is small, a vortex gets pinned in the immediate vicinity of the vortex nucleation site. The probability of such a pinning event is not too low if a large number of nanoscopic particles have settled in the micro-aperture. This vortex is pushed to and fro by the alternating flow about its pinning sites. It depletes the local velocity at the nucleation site for the flow direction which pushes closer to it. It unpins randomly. We have studied the lifetime of the pinned state and the velocity at which unpinning takes place at temperatures varying from 14 to 44 mK. The unpinning velocity is found to be independent of temperature over this temperature range but to display fluctuations which show that vortex unpinning is governed by a stochastic process. These fluctuations are also found to be T-independent.

The lack of T-dependence rules out thermal activation of the unpinning process, but is not by itself sufficient proof that quantum tunnelling takes place [18]. Such quantum processes have been invoked in other physical situations, notably for the pinning of vortices in high T_c materials, for the nucleation of vortices in superfluid ^{4}He [9,19], for the nucleation of cavitation bubbles in helium at low temperature [20], etc...

In the present situation, the pinned vortex does behave as a quantum object [21] and a rather strong case for its escape from its pinning site by quantum tunnelling can be made. It can indeed be deduced from the measurements [17] that the vortex under study is about 4 nanometres in length and that it sits in a potential well about 5 K in depth. The fundamental mode of vibration of such a vortex (the Kelvin mode) has an energy of 180 mK and a zero point fluctuation amplitude of 1 Å . At the temperature of the experiment, 44 mK and below, thermal activation is far less probable than tunnelling.

4. DISCUSSION

Although the foregoing discussion of phase slippage and pinning experiments in superfluids is very abbreviated, it can already be seen to imply that quantum mechanics plays a rôle at different levels in superfluid helium.

At the mesoscopic level, the nucleation of vortices has been found to take place by quantum tunnelling below $\sim$ 150 mK [9]. Unpinning of nanometric vortices also takes place by quantum tunnelling [17], as discussed here.

The quantisation of circulation, which implies that of vortices, constitutes experimental evidence for the existence of a complex order parameter which extends over macroscopic distances. The state of zero circulation in the inertial frame [22,23], the Landau state, is the equivalent of the Meissner effect in superconductors. The measurement of circulation by phase slippage described here relies on the quantisation of the states of circulation in the loop. This set of states is used as a quantum register to keep track of the sequence of phase slips, enabling the measuring process to be repeated any number of times.

Finally, as discussed in [6], the de Broglie relation comes into play to establish the link between the Sagnac effect and the purely hydrodynamical effect used in the superfluid gyrometer: the phase of the superfluid order parameter is shifted by the applied rotation in exactly the same manner as that of a coherent beam of particles. The rotation-sensing experiment described here yields an illustration of how the coherent-state representation of the superfluid order parameter really works in the laboratory and justifies further Uhlenbeck's saying that 'one must watch like a Hawk to see how Planck's constant comes into the hydrodynamics' [24].

REFERENCES

1. W. Zimmermann, Jr., Contemp. Phys. 37 (1996) 219.
2. P.W. Anderson, Rev. Mod. Phys. 38 (1966) 298.
3. For a review, see E. Varoquaux and O. Avenel, Physica, B197 (1994) 306.
4. M. Cerdonio and S. Vitale, Phys. Rev. B29 (1984) 484.
5. O. Avenel and E. Varoquaux, Czech. J. Phys. 46–S6 (1996) 3319.
6. O. Avenel, P. Hakonen and E. Varoquaux, Phys. Rev. Lett. 78 (1997) 3602.
7. K. Schwab, N. Bruckner and R.E. Packard, Nature, 386 (1997) 585.
8. R. Dolesi, M. Bonaldi et al., Czech. J. Phys. 46–S1 (1996) 33.
9. G.G. Ihas, O. Avenel et al., Phys. Rev. Lett. 69 (1992) 327.
10. Yu. Mukharsky, O. Avenel and E. Varoquaux, to be published in J. Low Temp. Phys..
11. O. Avenel, P. Hakonen and E. Varoquaux, J. Low Temp. Phys. 110 (1998) 709.
12. G.E. Stedman, Rep. Prog. Phys. 60 (1997) 615.
13. P. Hakonen, O. Avenel and E. Varoquaux, J. Low Temp. Phys. 110 (1998) 503.
14. W.I. Glaberson and R.J. Donnelly, Prog. Low Temp. Phys. IX. ed. D.F. Brewer, (North-Holland,Amsterdam, 1986) Ch. 1.
15. P.J. Hakonen, K.K. Nummila, J.T. Simola, L. Skrbek and G. Mamniashvili, Phys. Rev. Lett. 58 (1987) 678; M. Krusius, J.S. Korhonen, Y. Kondo and E.B. Sonin, Phys. Rev. 47 (1993) 15113.
16. O. Avenel, M. Bernard, S. Burkhart and E. Varoquaux, Physica, B 210 (1995) 215 and references therein; E. Varoquaux, O. Avenel, M. Bernard and S. Burkhart, J. Low Temp. Phys. 101 (1995) 821. These sudden decreases of the superflow in the micro-aperture are now thought to be due to vortices re-entring the aperture.
17. E. Varoquaux, O. Avenel, P. Hakonen and Yu. Mukharsky, to be published in Physica.
18. G.E. Volovik, JETP Lett. 65 (1997) 217.
19. U.R. Fischer, Phys. Rev. B 58 (1998) 105.
20. H. Lambaré, P. Roche et al., Eur. Phys. J. B2 (1998) 381.
21. A.L. Fetter, Phys. Rev. 138A (1965) 429.
22. J.D. Reppy and C.T. Lane, Phys. Rev. 140 (1965) A106.
23. G.B. Hess and W.M. Fairbank, Phys. Rev. Lett. 19 (1967) 216.
24. quoted by S.J. Putterman, Superfluid Hydrodynamics, North-Holland, Amsterdam, 1974, in the Preface.

Quantum Coherence and Decoherence - ISQM - Tokyo '98
Y.A. Ono and K. Fujikawa (Editors)
© 1999 Elsevier Science B.V. All rights reserved.

Real-time Observation of Dynamics of Vortices in High-temperature Superconductors by Lorentz Microscopy

A. Tonomura[a,c], T. Matsuda[a], H. Kasai[a], O. Kamimura[a], K. Harada[a,c]
J. Shimoyama[b,c], K. Kishio[b,c], and K. Kitazawa[b,c]

[a]Advanced Research Laboratory, Hitachi, Ltd.
Hatoyama, Saitama 350-0395, Japan

[b] Department of Applied Chemistry, University of Tokyo
Tokyo 113-0033, Japan

[c]CREST, Japan Science and Technology Corporation (JST)

The onset of vortex motion in high-temperature Bi-2212 superconductors was directly observed using the 300 kV field-emission transmission electron microscope. We found various forms of vortex motion depending on magnetic field H and temperature T: Below the transition temperature of dynamic vortex motion, T_t, which was around 20 K, the vortices migrated slowly. Above T_t, the vortices moved in various forms of plastic flow depending on H and T. We interpret that the migration below T_t is due to the collective pinning of a large number of oxygen defects, and that above T_t the effect of larger and sparser defects becomes dominant instead of oxygen defects.

1. INTRODUCTION

A dissipation-free current can only be achieved in a superconductor when the vortices, produced by the magnetic field H and penetrateing the superconductor, are pinned down against the current-induced force. In order to attain high critical current J_c, much effort has been made to clarify the microscopic mechanism of the vortex depinning, *i.e.*, how vortices are depinned and start moving. Simulations and macroscopic measurements have predicted that vortices in *random* disorder begin to flow in filamentary flows[1-3] and that vortices in *regular* disorder begin to flow in various plastic flow forms.[4]

Lorentz microscopy using a field-emission electron beam opened a way to directly observe the motion of individual vortices[5]: Plastic motions have been observed in Nb thin films having artificial disorder.[6,7] Vortices in high-temperature superconductors have also been recorded in Lorentz micrographs with exposure time as long as 30 seconds, but not on video, since the images had low contrast and were very dark due to the weak magnetic fields of the vortices.

This resulted from the fact that the magnetic radius of vortices, *i.e.*, penetration depth λ in high-temperature superconductors is much larger than that in Nb, and that the film thickness is smaller than λ. As another way to dynamically observe vortices, scanning Hall probe microscopy has been developed, but the time resolution is still limited to one second.[8]

In the present experiment[9], we improved Lorentz microscopy and directly observed vortex motion at the depinning threshold region through a video system.

2. EXPERIMENTS

Vortices in high-temperature superconductors became dynamically observable in the following way: The image contrast was enhanced by increasing a film thickness to shorten the effective λ, and by increasing the tilting angle α of the film (see Fig. 1) to obtain a larger phase shift[10] of electrons due to a vortex. Simultaneously, the image brightness was increased by using a beam intensity with one order of magnitude larger than previously used one so that the vortices would be seen more clearly.

Thin film samples of Bi-2212 ($T_c = 85$ K) were prepared by peeling them off a crystal grown by the floating-zone technique.[11] Cleaved samples less than 10 μm in thickness often had thin platelets attached to them at their edges. We selected the platelets with a thickness more than 0.2 μm and a uniform area over 100 μm $\times$ 100 μm for the observation. In addition, we selected special platelets in which bend contour lines due to the Bragg reflections of incident electrons would not hide the vortex images.

When a Lorentz force was applied to vortices by increasing (or decreasing) H, vortices located near the edge of a sample film moved towards the inside (or outside) of the film. We observed how the vortices moved in the regions, 0 G $\leq H$ $\leq$ 45 G and 7 K $\leq T \leq$ 50 K, where individual vortices were dynamically observable. We found that their behavior changed markedly depending on H and T.

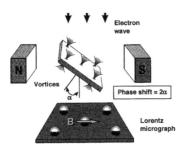

Figure 1. Experimental method to observe vortices by Lorentz microscopy. To obtain high-contrast and bright images, the film thickness, the tilting angle α, and the intensity of incident electrons were increased as much as possible.

A. Migration

The dynamic behavior of vortices above T_t was completely different from that below T_t which ranged from 17 K to 25 K depending on samples: at $T < T_t$ all the vortices migrated slowly almost at a uniform speed, maintaining the relative position of the vortices. Two video frames at 1s interval shown in Fig. 2a, b indicate that the vortex speed is 1.5 μm at $T = 20$ K and $dH / dt = 0.008$ G/s. The speed decreased rapidly as T decreased, indicating a microscopic image of the well-known strong pinning in Bi-2212 at low temperatures. The speed increased as dH / dt increased.

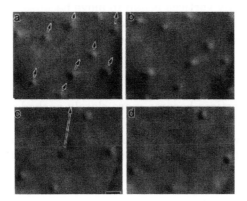

Figure 2. Lorentz micrographs showing the motion of extremely sparse vortices ($H < 1$ G) at the depinning threshold when the force F is exerted on vortices in a Bi-2212 film by changing H at $dH / dt = 0.008$G/s. The motion at low temperatures differs completely from that at high temperatures: Figures 2a and 2b are two Lorentz micrographs at 1s interval showing vortex migration almost in parallel with the speed 1.5 μm/s ($T = 20$ K). Figures 2c and 2d, before and after a sudden hopping of more than 50 μm, respectively, ($T = 30$ K) (Scale bar = 2 μm).

We conclude that this migration is due to the fact that in high-temperature superconductors vortices can be pinned at extremely small, densely-distributed oxygen defects. Since a single vortex penetrating a 2000 Å-thick film may be collectively pinned by more than 100 oxygen defects, vortices would move apparently smoothly when thermally activated.

B. Hopping

At $T > T_t$, pinning centers began to appear and vortices were no longer able to migrate (Fig. 2c). The vortices trapped at the pinning centers suddenly started to move away from them by hopping (Fig. 2d), and the vacated pinning centers were soon occupied by newly arriving hopping vortices. Since the hopping speed was so high that the vortex image on the video screen looked as if it were blinking on and off.

The reason why vortex motion changed at T_t is conjectured as follows: The collective pinning due to tiny oxygen defects decreases rapidly with the increase in T, and the pinning of larger and sparser defects, which have not yet been identified, becomes dominant instead.

Even when vortices became denser ($H > 1$ G) and the vortex-vortex interaction became stronger, vortices remained migrating until H became 45 G.

At $T > T_t$, we found several different forms of vortex motion when we increased H. We describe these forms mainly at $T = 30$ K in the following.

C. Random flow

At 1 G $< H < 5$ G, many vortices began to move because they were pushed by other hopping vortices, but they only moved for a short distance because they were interrupted by surrounding vortices. Overall, many vortices moved rather randomly.

D. Filamentary flow

At 5 G $< H < 7$ G, vortices became more closely packed and tended to move simultaneously along a filament (Fig. 3a, b). A filament, which usually consisted of a unit of several to a few tens of vortices, continued to flow for a few seconds to a few minutes. Vortices in the filament simultaneously hopped from one site to the adjacent one.

Such filamentary flow was predicted to occur at the depinning threshold in a two-dimensional (2D) film[1-3] where vortices favor a uniform distribution and the vortex lattice becomes easy to shear. Filamentary flow appeared only in the narrow regions of H and T, because our samples were considered to be between 2D and 3D films. We also found cases of two filaments joining each other.

E. River flow

At 7 G $< H < 45$ G, the vortex-vortex interaction became stronger, and vortices at rest formed domains of lattices approximately 10 μm in size. When vortices began to move, however, they did not maintain the lattice form, but some of them flowed in rivers[6] (Fig. 3c, d) though the rivers were not stationary but intermittently changing their locations and widths from time to time. It seemed that a single filament dragged adjacent lines forming rivers. A river often looked like consisting of a few lines of vortices and flowing along each line independently.

The observation of individually moving vortices at H higher than 45 G was difficult because of dimmer image contrast. To overcome this difficulty, we increased T instead of H so that the vortex-vortex force became larger than the pinning force.

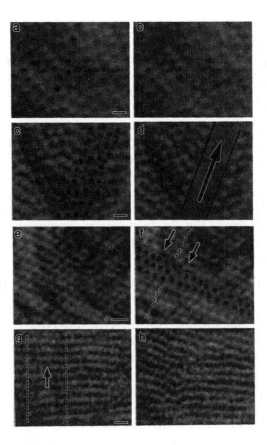

Figure 3. Lorentz micrographs showing various regimes of plastic flow of vortices at the depinning threshold when the force F is applied to vortices in a field-cooled film ($H > 1$ G). The flow forms vary greatly and depend on H and T: **a, b,** "red" vortices forming filamentary flow (before and during the flows, respectively) ($H =$ 6 G, $T = 40$ K). **c, d,** river flow consisting of intermittently flowing "red" vortices (before and during the flow, respectively) ($H = 10$ G, $T = 30$ K). **e, f,** before and after production of a dislocation due to a phase slip between two domains, respectively ($H = 20$ G, $T = 50$ K). **g, h,** before and after reorganization of vortex lattices due to phase slips between the displaced domain and the adjacent domains, respectively ($H = 20$ G, $T = 50$ K) (Scale bars = 2 μm).

F. Distorted-lattice flow

At $T = 50$ K, stationary vortices formed a single and rigid lattice over the entire field of view at $H > 15$ G. When a force was exerted on the lattice, it appeared to be divided into multiple domains each tending to move separately due to the spatial difference in pinning force. As a result, the lattice was deformed with time, changing its form in various ways: For example, two domains were

298

suddenly displaced by a short distance with each other (phase slips[12]) along the flow direction, producing edge dislocations between them at one time and annihilating them at a later time. An example is shown in Figs. 3e, f, where only the right domain was first displaced downwards producing an extra half plane in the left domain (the displacement was not the same but increased downwards from zero to one lattice spacing), and then the left domain was displaced downwards similarly returning to the perfect crystal form.

In other example (Fig. 3g), the lattice domain indicated by dotted lines tended to move upwards in the arrow direction, then it was suddenly displaced with other domains, and finally the whole vortex configuration was reorganized as shown in Fig. 3h, where the lattice orientation in the displaced domain changed. But after a few seconds, the other domains also moved, thus returning to the original perfect lattice form of Fig. 3g. In summary, vortices proceeded in units of domains, sometimes making slips between domains and sometimes changing lattice orientations in domains.

3. CONCLUSIONS

By using Lorentz microscopy, we directly observed the onset of vortex motion in high-temperature superconductors, which had been predicted only by simulations or by macroscopic experiments. We found the slow migration of vortices at low temperatures, which was considered to be due to the collective pinning effect of vortices by a large number of atomic defects such as oxygen defects. This pinning effect decreased rapidly with the rise in temperature, and vortices began to hop among larger and sparser defects above T_t, the transition temperature of dynamic vortex motion. In this region, we obtained an evidence of different types of plastic flow.

REFERENCES

1. H. J. Jensen, A. Brass, A. J. Berlinsky, Phys. Rev. Lett. **60** (1988) 1676.
2. N. Grønbech-Jensen, A. R. Bishop, D. Dominguez, Phys. Rev. Lett. **76** (1996) 2985.
3. C. J. Olson, C. Reichhardt, F. Nori, Phys. Rev. **B56** (1997) 6175.
4. C. Reichhardt, C. J. Olson, F. Nori, Phys. Rev. Lett. **78** (1997) 2648.
5. K. Harada et al., Nature **360** (5 November 1992) 51.
6. T. Matsuda et al., Science **271** (8 March, 1996) 1393.
7. K. Harada et al., Science **274** (15 November 1996) 1167.
8. A. Oral et al., Phys. Rev. Lett. **80** (1998) 3610.
9. A. Tonomura et al., to be published in Nature.
10. J. E. Bonevich et al., Phys. Rev. Lett. **70** (1993) 2952.
11. Y. Kotaka et al., Physica C **235-240** (1994) 1529.
12. L. Balents, M. P. A. Fisher, Phys. Rev. Lett. **75** (1995) 4270.

Quantum Coherence and Decoherence - ISQM - Tokyo '98
Y.A. Ono and K. Fujikawa (Editors)
© 1999 Elsevier Science B.V. All rights reserved.

Biprism interferences of He$^+$-ions

Franz Hasselbach and Uwe Maier

Universität Tübingen, Institut für Angewandte Physik, Auf der Morgenstelle 10,
D-72076 Tübingen, Germany

This paper is to describe the first realization of a biprism interferometer for ions. Due to the short wavelength of the ions an extremely high mechanical and electrical stability of its components was a crucial prerequisite. A 'supertip' field-ion source cooled down to 77 K illuminates an electrostatic biprism coherently. The primary interference pattern behind the biprism is magnified by a dual quadrupole lens, intensified by a channelplate image intensifier, fiberoptically transferred to a state of the art slow scan CCD-camera and evaluated by image processing in a personal computer. The magnification of the quadrupole is chosen very large ($\sim$5000 – 10 000) perpendicular to the fringe direction in order to resolve the fringes at the input channel plate. A relatively small magnification in fringe direction ($\sim$200) enhances the signal to noise ratio of the fringe pattern.

1. INTRODUCTION

While diffraction of atoms has been realized [1] shortly after L. de Broglie's hypothesis that matter is associated with a wave, the first atom interferometers were put into operation only at the beginning of this decade [2]. One cause for this long delay has been the missing technology indispensable to develop optical components for neutral particles. On the other hand, powerful optics for charged particles has been available for more than 50 years. But, due to the lack of a highly stable design no ion interferometer has been developed so far, in spite of the fact that ion interferometers were expected to solve similar important questions as atom interferometers do, namely fundamental tests of quantum mechanics and relativity, multiparticle interferometric effects [3], development of inertial sensors with unprecedented sensitivity [4]. Additionally, *charged atoms* couple to the electromagnetic field. Due to the internal structure of ions an ionic Aharonov-Bohm experiment will be the crucial test whether the phase shift caused by electromagnetic potentials depends on internal degrees of freedom of ions or is the same as for structureless 'point' particles as electrons.

2. EXPERIMENT

The experimental set-up of the interferometer for ions is presented in Fig. 1. It uses wavefront division by an ion optical biprism whereby a true physical separation of the coherent ion wave packets in space is achieved. In essence, it is a modified type of our electron interferometer [5], which has proved to be extraordinarily rugged, insensitive to vibrations, and electronically stable. The requirements which must be met by an ion interferometer with respect to mechanical stability and insensitivity to vibrations are even more stringent than in atom interferometers. While in atom interferometers thermal

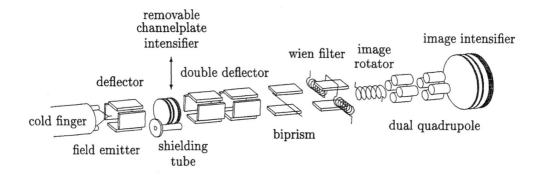

Figure 1. Schematical set-up of the ion interferometer

beams with wavelengths on the order of Å are used, the wavelengths of our ions in the energy range of some keV are a fraction of a pm only. An advantageous feature of the field-ion source incorporated in the instrument is, that it can be switched between ion and electron emission. Since all optical components of the interferometer are electrostatic, this gives us the possibility of a rough prealignment of the interferometer optics with a high intensity beam of field emitted electrons.

The very small part of the current of ions which is emitted into the tiny angle where the interference fringes are formed ($\sim 10^{-7}$), leads to exposure times as long as 30 minutes. The primary interference pattern is magnified by two cascaded electrostatic quadrupole lenses and intensified by two cascaded channel plates. Fringe pattern is transfered by a tapered fiberoptic from the fluorescent screen to a cooled slow scan CCD-camera.

2.1. Ion Source

In order to maximize the coherent flux of He^+-ions emitted by the field ion source, we use a specially treated 'supertip' [7], cooled down during operation to 77 K. The 'supertip', i.e., a protrusion consisting of a small number of tungsten atoms on a $\langle 111 \rangle$-oriented tungsten field emission tip, which has a relatively large radius of curvature of the apex, is prepared in-situ in the interferometer. During preparation, its emission pattern can be controlled on the screen of a channel plate image intensifier, which is inserted temporarily between ion source and interferometer (Fig. 1). After preparation, the source is emitting into a single spot of angular diameter of about 1° only. The brightness of our present source is rather low ($10^3 \ A/cm^2 \cdot sr$ at 77 K). More details about the ion source are given in [8]. In order to achieve a substantially improved brightness as a next step, we will lower the temperature of the tip to about 10 K.

Besides brightness, to achieve stable emission from one single site with atomic dimensions during the whole exposure time of about 30 minutes and a lifetime of the 'supertip' of at least 60 minutes turned out to be greatest problems. More than 100 differently processed $\langle 111 \rangle$-oriented and $\langle 100 \rangle$-oriented tungsten field emission tips were evaluated experimentally. The results given in Fig. 2 were obtained with a ⟨111⟩-oriented tip with a total lifetime of about 2 hours.

2.2. Image pickup and processing

An ideal detector for our purpose has high detection efficiency and detective quantum efficiency (DQE) for single ions. Exposure times of about 1 hour, live observation of incoming single ions and simultaneous presentation of the events integrated in time on the screen of the image processing system should be possible. The detective quantum efficiency describes the reduction of the signal-to-noise ratio of a detector system and characterizes its quality. For our present system of two cascaded channelplates the DQE is limited to ≤ 0.5 by the pulse-height distribution caused by the statistically varying scintillation intensities of the fluorescent screen. Standard acquisition-schemes, e.g., on-chip integration do not overcome this low value of the DQE.

Fortunately, for a small number of events per second (low ion current), a normalization technique can be applied which improves the DQE of our detection system [6] to about 0.8 for low mass ions with an energy of 2 keV. This image pick-up technique is applicable only if within the the exposure time of a frame, multiple events in a single pixel can be excluded (about 400 ms for our system). It works as follows: Due to the statistics of the multiplication process in the channelplates the intensity of single event scintillations varies. Usually it exposes not a single, but a small number of adjacent pixels of the CCD-chip with different intensities. The pixel intensities are digitized and stored in the frame grabber. The frames are corrected by subtraction of a mean dark current frame. Then, the intensity stored in every single pixel has to pass an adjustable threshold which suppresses remaining dark-current and read-out noise. Only real signals survive this process and contribute to the frame. For each single event, the pixel with maximum exposure is localized and normalized to the minimal discernible intensity step: one grey level. The localized and normalized pixels are written into the corresponding pixels of a second frame memory. This process is repeated, until a frame with an exposure time of, say 20 minutes is accumulated in this second memory. This processing completely supresses dark current noise, signal independent noise and pulse height statistics of the channel-plates for bright single-event scintillations (which can be achieved by high channel-plate gain). Therefore, in addition to the improved DQE of about 0.8, dynamic range and spatial resolution of the frames are superior to those taken with the standard procedure.

3. RESULTS/DISCUSSION

In order to find the best type of field ion emitter and to optimize the brightness and stability of the ion source more than 100 tips different in orientation were examined. We succeeded in recording diffraction at both edges of the biprism filament and biprism interference fringes after some additional tests of the stability of the interferometer. The results are shown in Fig. 2 below. The bright round area in the middle of the frames is a shadow of a diaphragm located near the field ion emitter and is either caused by beam ions neutralized near the field emitter tip or ultraviolet recombination radiation. In reality, this area is much brighter than is seen in the frames. Within the bright area ion interferences are impossible to be recovered even with our normalization technique. Only by aligning the interferometer in such a way that the ion fringes are formed outside of the

302

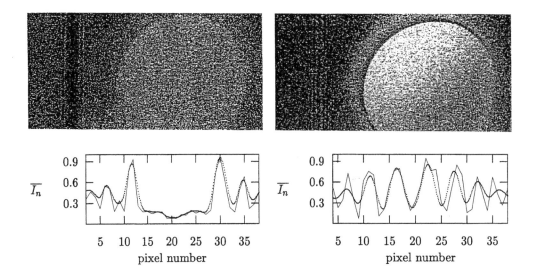

Figure 2. Edge-diffraction and biprism interferences with 3 keV-Heliumions. The lines are raw data summed-up over 50 pixel-lines, the dotted linescans are low-pass filtered and spline-interpolated data. $\overline{I_n}$ is the normalized mean intensity.

bright area they can be detected.

The first ion fringe patterns presented here are rather noisy. In order to develop this ion interferometer into a reliable research tool, the stability of the ion source and its brightness has to be improved substantially. We hope to reach this goal by going from emitter temperatures of 77 K down to about 10 K and by using sligthly higher beam voltage ($\sim$15kV).

Acknowledgement
We would like to thank the "Deutsche Forschungsgemeinschaft" for constant support.

REFERENCES

1. I. Esterman, O. Stern, Z. Phys. 61 (1930), 95.
2. O. Carnal, J. Mlynek, Phys. Rev. Lett. 66 (1991), 2689; D. W. Keith, Ch. R. Ekstrom, Q. A. Turchette, D. E. Prichard, Phys. Rev. Lett. 66 (1991), 2693.
3. F. Shimizu, in 'Atom Interferometry', P. R. Berman ed., Academic Press (1997), p.153.
4. J. F. Clauser, Physica B 151 (1988), 262-272; T. L. Gustavson, P. Bouyer, M. A. Kasevich, Phys. Rev. Lett. 78 (1997), 2046; A. Lenef et al., Phys. Rev. Lett. 78 (1997), 760.
5. F. Hasselbach, Z. Phys. B — Condensed Matter 71 (1988), 443-449; F. Hasselbach, M. Nicklaus, Phys. Rev. A 48 (1993), 143-151.
6. K.-H. Hermann, D. Krahl, Adv. Opt. El. Micr. 9(1984), 1-64.
7. P. Schwoebel, G. R. Hanson, J. de Physique Coll. C2 47 (1986), 59-66.
8. F. Hasselbach, U. Maier, in Quantum Coherence and Decoherence, K. Fujikawa, Y.A. Ono eds. 1996 Elsevier Science.B.V., p. 69-72.

Quantum Coherence and Decoherence - ISQM - Tokyo '98
Y.A. Ono and K. Fujikawa (Editors)
© 1999 Elsevier Science B.V. All rights reserved.

Neutron Spin Precessions on Quasibound States through Magnetic Fabry-Pérot Films in Tunneling Region

N. Achiwa[a], M. Hino[b], T. Ebisawa[b], S. Tasaki[b], T. Kawai[b], D. Yamazaki[c]

[a]Department of Physics, Kyushu University, Fukuoka, 812-8581, Japan

[b]Research Reactor Institute, Kyoto University, Osaka, 590-0494, Japan

[c]Department of Nuclear Engineering, Kyoto University, Kyoto, 606-8501, Japan

The shift of Larmor precessions of neutron spin transmitted through multiply coupled magnetized [permalloy45(PA)-germanium(Ge)]^n-PA Fabry-Pérot resonators for n=2 and 10 in tunneling region, has been measured by means of a neutron spin interferometer. Across the first bound state in tunneling region as a function of the incident neutron angle an amplitude of sinusoidal shift of Larmor precessions through the coupled Fabry-Pérot resonator seems to be proportional to the number of coupled resonator, though the tunneling probability and the transmission half width of the bound state seems to keep constant values with increasing it. The shift number of Larmor precession for n=10, is 1.64 turns which is well reproduced by the theoretical phase difference of ↑ and ↓ spin neutron wave functions based on one-dimensional Schrödinger equation.

1. INTRODUCTION

We have established neutron spin echo interferometry[1] using a neutron spin echo interferometer in KURRI and JAERI, in order to observe a quantum spin precession which is given by the phase difference of ↑ and ↓ neutron wave function. We have measured tunneling phase shift of neutron wave function by a spin precession of neutron through magnetic thin films of Permalloy45($Fe_{55}Ni_{45}$) abbreviated as PA with the thicknesses of 200, 300, 400Å[2] by Larmor precession, as well as the phase shift due to the bound state of ↑ spin function through magnetic Fabry-Pérot thin films of PA(150Å) -Ge(800Å)-PA(150Å) and [PA(100Å) -Ge(800Å)]^2-PA(100Å)[3], in tunneling region by the spin precession. The bound time of ↑ neutron spin in the tunneling potential can be estimated from the phase shift of ↑ neutron, $\Delta\phi$ across the bound state, which corresponds about 9.3×10^{-9}sec using the relation $\delta t=\Delta\phi\lambda_\perp/v_\perp=\frac{\Delta\phi hm}{\hbar^2 k_{Ge}^2}$[3]. Here the wave number k_{Ge} is the perpendicular wave number component to the surface plane in the Ge layer as $k_{Ge} = \frac{\sqrt{E-2mU_{Ge}}}{\hbar}$, where $E = \frac{(\hbar k_\perp)^2}{2m}$ the perpendicular component of the incident neutron wave vector, U_{Ge} the nuclear potential 94neV. The tunneling potential for PA, U_{PA}+B for ↑ spin is 308neV, here the magnetic induction, B is 1.4T(84neV) in the PA layer.

On the other hand, the phase shift through a various type of magnetic multilayer film was interpreted as a phase difference due to the dynamical diffraction effect[4]. These extra spin precessions through multilayer magnetic films are well reproduced by the sim-

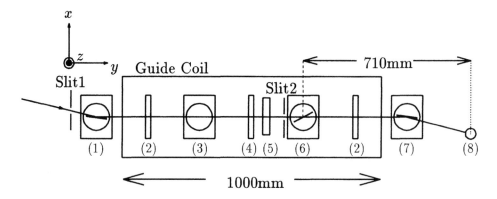

Figure 1. Schematic layout of the neutron spin interferometer at JRR-3M. (1) Polarizer, (2) $\pi/2$ spin flipper coil, (3) Precession coil I(PC1), (4) π spin flipper coil, (5) Accelerator coil(sub-coil), (6) Precession coil II(PC2)& a Fabry-Pérot magnetic film(sample) on a goniometer, (7) Analyzer, (8) ^{3}He detector.

ulation between the phase difference of ↑ and ↓ spin wave function by solving Schrödinger equation based on one dimensional multi-well potentials, whose wave vectors are the perpendicular component of the incident neutrons against the layer surface. But it is still not clear whether the measured neutron tunneling phase can be interpreted as the time measured by the group velocity or not. If we could get a longer bound time of $\sim 10^{-7}$sec in the tunneling barrier, we could have a chance to observe it by a Time-Of-Flight(TOF) method. Let us consider a magnetic Fabry-Pérot film of a coupled multi-well type potential for ↑ neutron spin and a smaller flat potential for ↓ spin.

2. EXPERIMENTS AND RESULTS

The experiments were carried out with transverse NSE interferometer inserting the magnetic Fabry-Pérot multilayers on a goniometer in one of the Larmor precession fields as shown in Fig. 1. The wavelength resolution with polarizer was 12.6±0.44(FWHM)Å and the divergent angle was 0.7×10^{-3} rad. The shift of neutron spin echo precession through the magnetic Fabry-Pérot multilayer was measured as a function of the incident angle of the precessing neutron.

As shown in Fig. 2 the transmission probabilities through [PA(200Å)-Ge(400Å)]^{10}-PA(200Å) the multiple Fabry-Pérot film for ↑ and ↓ spin of neutrons were measured where the bound state angle of ↑ spin is higher than that for the ↓ spin. This is due to the difference of wave number $k_\pm$ for them. The observed transmission probabilities are well reproduced by the calculated ones, $T_\pm$ in eq.(2) taking into account the effect of the wavelength resolution. The shift of Larmor precession versus incident angle was measured across the bound state condition. Around the bound state angle an anomalous oscillatory phase shift of Larmor precession was observed as shown in Fig. 3.

At the bound condition the ↑ spin neutron is anomalously transmitted through the tunneling potential and recombines with the transmitted ↓ spin giving extra-Larmor precession. The extra Larmor precession can be calculated by the equ.(3) as follows. Then,

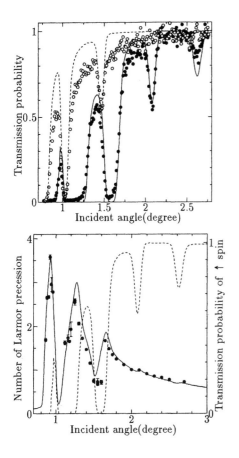

Figure 2. Transmission probabilities of ↑ (•) and ↓ (○) spin neutrons through [PA(200Å)-Ge(400Å)]¹⁰-PA(200Å) Fabry-Pérot multilayer as a function of incident angle for the wavelength of 12.6Å. The solid and broken lines are the calculated transmission probabilities based on one-dimensional Schrödinger equation with periodic rectangular potentials for ↑ and ↓ spins, respectively.

Figure 3. The shift of NSE signals as well as transmission probability through [PA(200Å)-Ge(400Å)]¹⁰ -PA(400Å Fabry-Pérot multilayer as a function of incident angle for the wavelength of 12.6Å. The solid line is the calculated Larmor precession as well as the broken line the calculated transmission probability across the bound state angle as a function of incident angle, on the basis of solving Shrödinger equations for one dimensional periodic potential of rectangular type for ↑ and ↓ spins.

the extra shift of Larmor precession as a function of the incident angle are well reproduced as shown in Fig. 3.

3. SIMULATION AND CONCLUSION

We have carried out the simulation of the spin precession through magnetic Fabry-Pérot film of multi-well type by changing the incident neutron angle across the critical angle of total reflection with various thickness and various number of well. By solving Schrödinger equation on the one dimensional periodic rectangular potential for ↑ and ↓ spin neutrons, the wave function in the j-th multilayer in the state of Larmor precession is given by

$$\psi_{j\pm}(r) = A_{j\pm} \cdot e^{ik_{j\pm}r} + B_{j\pm} \cdot e^{-ik_{j\pm}r}. \tag{1}$$

If we represent $\psi_{j\pm}(r)$ by a vector $\begin{pmatrix} A_{j\pm} \\ B_{j\pm} \end{pmatrix}$, at the first interface $\psi_{0\pm}(r_0) = \begin{pmatrix} 1 \\ r_\pm \end{pmatrix}$ and at the last interface $\psi_{j\pm}(r_j) = \begin{pmatrix} t_\pm \\ 0 \end{pmatrix}$, using the reflection and the transmission coefficients $r_\pm$ and

$t_\pm$. These are related by a transfer matrix $\boldsymbol{M}_{j\pm}$[5] as

$$\begin{pmatrix} 1 \\ r_\pm \end{pmatrix} = \boldsymbol{M}_{1\pm}\cdots\boldsymbol{M}_{j\pm}\begin{pmatrix} t_\pm \\ 0 \end{pmatrix} = \begin{pmatrix} M_{11\pm} & M_{12\pm} \\ M_{21\pm} & M_{22\pm} \end{pmatrix}\begin{pmatrix} t_\pm \\ 0 \end{pmatrix}. \tag{2}$$

Solving Eq. (2), one can obtain transmission coefficients as $t_\pm = \sqrt{T_\pm}e^{i\Delta\phi_\pm}$, where $T_\pm$ are the transmission probabilities. Then we get the extra Larmor precession, $\Delta\phi = \Delta\phi_+ - \Delta\phi_-$ through a magnetic multilayer.

$$\langle S_x \rangle = \hbar\cos(\Delta\phi_+ - \Delta\phi_-)\frac{\sqrt{T_+T_-}}{T_+ + T_-}, \tag{3}$$

$$\langle S_y \rangle = -\hbar\sin(\Delta\phi_+ - \Delta\phi_-)\frac{\sqrt{T_+T_-}}{T_+ + T_-}, \tag{4}$$

$$\langle S_z \rangle = \frac{\hbar}{2}\frac{T_+ - T_-}{T_+ + T_-}. \tag{5}$$

Then, the amplitude of the NSE signal, P can be expressed by the transmission probabilities $T_\pm$ as

$$P = P_0\frac{|\langle S_{x,y} \rangle|}{|\langle S_{x,y} \rangle| + \frac{1}{2}|\langle S_z \rangle|} = P_0\frac{4\sqrt{T_+T_-}}{4\sqrt{T_+T_-} + |T_+ - T_-|}, \tag{6}$$

where P_0 is the NSE signal in the absence of the magnetic layer. The simulations of the transmission probability $T_\pm$ and the extra Larmor precession, $\Delta\phi = \Delta\phi_+ - \Delta\phi_-$ were carried out by taking into account of the resolution of the wavelength. It is concluded that the number of spin precession for the first order bound state in a tunneling barrier is proportional to the number of well, while the transmission probability as well as the transmission wavelength width for the first order bound state in a tunneling region keeps constant values with increasing the number of well. When the number of well is 5 and 10, with the gap of Ge film 400Å, and with the barrier thickness of PA 200Å, the shift number of the spin precession across the first order bound state in the tunneling barrier, is about 0.85 and 1.64 turns, respectively which corresponds to the bound time of 4.6 and 9.0×10^{-8}sec, respectively. For the n=10 film, the observed shift spin precession is well reproduced by the simulation. The estimated bound time of $\sim10^{-7}$sec in the Fabry-Pérot film could be measured by the TOF neutron spectrometer near future.

REFERENCES

1. N. Achiwa, M. Hino, S. Tasaki, T. Ebisawa, T. Akiyoshi and T. Kawai, J. Phys. Soc. Jpn. **65** (1996) Suppl. A 183. T. Ebisawa *et al.*, *ibid.* 66.
2. M. Hino, N. Achiwa, S. Tasaki, T. Ebisawa, T. Akiyoshi and T. Kawai, J. Phys. Soc. Jpn. **65** (1996) Suppl. A 281.
3. M. Hino, N. Achiwa, S. Tasaki, T. Ebisawa, T. Akiyoshi and T. Kawai, Physica B **241-243** (1998) 1083.
4. N. Achiwa, M. Hino, S. Tasaki, T. Ebisawa, T. Akiyoshi and T. Kawai, Physica B **241-143** (1998) 1068.
5. S.Yamada, T.Ebisawa, N.Achiwa, T.Akiyoshi and S.Okamoto, Annu. Rep. Res. Reactor Inst. Kyoto Univ., **2**(1978)8.

Quantum Coherence and Decoherence - ISQM - Tokyo '98
Y.A. Ono and K. Fujikawa (Editors)
© 1999 Elsevier Science B.V. All rights reserved.

Measurement of the transverse coherency of spin precessing neutron using multilayer spin splitters

M. Hino[a], T. Ebisawa[a], S. Tasaki[a], T. Kawai[a], N. Achiwa[b], H. Tahata[c],
D. Yamazaki[c] and Y. Otake[d]

[a]Research Reactor Institute, Kyoto University, Osaka, 590-0494, Japan

[b]Department of Physics, Kyushu University, Fukuoka, 812-8581, Japan

[c]Department of Nuclear Engineering, Kyoto University, Kyoto, 606-8501, Japan

[d]The Institute of Physical and Chemical Research(RIKEN), Hyogo, 679-5143, Japan

Authors propose a new method to measure interference as a visibility of spin precessing neutron which consisted of ↑ and ↓ spin components are separated by a pair of multilayer mirrors("spin splitter"). We have measured the transverse coherency up to the separation of 4.0μm using the visibility of spin precession for neutron wavelength of 12.6Å.

1. INTRODUCTION

In quantum mechanical description, Larmor precessing neutron state is represented as a coherent superposition of ↑ and ↓ spin states[1]. In other words, coherency of ↑ and ↓ spin waves of neutron can be estimated by whether or not one can observe the spin precession. Let us consider spin precessing neutron reflected by a pair of spin splitters as shown in

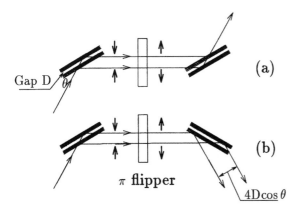

Figure 1. Schematic view of transverse separation and convergence of ↑ and ↓ spin neutron wave by a pair of spin splitters with (a) (+,−) and (b) (+,+) arrangements. The spin splitter consists of a magnetic layer on top, followed by a gap layer and a nonmagnetic layer.

Fig.1. The spin splitter consists of a magnetic layer on top, followed by a gap layer and a nonmagnetic layer[2,3]. At the incident angle θ, the ↑ and ↓ spin neutron waves reflected by the first spin splitter are separated transversely with the distance of $2D\cos\theta$, and the path difference gives the phase difference of ↑ and ↓ spin, $\Delta\phi_1 = 2\pi\frac{2D\sin\theta}{\lambda}$. The ↑ and ↓

308

spin states of the spin precessing neutron are exchanged at π flipper, and the $\uparrow$ and $\downarrow$ spin neutron waves are reflected by the second spin splitter with the arrangements of $(+,-)$ or $(+,+)$ as shown in Fig.1. The transverse separation with the $(+,-)$ arrangement is canceled out. On the other hand the separation with $(+,+)$ one is doubled. For each incident neutron wavelengths, $\Delta\phi_1$ is compensated with the phase difference $\Delta\phi_2$ due the second spin splitter, then we can observe the spin interference for less monochromatic incident beam. The transverse coherency of $\uparrow$ and $\downarrow$ spin neutron waves can be estimated from loss of visibility of the spin precession as a function of the gap thickness. We would like to emphasize that the $\uparrow$ and $\downarrow$ spin neutron waves propagate parallel each other. We can predict that the spin precession can not be observed if the distance $4D\cos\theta$ is more than a maximum length of the coherent separation which does not means coherence length well established in classical optics. In this paper, we try to measure the maximum length for transverse coherent separation of $\uparrow$ and $\downarrow$ spin neutron waves separated by a pair of multilayer spin splitters.

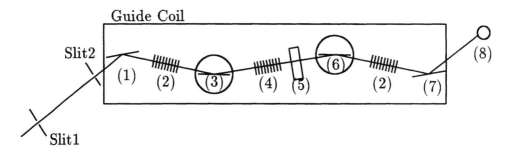

Figure 2. Schematic view of the neutron spin interferometer installed at MINE at JRR-3M. (1) polarizer, (2) $\pi/2$ RF(Radio Frequency) coil, (3) first spin splitter, (4) π RF coil, (5) Accelerator coil, (6) second spin splitter, (7) analyzer, (8) ^{3}He detector

2. EXPERIMENTAL RESULTS AND DISCUSSION

Figure 2 shows a neutron spin interferometer installed at MINE at JRR-3M in JAERI. MINE provides the cold neutron beam with a wavelength of 12.6 Å and the distribution of 3.5 % with full width at half maximum(FWHM). The interferometer is similar to neutron spin echo(NSE) spectrometer[4], and the amplitude of spin precession corresponds to polarization(visibility) of NSE signal measured as a function of the accelerator coil current. Comparing these polarizations of NSE signals measured with $(+,-)$ and $(+,+)$ arrangements, we can observe the transverse coherency as a function of the distance $4D\cos\theta$ without the effects of incident wavelength distribution and some experimental errors occurred from the effect of imperfectness of the spin splitter, such as roughness and inhomogeneity of the gap thickness. However, we can not cancel the effect of incident divergent angle with $(+,+)$ arrangement. Therefore, the experiment were carried out with various divergent angle of incident beam, and we extrapolated the polarization of

NSE signal when the divergent angle was 0. Here the divergent angle was given by these width of the slit1 and slit2, and the distance between the slit1 and slit2 was 2.0m.

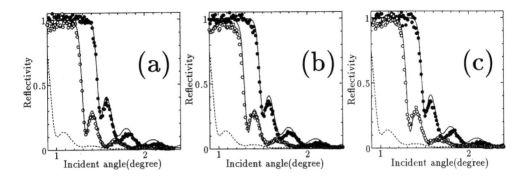

Figure 3. Reflectivity of ↑ (•) and ↓ (○) spin neutrons reflected by (a) the first spin splitter, (b) the second spin splitter with $(+, -)$ arrangement, (c) the second spin splitter with $(+, +)$ arrangement.

The spin splitter is deposited on a well polished silicon substrate with vacuum evaporation technique, and composed of Permalloy45($Fe_{55}Ni_{45}$) magnetic layer with 80nm thickness, germanium layer with thickness D, and nickel layer with 80nm thickness. Figure3(a) shows reflectivity of ↑ and ↓ spin neutrons reflected by the first composite mirror. The closed and open circles indicate the reflectivity of ↑ and ↓ spin neutrons, respectively, and are well reproduced by the theoretical(solid) lines calculated from solving one-dimensional Schrödinger equation for the optical potentials in the mirror. For the magnetic layer, the average nuclear and magnetic potential values are 220neV and 96.5neV, respectively, and the magnetic potential corresponds to 1.6T. The values of average nuclear potential for the germanium layer and nickel layer are 94neV and 243neV, respectively. The dotted line indicates theoretical reflectivity of ↓ spin neutrons reflected by the top(magnetic) layer of composite mirror. Therefore, we adopted the incident angle of 1.15° to function the mirror as ideal spin splitter. Figure3(b) shows reflectivity of ↑ and ↓ spin neutron reflected by the second spin splitter with $(+, -)$ arrangement, and Fig.3(c) shows those with $(+, +)$ arrangement. In the same way, we aligned the second splitter. The spin splitters for all conditions work without loss of the reflectivity, we were successful in realization of ideal conditions shown in Fig.1.

As shown in Fig.4, the polarization of NSE signal with $(+, -)$ arrangement , $P_{NSE}(+, -)$, is almost independent of the divergent angle of incident beam, on the contrary that with $(+, +)$ arrangement, $P_{NSE}(+, +)$, decays as a function of the divergent angle. The roughness on surface of the mirror increase with the gap thickness. Although the ideal expectation of $P_{NSE}(+, -)$ is 1, the experimental expectations of $P_{NSE}(+, -)$ at divergent angle of 0 for the gap thickness of 0.5 and 1.0 μm are 0.824±0.016 and 0.777±0.016, respectively. In case of gap thickness 0.5μm at the extrapolative incident divergent angle of 0, the ratio of the P_{NSE} with $(+, -)$ and $(+, +)$ arrangements, $P_{ratio}=P_{NSE}(+, +)/P_{NSE}(+, -)$, is 0.977±0.023. On the other hand, the ratio of gap thickness 1.0μm, P_{ratio}, is 0.849±0.023.

310

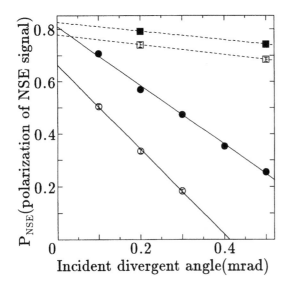

Figure 4. The changes of polarization of NSE signal(P_{NSE}) reflected by a pair of spin splitter as a function of incident divergent angle. The open and closed signs(circles and squares) indicate P_{NSE} reflected by a pair of spin splitter whose gap thickness are $0.5\mu m$ and $1.0\mu m$, respectively. The circles and squares indicate P_{NSE} with $(+,+)$ and $(+,-)$ arrangements, respectively.

Here the distance between the second spin splitter and detector is 730mm. To consider the effect of spherical wave from the spin splitter, the optical path difference between $\uparrow$ and $\downarrow$ spin neutron waves is given by $\Delta L_{\uparrow\downarrow} \simeq 0.73 \times 10^{10} \times (4.0/0.73 \times 10^{-6})^2/2 \simeq 0.13 \ll 12.6$ Å, and then the effect of spherical wave is negligible. Thus, we can interpret that the loss visibility is caused from less overlap of coherent probability distribution of $\uparrow$ and $\downarrow$ spin neutron waves. Applying the Gaussian distribution of the transverse separation to the P_{ratio} values of 0.977 and 0.849, respectively, the standard deviation is roughly estimated to be more than 10 μm and the width(FWHM) of the Gaussian distribution is twenty thousand times larger than the incident wavelength of 12.6 Å.

This work was supported by the inter-university program for the common use KURRI and JAERI facility, and financially by a Grant-in-Aid for Scientific Research from the Japanese Ministry of Education, Science and Culture(No.07209202, No.08874019 and No.10440122). One of the authors(M.H.) was supported by JSPS.

REFERENCES

1. J. Summhammer, G. Badurek, H. Rauch, U. Kischko and A. Zeilinger, Phys. Rev. A**27**, 2523(1983)
2. H. Funahashi, T. Ebisawa, T. Haseyama, M. Hino, A. Masaike, Y.Otake, T. Tabaru, S.Tasaki Phys. Rev. A**54**, 649(1996)
3. T. Ebisawa, S.Tasaki, T. Kawai, M. Hino, N. Achiwa Y.Otake, H. Funahashi, D. Yamazaki and T. Akiyoshi, Phys. Rev. A**57**, 4720(1998)
4. F. Mezei, ed., *Neutron Spin Echo, Lecture Notes in Physics* **128** (Springer, Heidelberg, 1980).

Quantum Coherence and Decoherence - ISQM - Tokyo '98
Y.A. Ono and K. Fujikawa (Editors)
© 1999 Elsevier Science B.V. All rights reserved.

A delayed choice experiment using a multilayer cold neutron pulser

T. Kawai[a], T. Ebisawa[a], S. Tasaki[a], M. Hino[a], D. Yamazaki[b], Y. Otake[c], N. Achiwa[d]

[a]Research Reactor Institute, Kyoto University, Osaka, 590-0494, Japan

[b]Department of Nuclear Engineering, Kyoto University, Kyoto, 606-8501, Japan

[c]The Institute of Physical and Chemical Research(RIKEN), Hyogo, 679-5143, Japan

[d]Department of Physics, Kyushu University, Fukuoka, 812-8581, Japan

A delayed choice experiment was done using the cold neutron spin interferometer of Jamin type arrangement which is installed at the JRR-3M of JAERI. A pair of identical composite neutron mirrors composes the pair of the wave splitters of the interferometer. The second wave splitter was placed in the pulsed magnetic field for creating the conditions of whether the second wave splitter is introduced or not when a polarized neutron reaches at the position of the second wave splitter. It was shown that whether the interference fringe is obtained or not depends only on whether the second wave splitter is introduced or not, after the neutron passed through the first wave splitter.

1. INTRODUCTION

The neutron is a fermion with a spin angular momentum of quantum number 1/2 and thus interferes only with itself. A multilayer neutron mirror is an artificial lattice with a large interplanar spacing and gives a simple one-dimensional neutron optical potential including the magnetic one. The marked advantage of an artificial multilayer neutron mirror is to enable us to make a composite mirror which consists of layers of different substances. Multilayer neutron polarizer working under a very low applied field around zero gauss (low magnetic-field neutron polarizer) [1] are especially useful for cold neutron interferometry.

In 1978, J.A.Wheeler presented the delayed choice thought-experiment to verify Bohr's view concerning the propagating behaviour of a photon after being split through a semi-reflecting mirror [2]. In that experiment, it is required to establish the condition of whether the second wave splitter is introduced or not at the point of intersection of the two partial waves split by the first wave splitter. The switching function of a cold neutron pulser could be used for that purpose. We call "a low magnetic-field neutron polarizer placed in the pulsed magnetic field" a cold neutron pulser [3].

The Jamin type neutron interferometer has been developed by H.Funahashi [4] and the neutron spin interferometer of Jamin type arrangement by T.Ebisawa [5]. We have done a delayed choice experiment using the neutron spin interferometer which is installed at the cold neutron guide tube of JRR-3M in the Japan Atomic Energy Research Institute(JAERI). We report the results of the delayed choice experiment using a cold neutron pulser and the Jamin type neutron spin interferometer [6].

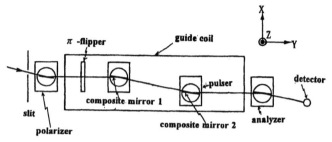

Figure 1. Schematic arrangement of the delayed choice experiment. The pair of identical composite neutron mirrors composes a Jamin type neutron spin interferometer.

2. EXPERIMENTS AND DISCUSSIONS

The composite neutron mirror consisting of a Ni/Ti multilayer ((Ni:85 Å/Ti:85 Å)× 5(layers)), a Ge gap layer(4000 Å) and a Permalloy ($Fe_{55}Ni_{45}$)/Ge multilayer ((Permalloy:85 Å/Ge:85 Å)×5(layers))(PGM) was fabricated by evaporating in series on the single Si wafer in a magnetic field of 130 gauss. The calculated reflectivity of the PGM of 5 layers is about 10 %[7]. The outmost PGM functions as a wave splitter for neutrons polarized in the direction parallel to the magnetic field(quantization axis) and the Ge gap layer plays the role of a phase shifter in the interferometer. The schematic experimental arrangement is shown in Fig. 1.

The pair of identical composite neutron mirrors(wave splitters) was set at positions about 44 cm apart. Polarized 12.6 Å neutrons with FWHM bandwidth of 3.5 % are incident on the first wave splitter and divided into two coherent partial waves. The flipping ratio of the polarized neutrons was about 14/1. The phase difference between the two partial waves induced through the first wave splitter is compensated by the second one because the two partial waves are superposed coherently upon each other at the second wave splitter. When the second mirror rotates by a small angle $\delta\theta$ on a vertical axis parallel to the mirror surface, the phase difference changes, and interference fringes are observed by varying $\delta\theta$. This is a Jamin type interference fringe. In the case of the delayed choice experiment, the second wave splitter is set in the pulsed magnetic field of 20 gauss to make a condition of whether the PGM of the second wave splitter is introduced or not after a neutron passed through the first wave splitter. The cold neutron pulser of PGM functions as a neutron optical switch.

Two delayed-choice modes are established: (1) the PGM of the second mirror is not placed when the neutron reached the first mirror, and it is introduced after the neutron passed through the first mirror, (2) the second mirror was put when the neutron has reached the first mirror, and it is removed after the neutron passed through the first mirror. Whether the second wave splitter is put or removed is decided after the neutron passed through the first mirror. This is the meaning of "delaying choice". The two-channel

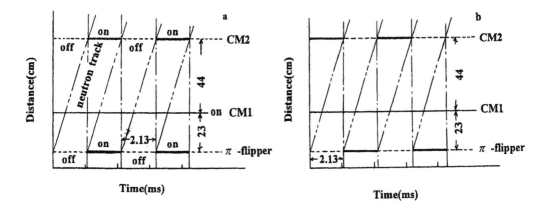

Figure 2. The time evolution in the delayed choice experiment. In case **a**, the PGM of second mirror(CM2) is not placed when the neutron reached the first mirror(CM1), and it is introduced after the neutron passed through CM1. In case **b**, the PGM of CM2 was placed when the neutron has reached CM1, and it is removed after the neutron passed through CM1.

oscillator is used to provide the accurate synchronization and timing for the π-flipper and the pulsed magnetic coil. Two modes were discriminated clearly by adjusting trigger pulse frequencies of the π-flipper and the current flowing into the pulsing coils, and the phase between them. The trigger pulse frequencies of the π-flipper and the current flowing into the pulsing coil for the second mirror are the same, that is 234 Hz, of which a half period corresponds to the neutron flight time from the π-flipper to the second mirror. The phase between two trigger pulses from the two outputs of the oscillator is zero in case of the first mode. In case of the second mode it is π as shown in Fig. 2. The neutron is polarized in the direction parallel to the guide field by the polarizer. Therefore, the neutron polarized

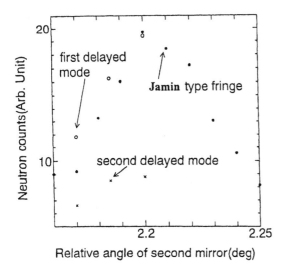

Figure 3. The interference fringe observed by changing the relative angle between the first and the second mirrors. Interference fringe appeared in the first delayed mode, and disappeared in the second delayed mode.

in the direction parallel to the guide field could reach the first composite mirror and split into two partial waves through it, if the π-flipper is off when the polarized neutron reached the π-flipper. The two partial waves interferes through the second composite mirror, if the magnetic field applied to the PGM of the second composite mirror is parallel to the guide field when the two partial waves reached the second composite mirror.

In the first mode, the interference fringe was obtained and in the second mode the interference fringe disappeared as shown in Fig. 3.

These results show that whether the interference fringe is obtained or not depends only on the condition of whether the second wave splitter is introduced or not after the neutron passed through the first wave splitter. Delaying choice has no effect on the interference. From this result, we could deduce that the neutron wave propagates both ways with equal probability after being split by the first wave splitter and does not choose one of the two ways, as explained by the quantum mechanics.

Acknowledgements

We would like to thank Prof. A.Masaike for his kind support and encouragement. This work was supported by inter-university collaboration programs of KUR and JAERI, and financially by the Grant-in-Aid for Scientific Research of the Ministry of Educations, Science and Culture of Japan under the program numbers 04244103, 08874019.

REFERENCES

1. T.Kawai, Proc. of *International School and Symposium on Physics in Materials Science*, Inst. Atomic Energy, Swierk, Poland, Sep. 13-19, 1998.
2. J.A.Wheeler, *Mathematical Foundations of Quantum Theory*, ed., A.R.Marlow, (1978) 9-48, Academic Press.
3. T.Kawai, T.Ebisawa, S.Tasaki, M.Hino, D.Yamazaki, H.Tahata, T.Akiyoshi, Y.Matsumoto, N.Achiwa, Y.Otake, Physica **B241-243** (1998) 133.
4. H.Funahashi, T.Ebisawa, T.Haseyama, M.Hino, A.Masaike, Y.Otake, T.Tabaru, S.Tasaki, Phys.Rev., **A54** (1996) 649.
5. T. Ebisawa, S.Tasaki, T. Kawai, M. Hino, N. Achiwa Y.Otake, H. Funahashi, D. Yamazaki and T. Akiyoshi, Phys.Rev., **A57** (1998) 4720.
6. T.Kawai, T.Ebisawa, S.Tasaki, M.Hino, D.Yamazaki, T.Akiyoshi, Y.Matsumoto, N.Achiwa, Y.Otake, Nucl.Instrum. Methods, **A410** (1998) 259.
7. S.Tasaki, J.Appl.Phys., **71** (1992) 2375.

Quantum Coherence and Decoherence - ISQM - Tokyo '98
Y.A. Ono and K. Fujikawa (Editors)
© 1999 Elsevier Science B.V. All rights reserved.

Fluctuation-dissipation theorem and quantum decoherence

Kazuo Fujikawa

Department of Physics, University of Tokyo
Bunkyo-ku, Tokyo 113, Japan

The fluctuation-dissipation theorem of Callen and Welton is reformulated in such a way that the presence of dissipation with a well-defined dissipative coefficient, which little depends on the temperature in the region we are interested in, necessarily leads to the presence of effective oscillator freedom. The basic idea of Feynman-Vernon and Caldeira-Leggett is thus realized without introducing an infinite number of oscillators. Combined with dispersion relations, the quantum tunneling (coherence) with dissipation is formulated in a model independent manner both at vanishing and finite temperatures.

1. FLUCTUATION-DISSIPATION THEOREM OF CALLEN AND WELTON

We present a reformulation of the fluctuation-dissipation theorem of Callen and Welton[1] in such a manner that the basic idea of Feynman-Vernon[2] and Caldeira -Leggett[3][4], which simulates the dissipative medium by an infinite number of oscillators, becomes manifest without actually introducing oscillators.

We start with a Hamiltonian

$$H = H_0(Q) + Qq \cos \omega t, \tag{1}$$

where $H_0(Q)$ is the unperturbed Hamiltonian for the dissipative medium, which has eigenstates $H_0|E_n\rangle = E_n|E_n\rangle$. The variable Q is a (generally very complicated) bosonic composite operator and $q \cos \omega t$ describes an (infinitesimal) external dynamical freedom which perturbs the dissipative medium. The variable q is treated as a classical variable for the moment.

If the dissipative medium is initially in the state $|E_n\rangle$, the lowest order transition rate is given by the Fermi's golden rule by treating the last term in (1) as small perturbation,

$$w = \frac{\pi q^2}{2\hbar}\left[|\langle E_n + \hbar w|Q|E_n\rangle|^2 \; \rho(E_n + \hbar w) + |\langle E_n - \hbar w|Q|E_n\rangle|^2 \; \rho(E_n - \hbar w)\right]. \tag{2}$$

Since the first term stands for the absorption of energy $\hbar w$ and the second term for the emission of energy $\hbar w$, the energy absorption rate by the dissipative medium is given by multiplying $\hbar \omega$ to the difference of these two terms. If the system is initially in thermal equilibrium at temperature T, we must average over all initial states, weighting the state

$|E_n\rangle$ by the (normalized) Boltzmann factor $f(E_n)$ which satisfies $f(E_n + \hbar w)/f(E_n) = e^{-\beta\hbar w}$ with $\beta = \frac{1}{kT}$. Then the energy dissipation per unit time is given by

$$P(\omega) = \frac{q^2\pi\omega}{2}(1 - e^{-\beta\hbar w}) \int_0^\infty dE\, \rho(E)f(E)\Big[|\langle E + \hbar w|Q|E\rangle|^2\, \rho(E + \hbar w)\Big] \tag{3}$$

where we replced the summation over n by an integration over energy and used $|\langle E - \hbar w|Q|E\rangle| = |\langle E|Q|E - \hbar w\rangle|$ together with a shift in energy.

We next define the phenomenological *macroscopic* dissipative coefficient (resistance) $R(\omega)$ on the basis of the following reasoning[5]. We first define an infinitesimal coordinate $q(t) = q\cos\omega t$ in H_I. The existence of the energy dissipation into the medium (3) induced by the external perturbation $q(t)$ suggests the presence of a dissipative force (reaction) $F = -R(\omega)\dot{q}(t)$ acting on the variable $q(t)$, which oscillates with frequency ω. Here $R(\omega)$ is a real function. The energy dissipation per unit time generated by this phenomenological reactive force is given by

$$P(\omega) = -\overline{F\dot{q}(t)} = \frac{\omega^2}{2}R(\omega)q^2 \tag{4}$$

where the overline indicates time averaging.

Combining (3) and (4), we obtain the microscopic expression for the dissipative coefficient (resistance) $R(\omega)$

$$R(\omega) = \frac{\pi}{\omega}(1 - e^{-\beta\hbar w}) \int_0^\infty dE\, \rho(E)f(E)\rho(E + \hbar w)\,|\langle E + \hbar w|Q|E\rangle|^2. \tag{5}$$

From this expression of $R(\omega)$, we find the basic relations[5],

$$\frac{2}{\pi}\frac{\hbar w}{2}\left[1 + \frac{1}{e^{\beta\hbar w} - 1}\right]R(\omega) = \hbar \int_0^\infty dE\, \rho(E)f(E)\rho(E + \hbar w)\,|\langle E + \hbar w|Q|E\rangle|^2, \tag{6}$$

$$\frac{2}{\pi}\frac{\hbar w}{2}\left[\frac{1}{e^{\beta\hbar w} - 1}\right]R(\omega) = \hbar \int_0^\infty dE\, \rho(E)f(E)\rho(E - \hbar w)\,|\langle E - \hbar w|Q|E\rangle|^2. \tag{7}$$

These relations stand for a proto-type of the fluctuation-dissipation theorem of Callen and Welton. The fluctuation-dissipation theorem as it stands is a mathematical identity and contains no physical contents by itself. What is remarkable is that we obtain highly non-trivial relations in (6) and (7) if one *assumes* that the dissipative coefficient $R(\omega)$ little depends on the temperature in the region we are interested in. We can recognize the left-hand side of (6) as standing for the spontaneous and induced emissions of the second quantized bosonic oscillator with frequency ω into the dissipative medium, whereas the left-hand side of (7) is recognized as the (induced) absorption of these oscillators from the dissipative medium at temperature T. Moreover, the spectrum of these effective oscillators is precisely specified by the dissipative coefficient $R(\omega)$. These properties realize the basic physical idea of Feynman-Vernon[2] and Caldeira-Leggett[3].

2. QUANTUN TUNNELING (COHERENCE) WITH DISSIPATION

We consider the total Hamiltonian

$$H = H_0(Q) + H_0(q) + H_I(q, Q) \tag{8}$$

where $H_0(q) = \frac{1}{2M}p^2 + V(q)$ describes the unperturbed Hamiltonian of the quantum system we are interested in, and $H_I(q,Q) = qQ$ stands for the interaction Hamiltonian in (1). But now the variable q is promoted to a hermitian quantum operator; the explicit time dependence disappears in the Schroedinger picture. In the context of macroscopic quantum tunneling (or quantum coherence), we choose $V(q)$ as a symmetric double well potential. $H_0(Q)$, which describes the dissipative medium , is indirectly specified by our fluctuation-dissipation theorem (6) and (7).

We now start with the eigenstates of $H_0(q)$, $H_0(q)|n\rangle = E_n|n\rangle$, and consider the transition probability for $n \rightarrow m + \hbar\omega$ by emitting energy $\hbar\omega$ to the dissipative medium, which is assumed to be in thermal equilibrium with temperature T. The transition probability for this process is given by the lowest order perturbation of $H_I(q,Q)$ as

$$w(n \rightarrow m + \hbar\omega)$$
$$= \frac{2\pi}{\hbar}|\langle m|q|n\rangle|^2 \int_0^{\hbar\Lambda} \int_0^\infty \rho(E + \hbar\omega)|\langle E + \hbar\omega|Q|E\rangle|^2\rho(E)f(E)dE\delta(E_n - E_m - \hbar\omega)d(\hbar\omega)$$
$$= \frac{2\pi}{\hbar}|\langle m|q|n\rangle|^2 \frac{2}{\pi}\int_0^{\hbar\Lambda} \frac{\hbar\omega}{2}[1 + \frac{1}{e^{\hbar\omega/kT} - 1}]\frac{R(\omega)}{\hbar}\delta(E_n - E_m - \hbar\omega)d(\hbar\omega) \qquad (9)$$

where we used eq.(6) and introduced an explicit cut-off $\hbar\Lambda$ of effective frequency of the dissipative medium. At finite temperature we also have an absorption probability by using (7), or from (9) by simply replacing $\hbar\omega$ by $-\hbar\omega$ as

$$w(n + \hbar\omega \rightarrow m) = \frac{2\pi}{\hbar}|\langle m|q|n\rangle|^2 \frac{2}{\pi}\int_0^{\hbar\Lambda} \frac{\hbar\omega}{2}[\frac{1}{e^{\hbar\omega/kT} - 1}]\frac{R(\omega)}{\hbar}\delta(E_n - E_m + \hbar\omega)d(\hbar\omega) \qquad (10)$$

It is interesting that these formulas satisfy the detailed balancing relation, $e^{\hbar\omega/kT}w(n \rightarrow m + \hbar\omega) = w(m + \hbar\omega \rightarrow n)$, with $\hbar\omega = E_n - E_m$.

We next define the half-width of the state $|n\rangle$ for emission(when we prepare the state $|n\rangle$ at $t = 0$) $1/2\Gamma_n^{(+)} = \hbar/2\sum_m w(n \rightarrow m + \hbar\omega)$, and the corresponding one for absorption $1/2\Gamma_n^{(-)} = 1/2\hbar\sum_m w(n + \hbar\omega \rightarrow m)$. It can be confirmed that eq. (9) gives $(1/2)\Gamma_n = (1/2)\hbar\eta/M$ at $T = 0$ for a simple harmonic oscillator $H_0(q) = (1/2M)p^2 + (M\omega^2/2)q^2$ and Ohmic dissipation $R(\omega) = \eta$[3]; this expression of Γ_n is consistent with a damped oscillator $M\ddot{q} + \eta\dot{q} + M\omega^2q = 0$, which in turn justifies the normalization of H_I in (8).

The basic idea in our attempt to reproduce the results of the Caldeira- Leggett model[3] without introducing an auxiliary infinite number of oscillators is to write dispersion relations(or second order perturbation formula in the linear response approximation), which relate the imaginary part of self-energy correction to the corresponding real part. The imaginary parts are evaluated by means of the fluctuation-dissipation theorem as in (9) and (10).

We thus write (a generalization of) the dispersion relation for the self-energy correction $\Sigma_n(E)$ to the energy eigenvalue E_n as

$$\Sigma_n(E) = \frac{1}{\pi}\int_0^\infty \frac{Im\Sigma_n^{(+)}(E')dE'}{E' - E - i\epsilon} + \frac{1}{\pi}\int_{-\hbar\Lambda}^\infty \frac{Im\Sigma_n^{(-)}(E')dE'}{E' - E - i\epsilon} \qquad (11)$$

where $Im\Sigma_n^{(+)}(E) \equiv \frac{1}{2}\Gamma_n^{(+)}(E) = \hbar/2\sum_m w(n \rightarrow m + \hbar\omega)|_{E_n=E}$ and similarly $Im\Sigma_n^{(-)}(E) \equiv \frac{1}{2}\Gamma_n^{(-)}(E) = 1/2\hbar\sum_m w(n + \hbar\omega \rightarrow m)|_{E_n=E}$. Note that the lower bound of the integration range in the second term in eq.(11) starts at $-\hbar\Lambda$ due to the definition in (10).

We thus obtain

$$
\begin{aligned}
\Sigma_n(E) &= \sum_m |\langle m|q|n\rangle|^2 \frac{1}{\pi} \int_0^{\hbar\Lambda} \{ \frac{\hbar\omega}{E_m + \hbar\omega - E - i\epsilon}[1 + \frac{1}{e^{\hbar\omega/kT} - 1}] \\
&+ \frac{\hbar\omega}{E_m - \hbar\omega - E - i\epsilon}[\frac{1}{e^{\hbar\omega/kT} - 1}] \} \frac{R(\omega)}{\hbar} d(\hbar\omega)
\end{aligned}
\tag{12}
$$

For the Ohmic dissipation, $R(\omega) \equiv \eta = constant$, for example, this expression of $\Sigma_n(E)$ agrees with the result of the field theoretical formulation of the Caldeira-Leggett model[6]. It should be noted that the vacuum fluctuation term (spontaneous emission) in (6) plays a central role in our application, unlike the conventional applications of the fluctuation-dissipation theorem where the vacuum fluctuation is usually subtracted.

For the *two-level* approximation, which is valid for the lowest two levels in a deep double-well potential, we have the result from (12)[6], $Re\Sigma_2(E_2) - Re\Sigma_1(E_1) \simeq \epsilon\bar{\eta}\ln(e^{-2}\beta\hbar\Lambda)$, with the zeroth order energy difference $\epsilon \equiv E_2 - E_1$ and $\bar{\eta} = \frac{2\eta}{\pi\hbar}|\langle 0|q|1\rangle|^2$ for the temperature $\epsilon \ll 1/\beta \ll \hbar\Lambda$. We also have $Re\Sigma_2(E_2) - Re\Sigma_1(E_1) \simeq \epsilon\bar{\eta}\ln(\hbar\omega_0/\epsilon)$ for $T = 0$.

The energy splitting (*order parameter* of quantum coherence) corrected by the dissipation, $\epsilon_{(1)} \equiv (E_2 - Re\Sigma_2(E_2)) - (E_1 - Re\Sigma_1(E_1))$, is then given by

$$
\begin{aligned}
\epsilon_{(1)} &\simeq \epsilon[1 - \bar{\eta}\ln(e^{-2}\beta\hbar\Lambda)] \simeq \epsilon[(e^{-2})\beta\hbar\Lambda]^{-\bar{\eta}} \quad, \quad T \neq 0 \\
\epsilon_{(1)} &\simeq \epsilon[\hbar\omega_0/\epsilon]^{-\bar{\eta}} \simeq \epsilon[\hbar\omega_0/\epsilon]^{-\bar{\eta}/(1-\bar{\eta})} \quad, \quad T = 0
\end{aligned}
\tag{13}
$$

after the renormalization group improvement;ω_0 is the curvature at the bottom of the double-well potential. These results are in agreement with the dilute instanton analysis for the case of Ohmic dissipation[3][4] and suggest the suppression of quantum coherence.

In conclusion, we can analyze the quantum coherence on the basis of general principles only, namely, the fluctuation-dissipation theorem of Callen and Welton and dispersion relations (or unitarity and causality), without referring to an explicit model Lagrangian which simulates dissipative effects. It should however be noted that the explicit model Lagrangian of Caldeira and Leggett[3] is more flexible amd covers a wider range of phenomena.

REFERENCES

1. H. B. Callen and T. A. Welton, Phys. Rev. 83(1951) 34.
2. R. P. Feynman and F. L. Vernon, Ann. of Phys. (N.Y.), 24(1963) 118.
3. A. O. Caldeira and A. J. Leggett, Ann. of Phys. 149(1983) 374;Phys. Rev. Lett. 46(1981) 211.
4. A. J. Bray and M. A. Moore, Phys. Rev. Lett. 49(1982) 1545; S. Chakravarty, Phys. Rev. Lett. 49(1982) 681; G. Schoen and A. D. Zaikin, Phys. Rep. 198(1990) 237 and references therein.
5. K. Fujikawa, Phys. Rev. E57 (1998)5023; K. Fujikawa and H. Terashima, Tokyo report, UT-819 (1998).
6. K. Fujikawa, S. Iso, M. Sasaki, and H. Suzuki, Phys. Rev. Lett. 68(1992) 1093; Phys. Rev. B46(1992) 10295 .

Quantum Coherence and Decoherence - ISQM - Tokyo '98
Y.A. Ono and K. Fujikawa (Editors)
© 1999 Elsevier Science B.V. All rights reserved.

Non-Hermitian quantum mechanics and localization in physical systems

Naomichi Hatano[a]

[a]MS-B262, Theoretical Division, Los Alamos National Laboratory
Los Alamos, NM 87545, USA

Recent study on a delocalization phenomenon of a non-Hermitian random system is reviewed. A complex energy spectrum of the system indicates delocalization transition of its eigenfunctions. It is demonstrated that, using this property, one can calculate the localization length of the *Hermitian* Anderson model by inspecting the complex energy spectrum of its non-Hermitian version.

1. INTRODUCTION

In the last few years, there has been a novel development in the study of localization phenomena, namely non-Hermitian localization and delocalization [1–24]. A new type of delocalization transition was found in a simple quantum-mechanical model with a non-Hermite Hamiltonian. In this paper, I describe interesting properties of the non-Hermitian system, a few physical motivations of studying it, and new results on the Anderson localization.

The model is defined by the Hamiltonian [1,2]

$$\mathcal{H} \equiv \frac{1}{2m}(\vec{p} + i\vec{g})^2 + V(\vec{x}), \tag{1}$$

where a constant $\vec{g}$ makes the Hamiltonian non-Hermite. The operator $\vec{p}$ is the momentum and $V(\vec{x})$ is a random potential. In the Hermitian case $\vec{g} = \vec{0}$, the system is reduced to a fundamental model of the Anderson localization. In this limit, it is well accepted that the eigenfunctions are localized in one and two dimensions and that there are energy ranges of localized eigenfunctions in three dimensions and higher. It was discovered [1,2] that, in any dimensions, as the non-Hermitian field $\vec{g}$ is introduced and increased,

  (i) Each of the originally localized states is delocalized at its own critical point $\vec{g}_c$;

  (ii) The delocalization transition coincides with the instance where the corresponding eigenvalue becomes complex (Up to this point, the eigenvalue is fixed to the real value for $\vec{g} = \vec{0}$.);

  (iii) The inverse localization length κ of the original eigenfunction (for $\vec{g} = \vec{0}$) is equal to $|\vec{g}_c|$.

It is a remarkable fact to have delocalization even in one dimension. It is also a very characteristic and new feature of this system that one can detect the delocalization transition by calculating the spectrum of the system.

(a) (b)

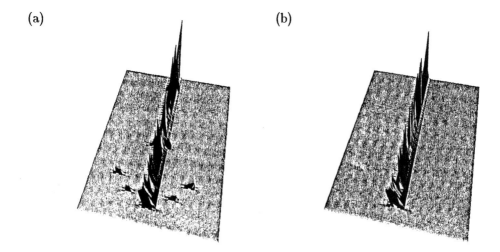

Figure 1. The density of states of $\mathcal{H}^2$, where $\mathcal{H}$ is the Hamiltonian (2), was calculated at the interval $1/200$ of Re $(\varepsilon/t)^2$ and Im $(\varepsilon/t)^2$. The real axis is plotted toward the far side of the figures, while the imaginary axis is plotted toward the left. The portion $-0.1 < \text{Re } (\varepsilon/t)^2 < 0.9$ and $-0.1 < \text{Im } (\varepsilon/t)^2 < 0.1$ is shown here for $\vec{g} = (0.2, 0, 0)$ with (a) $\Delta/t = 3.8$ and (b) $\Delta/t = 4.2$. The former has complex eigenvalues near the band center $\varepsilon \simeq 0$, whereas the latter does not.

2. LOCALIZATION LENGTH OF THE HERMITIAN ANDERSON MODEL

Combining the three properties described in Introduction, we can estimate the localization length of the Hermitian Anderson model by calculating the spectrum of the non-Hermitian model [2,15,16,23]. Suppose that we try to calculate the localization length of an eigenstate of the Hermitian case $\vec{g} = \vec{0}$. We increase $\vec{g}$, keeping observing the eigenvalue of the state. The eigenvalue is fixed to a value until $|\vec{g}|$ exceeds a critical point g_c. Then $1/g_c$ is our estimate of the localization length. For example, the exact solution [25] of the localization length of the Lloyd model (the one-dimensional Anderson model with a Lorentzian potential randomness) is reproduced by the above procedure from the exact solution [10] of the energy spectrum of the non-Hermitian Lloyd model [16].

Note that the conventional method of estimating the localization length is to calculate the Lyapunov exponent of the random transfer matrix of a long stripe of the Anderson model [26]. The method using the non-Hermitian Anderson model provides a new independent way of estimating the localization length. It is also noted that the above procedure can yield the localization length in any direction, which would be difficult in the transfer matrix approach. The present procedure will be convenient in studying lattice anisotropy of localization.

We here demonstrate the above procedure for the three-dimensional Anderson model. The (Hermitian) Anderson model in three dimensions can have an energy region of delocalized states around the band center. The width of the delocalized region depends on the randomness of the potential. The phase diagram in the (ε, Δ) plane (where ε is the energy and Δ is the amplitude of the randomness) is of continuing interest [26].

If there is a delocalized state for $\vec{g} = \vec{0}$, its eigenvalue will become complex as soon as we turn on $\vec{g}$. Thus we can detect the delocalized region of the Hermitian Anderson model by putting a weak $\vec{g}$ to the system. (In actual calculations, not all delocalized states will become complex because of anisotropic localization [2].) We carried out a numerical calculation for a lattice version of the non-Hermitian Anderson model

$$\mathcal{H} \equiv \sum_{\vec{x}} \left[-\frac{t}{2} \sum_{\nu=1}^{d} \left(e^{\vec{g}\cdot\vec{e}_\nu} |\vec{x} + \vec{e}_\nu\rangle \langle\vec{x}| + e^{-\vec{g}\cdot\vec{e}_\nu} |\vec{x}\rangle \langle\vec{x} + \vec{e}_\nu| \right) + V_{\vec{x}} |\vec{x}\rangle \langle\vec{x}| \right], \qquad (2)$$

where $\vec{e}_\nu$ is the lattice vector in each dimension and $V_{\vec{x}}$ is the random potential chosen for each site uniformly from $[-\Delta, \Delta]$. The lattice size was 100^3.

We developed a new method of calculating the density of states of non-Hermitian matrices, on the basis of the Lanczos algorithm. (The details of the method will be reported elsewhere [27].) Figure 1 shows the density of states of $\mathcal{H}^2$ in the complex energy plane. Since the numerical convergence is better for extremal eigenvalues, we can focus on the band center $\varepsilon \sim 0$ by treating $\mathcal{H}^2$ rather than $\mathcal{H}$ itself. The result indicates that the phase boundary between the localized phase and the delocalized phase lies within $3.8 < \Delta/t < 4.2$ at $\varepsilon = 0$, which is consistent with a result by the conventional transfer-matrix method [26].

3. OTHER PHYSICAL MOTIVATIONS

Let us conclude the present paper with other problems in physics that are related to the present non-Hermitian model. Flux line (or vortex line) in high-T_c materials has spawned a new branch of physics [28]. One of the interesting topics is flux-line pinning due to columnar and planar defects in superconductors. When the magnetic field applied to the superconductor is parallel to these extended defects, the flux lines generated by the field are easily pinned by the defects, which stabilizes the superconductivity. A depinning transition has been observed when the field is tilted away from the axis of the defects [29].

It has been argued that the depinning due to the field tilt is equivalent to the delocalization of the non-Hermitian system described in the previous section [1,2,30]. The depinning point is hence estimated by calculating the spectrum of the Hamiltonian (1).

In yet another motivation, the imaginary-time Schrödinger equation for the Hamiltonian (1) may be regarded as a Fokker-Planck equation. This equation can describe bacteria population in a constant flow with random distribution of nutrients [12,24] and chemical reaction system with random catalyst distribution. Mudry et al. [14] pointed out a relation between the non-Hermitian Anderson model (1) and the Dirac fermion in a weak random gauge field. As a related problem, non-Hermitian random matrix theory has been brought into focus lately. There are various physical motivations for this theory, from a model of chiral symmetry breaking in QCD [31,32,13] to neural networks [33]. A recent paper [34] studied a non-Hermitian $\mathcal{PT}$-symmetric oscillator from the point of view that the Hermiticity of Hamiltonian is not necessarily a physical requirement but the $\mathcal{PT}$ symmetry is.

Many physicists have thought that non-Hermite Hamiltonians have no physical meaning. In fact, the last few years saw a growing number of physical non-Hermitian models.

More intensive studies of these models are awaited.

REFERENCES

1. N. Hatano and D.R. Nelson, Phys. Rev. Lett. **77** (1996) 570.
2. N. Hatano and D.R. Nelson, Phys. Rev. B **56** (1997) 8651.
3. L.-W. Chen, L. Balents, M.P.A. Fisher and M.C. Marchetti, Phys. Rev. B **54** (1996) 12798.
4. N. Shnerb, Phys. Rev. B **55** (1997) R3382; Phys. Rev. B **57** (1998) 8571.
5. K.B. Efetov, Phys. Rev. Lett. **79** (1997) 491; Phys. Rev. B **56** (1997) 9630; Phil. Mag. B **77** (1998) 1135.
6. J. Feinberg and A. Zee, Nucl. Phys. B **504** [FS] (1997) 579; Report No. cond-mat/9706218, to appear in Phys. Rev. E; Report No. cond-mat/9710040.
7. R.A. Janik, M.A. Nowak, G. Papp and I. Zahed, Report No. cond-mat/9705098.
8. P.W. Brouwer, P.G. Silvestrov and C.W.J. Beenakker, Phys. Rev. B **56** (1997) R4333.
9. I.Ya. Goldsheid and B.A. Khoruzhenko, Phys. Rev. Lett. **80** (1998) 2897.
10. E. Brezin and A. Zee, Nucl. Phys. B **509** [FS] (1998) 599.
11. N. Zekri, H. Bahlouli and A.K. Sen, J. Phys.: Condens. Matter **10** (1998) 2405.
12. D.R. Nelson and N. Shnerb, Phys. Rev. E **58** (1998) 1383.
13. R.A. Janik, M.A. Nowak, G. Papp and I. Zahed, Acta Phys. Pol. B **28** (1997) 2949.
14. C. Mudry, B.D. Simons and A. Altland, Phys. Rev. Lett. **80** (1998) 4257.
15. A. Zee, Physica A **254** (1998) 317.
16. N. Hatano, Physica A **254** (1998) 300.
17. N.M. Shnerb and D.R. Nelson, Phys. Rev. Lett. **80** (1998) 5172.
18. Y.V. Fyodorov, M. Titov and H.-J. Sommers, Phys. Rev. E **58** (1998) R1195.
19. P.G. Silvestrov, Report No. cond-mat/9802219v2; Report No. cond-mat/9804093.
20. N. Hatano and D.R. Nelson, Phys. Rev. B **58** (1998) No. 13.
21. R.A. Lehrer and D.R. Nelson, Report No. cond-mat/9806016.
22. P.W. Brouwer, C. Mudry, B.D. Simons and A. Altland, Report No. cond-mat/9807189.
23. C. Mudry, P.W. Brouwer, B. I. Halperin, V. Gurarie and A. Zee, Report No. cond-mat/9807391.
24. K.A. Dahmen, D.R. Nelson and N.M. Shnerb, Report No. cond-mat/9807394.
25. T. Hirota and K. Ishii, Prog. Theor. Phys. **45** (1971) 1713.
26. B. Kramer and A. MacKinnon, Rep. Prog. Phys. **56** (1993) 1469.
27. N. Hatano and A. Zee, in preparation.
28. G.W. Crabtree and D.R. Nelson, Physics Today **50** No. 4 (1998) 38.
29. I.M. Obaidat, S.J. Park, H. Safar and J.S. Kouvel, Phys. Rev. B **56** (1997) R5774.
30. D.R. Nelson and V. Vinokur, Phys. Rev. B **48** (1993) 13060.
31. M.A. Stephanov, Phys. Rev. Lett. **76** (1996) 4472.
32. M.A. Halasz, J.C. Osborn and J.J.M. Verbaarschot, Phys. Rev. D **56** (1997) 7059.
33. H.J. Sommers, A. Crisanti, H. Sompolinsky and Y. Stein, Phys. Rev. Lett. **60** (1988) 1895.
34. C.M. Bender and S. Boettcher, Phys. Rev. Lett. **80** (1998) 5243.

Quantum Coherence and Decoherence - ISQM - Tokyo '98
Y.A. Ono and K. Fujikawa (Editors)
© 1999 Elsevier Science B.V. All rights reserved.

The double Stern-Gerlach experiment with multilayer cold neutron interferometry to test some quantum measurement theories

Y.Otake, H.Funahashi[a], S.Tasaki[b], T.Ebisawa[b], and T.Kawai[b]

The Institute Physical and Chemical Research, Mikazuki, Sayo-gun, Hyogo 679-5143, Japan,
[a] Dept. of Phys., Kyoto Univ., Sakyo-ku, Kyoto 606-8502, Japan
[b] KURRI, Kumatori, Sennan-gun, Osaka 590-0494, Japan,

The success of multilayer mirror interferometer for cold neutrons enabled us to perform double Stern Gerlach experiment, which was proposed by Machida and Namiki to test some quantum mechanical measurement theories. We have observed interference fringe as they predicted. This experimental result excludes measurement theories which derive decoherence with a direct link to the orthogonal decomposition in quantum mechanical measurement process.

1. Introduction

Quantum mechanical coherence and decoherence phenomena are now studied widely in theoretical and experimental approaches. Many experiments with thermal neutrons[1], electrons[2], and atoms[3] etc. have been performed so far. Since the first success of multilayer interferometer for cold neutrons in 1995[4], cold neutron interferometry using multilayer mirrors is recognized as a powerful tool to reveal quantum coherent and decoherent nature clearly. One of the distinguished points of this interferometer is its wide application by using magnetic multilayer mirrors. The feasibility of phase-spin-echo interferometer using multilayer spin splitter was demonstrated by T.Ebisawa et al.[5]. There are remarkable experiments with this kind of interferometer, such as a measurement of transverse coherence length of cold neutrons by M.Hino et al.[6]. A magnetic multilayer mirror functions either as a reflective mirror or as a transparent mirror for a beam of polarized neutrons depending on polarity of external magnetic field. Kawai et al. developed new magnetic mirrors of very quick response and have performed "Delayed choice experiment with cold neutrons"[7]. In this article, we report experimental result of double Stern-Gerlach experiment (dSG) by means of cold neutron interferometry using multilayer mirrors. Machida and Namiki proposed the dSG[8] to test some measurement theories especially Many-Worlds Interpretation,[9] Environment Theory[10] and Many-Hilbert-Space Theory (MHS)[11] by themselves.

2. Double Stern-Gerlach experiment and coherency

A transition from pure state into mixture state in a quantum mechanical measurement process is often expressed as follows.

$$\hat{\Xi}_I^{tot} = \hat{\rho}_I^Q \otimes \hat{\sigma}_I^D = \sum_i c_i c_j^* |u_i \rangle \langle u_j | \otimes \hat{\sigma}_I^D$$

$$\rightarrow \hat{\Xi}_F^{tot} = \hat{\rho}_F^Q \otimes \hat{\sigma}_F^D = \sum_k |c_k|^2 |u_k \rangle \langle u_k | \otimes \hat{\sigma}_{Fk,t}^D$$

Here, the total system Ξ^{tot} consists of a quantum system ρ^Q and an apparatus system σ^D. In the equations above, suffix I and F denote the initial and final state respectively. The initial state of the quantum system is a superposed state of a complete set of an eigenfunction $|u_k\rangle$. After the quantum measurement, there is no coherence, in other words, lack of interference terms, between eigenstates as is seen in the second equation above. This transition is a dephasing process. The mechanism of this decoherence is not solved yet. Many measurement theories have tried to explain decoherence process which is not governed by linear evolution such as Schrödinger equation. Now we make one point clear, we focus to understand which kind of physical processes take place in the quantum mechanical measurement. In the quantum measurement process, the superposed state is often orthogonally (spectrally) decomposed into each eigenstate of an observable before its detection. Many theorists don't distinguish these two essential steps, orthogonal decomposition step and detection step, in quantum measurement so that they derive decoherence with a direct link to the orthogonal decomposition. The dSG is the experiment that clearly gives an answer to the question whether the coherence between eigenstates preserves or not after orthogonal decomposition in quantum measurement process.

We explain dSG experiment of cold neutrons briefly. An incident neutron wave is a superposed state of spin up and spin down. It is simply divided into two waves keeping its spin unchanged by the first semi-transparent multilayer mirror. Each wave is spectrally decomposed into each orthogonal spin state by a magnetic mirror. Then, the second semi-transparent mirror recombines two waves. The point is whether the waves coming to this final channel after orthogonal spin decomposition keep their coherence or not.

3.dSG experiment and results

The experiment was performed using the multiyear interferometer for cold neutrons at the cold neutron guide tube of the reactor JRR-3M in JAERI. The beam channel provides the cold neutron beam with a wavelength of 12.6Å and a bandwidth of 3.5% at FWHM. Two pairs of multilayer mirrors were fabricated simultaneously in a magnetic field by successive vacuum evaporations of magnetic multilayer mirror A, the intermediate layer of 1 micron meter thickness, and multilayer mirror B on polished silicon substrates as is shown in Fig. 1. The magnetic multilayer mirror A had five layers made of Permalloy-germanium, whose bilayer thickness is 170Å. The multilayer mirror B was consist of five layers of nickel-titanium with the same bilayer thickness as mirror A. The magnetic mirror A spectrally decomposed the neutron wave into two spin eigenstates waves. The mirror B divided one wave into two or combined two waves keeping spin states unchanged. The measured reflectivity of the mirror A was 30% for parallel-polarized neutrons, and the reflectivity of mirror B was 20%.

Two pair mirrors were kept about 30cm apart each other. Each of them was placed in twenty gauss magnetic field, respectively, while the whole system is in the magnetic guide field of four gauss. The incident neutrons were polarized more than 96 %. In front of the interferometer of two pair mirrors we set Mezei coil to change neutrons polarization from parallel to perpendicular or anti-parallel to the polarity of the magnetic mirror A. When the glancing angle θ in Fig. 1 changes, the interference fringe was observed with parallel-polarized neutrons (parallel mode) as is shown in Fig. 2.

The dSG experiment was performed with incident neutrons perpendicular-polarized to the magnetic mirror (dSG mode). From now on we call the spin direction along the polarity of the mirror A as spin up. The superposed state of spin up and down neutrons was simply divided into two waves by mirror B1 in Fig. 1. Each mirror A1 or A2 of two pair mirrors spectrally decomposed incident superposed state of these two waves into spin up and down respectively. A spin up wave is separated from spin down through mirror A1, or A2. Each spin state now goes along different path, in other words, A1 and A2 spectrally decompose superposed waves, respectively. The final channel of reflected wave off the second pair mirror was consists of four components as is drawn in Fig. 1. One is reflected from mirror B1 and A2, the other is reflected from mirror A1 andB2, the other from A1 and A2, and another is from B1 and B2.

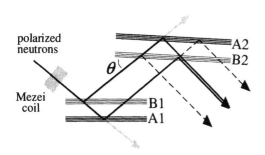

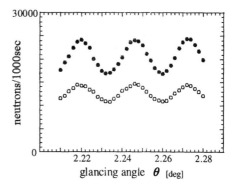

Fig.1 Schematic diagram of dSG experiment with multilayer interferometer. Magnetic mirrors of two pair mirrors are A1 and A2; semi-transparent mirrors of them are B1 and B2. θ is the glancing angle off the second pair mirror.

Fig.2 Intensity of neutrons reflected off the second pair mirror. Solid dots denote the interference fringe with parallel mode and open dots are that of dSG mode.

The clear interference pattern has been observed as is shown in Fig. 2. The upper data of solid dot is the interference fringe of parallel mode; the lower curve of open

326

dot is the data of dSG mode. The data of dSG mode clarifies that two waves of A1-B2 and B1-A2, which are solid lines in Fig. 1, keep coherence between them. The period and the constant phase shift of interference pattern of dSG mode agree with those of the interference pattern of parallel mode. The ratio of the measured interference amplitude between dSG and parallel modes is nearly the same as the theoretical calculation of 1:2. This factor two comes from the difference of the incident polarization between dSG mode and parallel mode. There is no loss of coherence in dSG mode. We observe coherence in the recombined wave after spin decomposition as MHS predicts[8].

4. Conclusion

We have performed double Stern-Gerlach experiment using multilayer interferometer for cold neutrons, which was proposed by Machida and Namiki to test some measurement theories[8]. The interference fringe has been clearly observed as they predicted. This experiment clarifies that the disappearance of the coherence among the eigenstates never happens through the spectral decomposition step. We make it clear that decoherence in between the eigenstates never happens through orthogonal decomposition in the measurement process. This experimental result excludes measurement theories which derive decoherence with a direct link to the orthogonal decomposition.

We have shown here the high potential of multilayer interferometer for cold neutrons and its wide application for direct tests of quantum-mechanical lows and nature.

The authors acknowledge Prof. M.Namiki for fruitful discussion. This work was supported by ISSP/JAERI facility and financially by a Grant-in-Aid for Scientific Research from Japanese Ministry of Education, Science and Culture (No.04244103).

References

[1] H. Rauch; Proc. 1st ISQM, Eds. S. Kamefuchi et al. Phys. Soc. of Japan, (1984) pp.277

[2] M.A.Kastner: Phys.Today 46(1993)24; Rev. Mod. Phys. 64(1992)849

[3] C.S.Adams, M.Sigel, J.Mlynek Phys.Rep. 240(1994)143-210

[4] H.Funahashi, et al.; Phys. Rev., A54(1996)649

[5] T.Ebisawa et al. Phys. Rev. A57(1998)4720.

[6] M.Hino et al. in this proceedings

[7] T.Kawai et al. in this proceedings

[8] S. Machida and M.Namiki; Proc. 2nd ISQM, Eds. M.Namiki et al. Phys. Soc. of Japan, (1987) pp.355

[9] H.Everett III, Rev. Mod. Phys. 29(1957)454

[10] W. H. Zurek, Phys. Rev. 24D(1981)1516, 26D(1982), Phys. Today, 44(1991)36

[11] M.Namiki, et. al, Decoherence and Quantum Measurements, World Scientific, 1998

Quantum Coherence and Decoherence - ISQM - Tokyo '98
Y.A. Ono and K. Fujikawa (Editors)
© 1999 Elsevier Science B.V. All rights reserved.

Quantum dynamical correlation of protons in water at T = 298K: New Raman light scattering and neutron Compton scattering experiments

C. A. Chatzidimitriou-Dreismann[a,b,*], T. Abdul-Redah[a], and R. M. F. Streffer[a]

[a] Iwan N. Stranski Institute for Physical and Theoretical Chemistry, Technical University of Berlin, Str. des 17. Juni 112, D-10623 Berlin, Germany

[b] Department of Physics, Uppsala University, P.O. Box 530, S-75121 Uppsala, Sweden

Recent theoretical work indicated the possible experimental accessibility of short-lived quantum entanglement (QE) of adjacent protons or other particles in condensed matter, even at room temperature. It was predicted that quantum entangled objects do not interact "properly" with an applied external field, thus showing "anomalous" scattering behaviour. Experimental evidence for this novel effect has been provided using Raman light scattering (RLS) and neutron Compton scattering (NCS) on liquid H_2O/D_2O mixtures and other systems. The results of these experiments, which have found so far no conventional interpretation, suggest for the first time direct evidence for short-lived QE of adjacent protons (and deuterons) in condensed systems at ambient conditions.

1. INTRODUCTION

The basic importance of quantum dynamics of proton transfer and H-bonds in physics, chemistry and biology is well known. Besides quantum aspects of single particles (e.g. quantum tunnelling) there exist various quantum correlation (or entanglement, QE) effects between *two* or *more* particles which have been studied intensively during the last two decades [1]. The experiments presented here have been motivated by our qualitative theoretical investigations [2]. At ambient conditions, the lifetime of the QE, i.e. the so-called *decoherence time* τ_D, is very short [1] and therefore it is commonly believed to have no experimental significance. In contrast, we suggested [2] that entangled objects may cause an "anomalous" component in the scattered (light or neutron) field. Intuitively, one may appreciate the existence of this new effect in such experiments, in which the interaction time between the probe and the scattering center is of the order of, or less than, τ_D. Our Raman light scattering (RLS) experiment [3] on liquid H_2O/D_2O mixtures succeeded to measure the predicted "anomaly" in the scattered light, thus providing experimental evidence for the QE of

*This work has been supported, in parts, by the Fonds der Chemischen Industrie (Frankfurt a.M.) and by a grant from the Royal Swedish Academy of Sciences (Stockholm).

the fermionic OH (and bosonic OD) oscillators, which combine electronic with nuclear protonic and deuteronic degrees of freedom. Since the neutrons are scattered by the nuclei only, the neutron Compton scattering (NCS) experiment [4], revealing essentially the same "anomaly", provides for the first time direct evidence for QE of adjacent H (and D) nuclei in condensed systems at room temperature.

2. RAMAN LIGHT SCATTERING

The relevant quantity for this experiment is the ratio R of the cross sections S_{OH} and S_{OD} for the stretching vibrational Raman transition of the OH and OD oscillators, respectively. S_{OH} and S_{OD} are obtained by integrating the scattered intensities A_{OH} and A_{OD} after frequency reduction [3] and normalisation by the respective H and D mole fractions x_H (=1-x_D) and x_D of the H_2O/D_2O mixtures. This gives:

$$R = \frac{A_{OH}/x_H}{A_{OD}/x_D} = \frac{S_{OH}}{S_{OD}}. \tag{1}$$

R was determined for pure H_2O and pure D_2O giving R^p, and for H_2O/D_2O mixtures with various x_D giving R^m. For more experimental details see [3].

The result of these experiments show that the ratios R^m of the mixtures of H_2O/D_2O are much smaller than R^p (see Fig. 1). The most striking result, however, is the strong dependence of the ratio of the molar scattering cross sections S_{OH}/S_{OD} on the H/D compositions x_D of the mixtures.

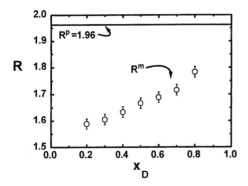

Figure 1. Dependence of the ratio $R^m = S_{OH}/S_{OD}$ of the integral Raman scattering cross sections of the OH stretching and the OD stretching vibration on the D content x_D of the H_2O/D_2O mixtures. R^p is the ratio of the cross sections of pure H_2O and D_2O (full horizontal line).

3. NEURTON COMPTON SCATTERING

These experiments have been performed at the eVS instrument using the spallation source of the ISIS Facility, Rutherford Appleton Laboratory (UK). Under the prevalent experimental conditions (energy transfer: 5-20 eV), the total scattered intensity I_i is always assumed to be proportional to the product of the number of particles N_i and the scattering cross section σ_i (of isotope i) [4]. Accordingly, the relation

$$\frac{I_H}{I_D} = \frac{N_H}{N_D} \cdot \frac{\sigma_H}{\sigma_D} = \frac{N_H}{N_D} \cdot Q \qquad \text{or equivalently} \qquad Q = \frac{I_H / x_H}{I_D / x_D} \qquad (2)$$

strictly holds. Since the ratio N_H/N_D is accurately known by sample preparation, the ratio I_H/I_D of the measured intensities can be used to determine the ratio of the scattering cross sections $Q=\sigma_H/\sigma_D$ of protons and deuterons

The main result of the NCS experiments on H_2O/D_2O mixtures is the strong dependence of Q on the D mole fraction x_D (see Fig. 2). This variation represents the new effect being inexplicable by standard neutron scattering theory, since the latter predicts $Q = 10.7$ independently of x_D. As can be seen in Fig. 2, these results were obtained repeatedly within different experimental periods (between March 1995 and June 1998) using different experimental setups. Whereas the data represented by the open circles were obtained using a uranium foil analyzer, all the other points were obtained using a standard gold foil analyzer. The slight difference between these two data sets might be due to the different energies the neutrons must have to be absorbed by the different nuclei of the two foils. This implies then that the magnitude of the considered QE effect may depend on the energy and/or momentum transfer the neutrons acquire during the scattering process (cf. the contribution of E. B. Karlsson et al. in this volume).

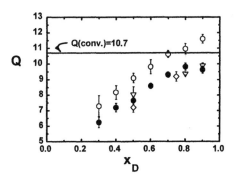

Figure 2. Dependence of the ratio $Q = \sigma_H/\sigma_D$ of the neutron scattering cross sections of H and D on x_D. The full line represents the conventionally expected value of $Q = \sigma_H/\sigma_D$ ($= 10.7$). Different symbols indicate results obtained in different experimental periods (between March 1995 and June 1998) involving different experimental setups (see text).

5. DISCUSSION

Two quite different, but conceptually related, inelastic scattering experiments are presented, the results of which clearly confirm the existence of the theoretically expected [2] protonic and deuteronic QE. Importantly, the NCS experiment on the H_2O/D_2O mixtures [4] provided for the first time direct experimental evidence for the new effect under consideration. This quantum correlation effect has found thus far no conventional (classical or quantal) interpretation. In this context, of particular interest is the fact that the relevant interactions in these two experiments are completely different - i.e., electromagnetic versus strong interaction. This fact corroborates the considered effect to be caused by QE, rather than by specific quantum interactions. Recall that QE may exist even in the absence of interactions between the entangled objects. It may be helpful to emphasize that the experimental results can alternatively be viewed to be related to an "effective change" of N_H and/or N_D being caused by QE [3,4].

In our previous theoretical work ([2] and the theoretical part of [3]) it has been shown that the intensity of photons or neutrons being scattered from entangled quantum systems may acquire an "anomalous" component. E.g., to be specific, in the case of the Raman experiment the scattering cross section depends on the intrinsic symmetry properties of the matrix element $\langle \Phi_{gr}(1,2)|H_{int}|\varphi_{gr}(1)\varphi_{ex}(2)\rangle$ with respect to parity transformation [3]. $\Phi_{gr}(1,2)$ represents the wavefunction of two entangled particles in their ground state before interaction with the light field and H_{int} is the relevant electromagnetic interaction Hamiltonian describing the Raman transitions. The two single-particle wavefunctions $\varphi_{gr}(1)$ and $\varphi_{ex}(2)$ describe the two oscillators after the Raman excitation of one of them (with wavefunction $\varphi_{ex}(2)$), which are now no more entangled. The decoherence of the two oscillators is related to the changed interaction with their environments. For more details, see [3]. Similar theoretical considerations motivated the NCS experiment [4], the results of which give direct evidence for the nuclear QE under consideration.

Our experimental investigations are presently extended to the study of liquid *urea*/H_2O/D_2O solutions and aqueous *electrolyte solutions*. Currently we study liquid C_6H_6/C_6D_6 mixtures (*benzene*) using RLS and NCS, in order to investigate to which extent the new protonic quantum correlation effect does also affect the covalent C-H bond. Very recently, we started to investigate the possibility to detect this effect with *coherent* neutron scattering using *neutron reflectivity* and *neutron interferometry*. Furthermore, we presently extend our work on QE into the topics of chemical reactions and surface phenomena by considering the *water splitting* process at the mercury electrode.

The considered new QE effect apparently affects certain important aspects of the structure and/or dynamics of proton transfer and chemical bonding in condensed systems. Therefore, it may have considerable applications in various physical, chemical and biological systems.

REFERENCES

1. (a) E.B. Karlsson and E. Brändas, Modern Studies of basic quantum concepts and phenomena. Proceedings of the Nodel Symposium No. 104, Physica Scripta (1998) in press; (b) B. d'Espagnat, Conceptual Foundations of Quantum Mechanics, 2nd edn, Addison-Wesley, Redwood City, 1989; (c) R. Omnès, The Interpretation of Quantum Mechanics, Princeton University Press, Princeton, 1994; (d) D. Giulini, et al. Decoherence and the Appearance of a Classical World in Quantum Theory, Springer, Berlin, 1996.
2. C.A. Chatzidimitriou-Dreismann, Adv. Chem. Phys. 80 (1991) 201; *ibid.* 99 (1997) 393.
3. C.A. Chatzidimitriou-Dreismann, U.K. Krieger, A. Möller, and M. Stern, Phys. Rev. Lett. 75 (1995) 3008.
4. C.A. Chatzidimitriou-Dreismann, T. Abdul-Redah, R.M.F. Streffer, and J. Mayers, Phys. Rev. Lett. 79 (1997) 2839.

Quantum Coherence and Decoherence - ISQM - Tokyo '98
Y.A. Ono and K. Fujikawa (Editors)
© 1999 Elsevier Science B.V. All rights reserved.

Evidence for anomalous correlations of protons in a metallic hydride

E.B.Karlsson[a], C.A.Chatzidimitriou-Dreismann[b], T.Abdul-Redah[b], R.M.F.Streffer[b], B.Hjörvarsson[c], J.Öhrmalm[c] and J.Mayers[d]

[a]Department of Physics, Uppsala University, P.O.Box 530, S-75121 Uppsala, Sweden

[b]I.N.Stranski-Institut für Physikalische und Theoretische Chemie, Technische Universität Berlin, Str.d.17.Juni 112, D-10623 Berlin, Germany

[c]Department of Physics, Royal Institute of Technology, S-10044 Stockholm, Sweden

[d]Rutherford-Appleton Laboratory, Chilton, OX 11 0QX, England

One of the consequences of the superposition principle of quantum mechnics is entanglement, i.e. the non-separability of wave functions for the individual particles involved. The existence of entanglement in real physical systems is now well established, not only for pairs of photons [1], but also for systems involving atoms and photons [2] or atoms and phonons [3] isolated in cavities or traps. Entanglement of particles in more strongly interacting environments is more difficult to observe because of the very fast loss of coherence and for particles in condensed matter it has long been considered inconceivable that any traces of entanglement could remain to be observable in experiments.

The present experiments deal with protons and deuterons in metallic hydrides, NbH_x and NbD_y and the mixed hydrides NbH_xD_y, where the protons (or deuterons) are situated with a separation of about 2.5 Å on a sublattice which forms strings throughout the crystal. At the concentrations used here (typically x = y = 0.85, or for the mixtures, x+y = 0.85, these strings are not completely filled so that H-(or D-) atoms can sit in pairs, triplets, etc, surrounded by vacancies. The main interactions of these protons (to be discussed in more detail below) are through the vibrations of the protons or deuterons themselves, the vibrations in the Nb lattice and the interaction with the conduction electrons. The frequencies and momentum transfers of the relevant collisions with the protons or deuterons are well known and the metallic hydride constitutes therefore a reasonably well characterized environment from the quantum point of view.

EXPERIMENTS

The experiments method used determines the time window over which the quantum system can be observed. It was known when this experiment was started that both Raman scattering [4] and scattering of neutrons with energies in the eV range [5] on water had show strong cross section anomalies, interpreted as entanglement effects on the protons (or deuterons) in the hydrogen bonds involved. Both these methods sample the situations at the 10^{-16}-10^{-15} s level. It was therefore of interest to find out whether similar local quantum effects would exist for the closely spaced protons in the metallic hydride environment. The scattering of eV neutrons is also well suited for this system.

The eV-neutron spectrometer at ISIS (Rutherford-Appleton Laboratory, UK) is a neutron Compton scattering instrument with so-called inversed geometry (i.e. the neutron energies are selected after the scattering). The time of flight spectra of the outgoing neutrons contain (usually well separated) speak due to the scattering of H,D and Nb, the areas A_i of which must be proportional to the product of the number of nuclei and the corresponding cross section, N_i σ_i, for each isotope if the atoms are considered as individual objects. With the concentrations x and y for the different hydrides one should then expect the area ratios (where $\sigma_H = 81.67$, $\sigma_D = 7.64$ and $\sigma_{Nb} = 6.255$ barns)

$A_H / A_{nb} = x \ \sigma_H / \ \sigma_{Nb} = 13.1$; $\qquad A_D / A_{nb} = y \ \sigma_D / \ \sigma_{Nb} = 1.22$;

$A_H / A_D = x \ \sigma_H / y \ \sigma_D = 10.7$ (1)

This is, however not what is observed and the cross section ratios σ_H / σ_{Nb} and σ_H / σ_D calculated from the measured area ratios and the concentrations x and y fall far below these values as seen in Table 1, while σ_D / σ_{Nb} comes closer to he "classically" expected value. The fact that the anomaly is largest when the lightest particle (the proton) is involved would not be unexpected if it were due to quantum effects.

Table 1
Angular averaged cross section determinations

x	y	A_H / A_{nb}	A_D/A_{nb}	σ_H / σ_{Nb}	σ_D / σ_{Nb}
0.73(3)	0.02(2)	8.7(4)	------	11.2(9)	------
0.61(3)	0.28(1)	6.3(2)	0.26(3)	10.3(8)	0.90(10)
0.39(2)	0.46(1)	3.9(1)	0.52(1)	10.0(8)	1.13(5)
0.16(2)	0.70(1)	1.54(6)	0.74(1)	10(2)	1.06(3)
0.03(2)	0.80(1)	------	0.87(1)	------	1.09(3)
				13.1	1.22

Another striking observation was that the mesured area ratios for neutrons scattered at different angles were not constant for the proton peaks in the time-of-flight spectra, but decreased with increasing scattering angles. This is not expected from the conventional scattering theory used for evaluating the data from the eV-instrument [6], and was also not observed in this experiment for the deuteron peaks, whose areas were independent of scattering angle (D-and H-peak areas were both normalized to the Nb-peak areas). Here it is important to note that the "scattering time" τ_S, which is the time window for the observation of the target nuclei, is actually shorter for the higher scattering angles Θ. A quantitative expression for τ_S has been given in a work by Watson [7] as $\tau_S \approx M\hbar / [q<p_M^2>^{1/2}]$, where M is the mass of the scatterer, $<p_M^2>^{1/2}$ the width of their momentum distribution (in their vibrational state), and q the momentum transfer during the neutron scattering. Since q is related to the scattering angle by $q = \sqrt{2m E_o}$ tg Θ, where E_o is the energy of the scattered neutrons (in the eV-instrument selected by filtering) and m the neutron mass, it is possible to express τ_S in terms of Θ as

$$\tau_S \approx M\hbar / 2[MmE_oE_{vibr}]^{1/2} \text{tg } \Theta \tag{2}$$

When the observed σ_H / σ_{Nb} cross section ratios for different scattering angles Θ are plotted as function of τ_S (see Fig. 1) it is realized that the anomalies occur only for the shortest observation times, while for times exceeding $0.6 \ 10^{-16}$ s the H/Nb cross section approaches the classical value 13.1. The anomaly can therefore be tentatively ascribed to a quantum entanglement effect that exist over times of the order of 10^{-16}s.
Most of the result mentioned here were obtained at room temperature (although data for 150K ae also included in Fig. 1). Experiments at temperatures down to 20K showed however very small changes, probably due to the effect that the H-atoms in this hydride are essentially in the zero-point vibrational state even at 300K and that quantum decoherence effects in this particular environment are only weakly T-dependent (see below)

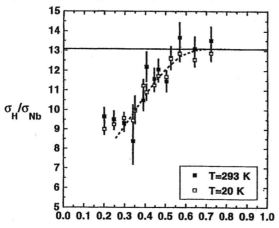

Figure 1

H/Nb cross section ratio as function of scattering time τ_S

334

DISCUSSION

At present, no theory has been developed for the cross section for neutron scattering on a proton in an entangled proton pair (or larger entangled structure), but there are reasons to expect hat interference terms will reduce it compared to that for scattering on an individual proton. Such effects may be different for proton (fermion) and deuteron (boson) pairs, etc. It remains to discuss whether τ eoh $= 10^{-16}$s would be a reasonable coherence time for a quantum entanglement of proton pairs (or larger local proton structures) in the actual environment. An order of magnitude estimate based on the formula by Joos and Zeh [8]

$$(\tau \text{ eoh })^{-1} \approx \kappa (ko)^2 n(x-x)^2 \tag{3}$$

with $\kappa \approx 1$ and a linear extension of a local object (x-x) = 2.5 Å which is hit by n quanta from the environment with momentum ko, actually gives a coherence time of this magnitude if the dominating perturbation is due to neighbouring protons hitting it when vibrating with frequency 1014s-1 and wavenumber $k_o =$ 6Å-1. The interaction with the Nb-phonons is weaker and the influence of the conduction electrons can be neglected, since only Fermi surface electrons can contribute to the decoherence [9].

The present experiments (together with the Raman [4] and neutron Compton scattering experiments [5] mentioned earlier) have therefore given evidence that quantum entanglement of protons can have measurable influences in condensed systems. Furthermore, they may be of importance in the initial stages of chemical reactions and therefore have far-reaching consequences, for instance for H-binding in chemistry and biology.

REFERENCES

1. A.Aspect, P.Grangier, and G.Roger, Phys.Rev.Lett.49(1982), 91
2. M.Brune, E.Hagley, J.Dreyer, X.Maitre, A.Maali, C.Wunderlich, J.M.Raimond, and S.Haroche, Phys.Rev.Lett.77 (1996), 4887
3. C.Monroe, D.M.Meekhof, B.E.King, and D.J.Wineland, Science 272(1996) 1131
4. C.A.Chatzidimitriou-Dreismann, U.K.Krieger, A.Möller, and M.Stern, Phys.Rev.Lett.75(1995) 3008
5. C.A.Chatzidimitriou-Dreismann, T.Abdul Redah, R.M.F.Streffer, and J.Mayers, Phys.Rev.Latt.79(1997), 2839
6. P.A.Seeger, A.d.Taylor, and R.M.Brugger, Nucl.Instr.Meth.Phys.Res.A 240(1985), 98
7. G.I.Watson, J.Phys.:Condens.Matter 8(1996), 5955
8. E.Joos and H.D.Zeh, Z.Phys. B59(1985), 233
9. E.Karlsson, R.Wäppling, S.W.Lidström, O.Hrtmann, R.Kadono, R.Keifl, R.Hempelmann and D.Richter, Phys.Rev.B52(1995) 6417

Quantum Coherence and Decoherence - ISQM - Tokyo '98
Y.A. Ono and K. Fujikawa (Editors)
© 1999 Elsevier Science B.V. All rights reserved.

The Wien filter as a tool to investigate the fundamentals of quantum mechanics

P. Sonnentag and F. Hasselbach

Institut für Angewandte Physik, Universität Tübingen,
Auf der Morgenstelle 10, D-72076 Tübingen, Germany

The Wien filter, i.e. a crossed electric and magnetic field, can be used in an electron interferometer for which-path experiments, as a quantum eraser, and to investigate decoherence.

1. COMPLEMENTARITY AND WAVE-PARTICLE DUALITY

1.1. Definition of complementarity

The notion of complementarity was introduced into physics by Niels Bohr in his famous Como lecture in 1927 [1]. He denoted thereby the logical relation between two descriptions or sets of concepts which, though *mutually exclusive*, are nevertheless *both necessary* for an *exhaustive* description of the situation [2, p. 363]. Already 19th century physics contained some kind of complementarity in the irreducibility of the second law of thermodynamics to mechanistic principles and in the mutual exclusiveness of the concept of temperature and a detailed description of the atoms. But in quantum mechanics, complementarity applies to one and the same object and not to different levels of description.

For Bohr, first of all space-time co-ordination on the one hand and the claim of causality or dynamical conservation laws on the other hand were complementary features of the description of atomic phenomena [1]. Another particular case of complementarity is wave-particle duality. This manifests itself especially in "welcher Weg" (which-path) experiments where observation of the 'path' taken by a particle destroys interference between the corresponding states. Bohr considered complementarity as resulting from the quantum postulate which implies that any observation will involve a non-negligible and 'uncontrollable' interaction with the agency of observation. In Heisenberg's uncertainty relations he saw a confirmation of his thesis that complementarity could not lead to logical contradictions, because the sharp exhibition of one of such complementary notions necessitates an experimental setup which differs totally from that required for the other [2, p. 363]. This again is illustrated clearly by the "welcher Weg" experiments.

But Bohr did not explicitly define complementarity. And Einstein [3] criticized that he, too, was not able to give the complementarity principle a sharp formulation. Therefore, already in the early days of quantum mechanics there was dissent among the leading physicists about the importance and the meaning of complementarity [2, p. 368 f.]. For example, Pauli [4] ascribed complementarity to two classical concepts which belong to the same classical mode of description (e.g., the particle picture) but are mutually exclusive in *quantum mechanics*, and not to two descriptions which are mutually exclusive already in *classical physics*. And even nowadays it is not clear if the wave picture is associated with the dynamic momentum-energy description and the particle picture with the kinematic space-time description, or vice versa [5]. Sen, Basu and Sengupta [6] make a distinciton between two kinds of complementarity, one being that of *variables*, the other that of *properties*. They also point out [7] that instead of talking about which-path information, one should better speak of which-*state* information because in quantum mechanics the notion of a path can be used only in a very restricted sense.

1.2. Mechanisms enforcing complementarity in "welcher Weg" experiments

In the literature, different *mechanisms* for the enforcement of complementarity are mentioned: (*possibility* of getting) "welcher Weg" information (distinguishability of paths), uncertainty relations, non-commuting dynamical variables (including canonically conjugate variables), pairs of variables obtained by a Fourier transformation of the state vector, change in the momentum spectrum (including random momentum transfer), entanglement with orthogonal states (either of the environment or of internal degrees of freedom or of an explicit which-path detector), randomization of phases, dephasing, mixed state, incoherent superposition of interference patterns, coherent superposition of probability amplitudes of interference patterns, difference in arrival times / longitudinal shift of the wave packets relative to each other, state reduction. Some authors claim that the mechanism *they* mention is responsible for complementarity in *all* experiments, whereas others state that in different experiments *different* mechanisms are at work while others do not play a role at all in these cases. We think instead that probably always *almost all* of the mechanisms mentioned are at work (though *one* of them would already be enough to ensure complementarity), but sometimes one of them can be considered the *cause* of the others. E.g., momentum transfer is always accompanied with entanglement because of conservation of momentum. And the equivalence of entanglement with orthogonal states of the environment and of dephasing has been shown by Stern, Aharonov and Imry [8].

2. THE WIEN FILTER AS A "WELCHER WEG" EXPERIMENT

The Wien filter, i.e., a crossed electric and magnetic field both perpendicular to the optical axis, is said to be in its matched state if the electric and magnetic force for the main energy component of the electron (or ion) beam cancel each other. In an interferometer, the Wien filter acts on a *wave packet* for which the matching condition is fulfilled merely as a wave packet shifter and leaves the phase difference unchanged (Fig. 1 a)). The wave packet shift is caused by different electric potentials on the two laterally separated paths in the Wien filter leading to different group velocities, the vanishing phase difference is due to opposite and equal phase shifts caused by the electric scalar and the magnetic vector potential. The decreasing overlap of the two parts of the wave packets with increasing excitation of the filter leads to decreasing fringe contrast.

It should be noticed that for the Wien filter phase relations between *different* energy components building up the wave packet do not matter – at least for *narrow* energy distributions. Therefore, e.g. for a mixed state which can be decomposed into monochromatic waves the contrast-reducing effect on the interference pattern turns out to be the same, provided that the energy spectrum equals that of the pure state of the wave packet, only

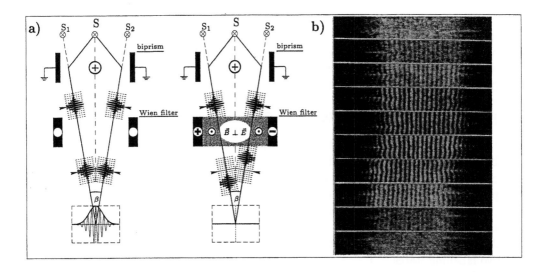

Figure 1. a) Electron biprism interferometer with Wien filter switched off (left) and Wien filter in its excited state (right). The wave packet shift exceeding the coherence length leads to the disappearance of interference. b) Restoration of contrast by a Wien filter. The longitudinal shift of the wave packets caused by electrostatic deflection elements (top) can be compensated (middle) and overcompensated (bottom) with the Wien filter [9].

the mechanism of the reduction of contrast being different: in case of the ensemble of monochromatic waves with statistically distributed phases it is the displacement of the incoherently superimposed intensity distributions relative to each other whereas for the wave packet it is the decrease in overlap or the non-stationarity of the phase difference between the two parts of the wave packet.

For the Wien filter, the reason for the loss of contrast in the interference pattern seems to differ from all the mechanisms mentioned above, but it is similar to that in a light interferometer with different lengths of the arms. And of course the loss of contrast can also be interpreted as being due to the increasing (*possibility* of getting) which-path information arising from the different times of flight of the two wave packets – though in the present state of our experiments we cannot really *get* this information because of the lack of knowledge of the instant of time when the electron was emitted.

The delay between the two wave packets can, loosely speaking, also be seen as a kind of entanglement: If we write the ordinary state space of the particle as a tensor product of the state space $\mathcal{X}$ of positions on the detection screen and of the state space $\mathcal{Z}$ of positions orthogonal to that screen (i.e., along the optical axis z), then because of the longitudinal distance Δz of the positions of the maxima of the partial wave packets we have entanglement with states in the space $\mathcal{Z}$. Yet this is only fully justified if the state of the split wave packet before entering the Wien filter is itself not an entangled state but a direct product state in $\mathcal{X} \otimes \mathcal{Z}$. In this case we really have entanglement with increasingly orthogonal states in $\mathcal{Z}$ when the excitation of the Wien filter is increased.

By increasing the excitation of the Wien filter we get a *continuous* transition from complete interference to complete which-path information (see Fig. 1 b)) (cf. [10–12]). This allows to test the inequality $\mathcal{V}^2 + \mathcal{D}^2 \leq 1$ between visibility $\mathcal{V}$ and distinguishability $\mathcal{D}$ derived by [13] (similar relations were found by [14,15] and [16] based on information theory resp. coherence considerations).

3. QUANTUM ERASER, DECOHERENCE

By combining two Wien filters with opposite excitation, the effect of the first of these can be undone. Again, this can be interpreted in different ways, either by the loss of "welcher Weg" information or by the fact that the z-coordinates of the maxima of the two parts of the wave packet are made equal again. The second Wien filter can be looked upon as a quantum eraser, but with the difference that we do not loose intensity by this process. This is because we do not project onto – or at least detect in correlation with – a 'labelling space' state (of which both packets have equal components), but instead equalize the two labelling states.

Furthermore, this demonstrates that coherence is not lost but only hidden after the first filter. Fig. 1 b) shows how interference which was lost due to the electrostatic fields of the deflection elements is restored by the Wien filter. This suggests that there is no decoherence emerging from the interaction of the electron with the electromagnetic fields or with the environment.

Acknowledgement
We thank Harald Kiesel for helpful discussions.

REFERENCES

1. N. Bohr, Nature, 121 (1928) 580.
2. M. Jammer, The Conceptual Development of Quantum Mechanics, Tomash Publishers, American Institute of Physics, 2. ed., 1989.
3. A. Einstein, in: P. A. Schilpp (ed.), Albert Einstein: Philosopher-Scientist, The Library of Living Philosophers, Evanston (Ill.), 1949.
4. W. Pauli, in: H. Geiger and K. Scheel (eds.), Handbuch der Physik, vol. 24/1, Springer, Berlin, 1933, p. 89.
5. W.G. Holladay, Am. J. Phys., 66 (1998) 27.
6. D. Sen, A.N. Basu, and S. Sengupta, Z. Naturforsch. A, 52 (1997) 398.
7. D. Sen, A.N. Basu, and S. Sengupta, Helv. Phys. Acta, 67 (1994) 785.
8. A. Stern, Y. Aharonov, and Y. Imry, Phys. Rev. A, 41 (1990) 3436.
9. M. Nicklaus and F. Hasselbach, Phys. Rev. A, 48 (1993) 152.
10. T.J. Herzog et al., Phys. Rev. Lett., 75 (1995) 3034.
11. B. Dopfer, H. Weinfurter, and A. Zeilinger, Verhandl. DPG, 33 (1998) 191.
12. S. Dürr, T. Nonn, and G. Rempe, submitted to Phys. Rev. Lett. (1998).
13. B.-G. Englert, Phys. Rev. Lett., 77 (1996) 2154.
14. W.K. Wootters and W.H. Zurek, Phys. Rev. D, 19 (1979) 473.
15. A. Zeilinger, Physica B, 137 (1986) 235.
16. L. Mandel, Opt. Lett., 16 (1991) 1882.

Quantum Coherence and Decoherence - ISQM - Tokyo '98
Y.A. Ono and K. Fujikawa (Editors)
© 1999 Elsevier Science B.V. All rights reserved.

Decoherence of wavepackets under inelastic scattering

Yoshimasa Murayama and Masaki Honda

Dept. Materials Sci. & Tech., Niigata University, Niigata 950-2102, Japan

Wavepacket (WP) dynamics in (1+1)-dimensions is discussed when it goes into a damping region for $t > 0$ and $x > 0$ [1]. The solution of a scattered WP is formulated first. The original WP is damped at the same time wavelets are generated. How the WP decoheres when it suffers inelastic scattering is discussed. A few simulation results are presented.

1. INTRODUCTION

Let us define an unperturbed WP as usual:

$$\varphi_0(x,t) = N^{-1/2} \sum_k A_k^p \chi_k(x) e^{-iE_k t/\hbar} \tag{1}$$

Here A_k^p is the Gaussian envelope function of k with a central wavevector p, and $N^{-1/2}$ is the normalization factor for the envelope. $\chi_k = L^{-1/2} e^{ikx}$, where L is the spatial extent for this problem. This WP goes into a damping region for $t > 0$ and $x > 0$, where inelastic scatterers exist and the damping rate in the lowest Born approximation is given by

$$\frac{1}{\tau_k} = \sum_{q\pm} \frac{1}{\tau_{k,k\pm q}} \qquad .$$

$$= \frac{2\pi}{\hbar} \sum_{q\pm} |\langle k|V|k\pm q\rangle|^2 \{\delta(E_k - E_{k\pm q} + \hbar\omega_q) + \delta(E_k - E_{k\pm q} - \hbar\omega_q)\}. \tag{2}$$

When WP suffers some inelastic scattering, the following Møller state

$$\chi_k(t) = e^{-itE_k/\hbar} \exp_+\{-i\int_0^t V(t')(e^{i\omega t'} + e^{-it\omega t'})dt'\}\chi_k(0) \approx e^{-it(E_k + \Delta E_k)/\hbar - t/\tau_k}\chi_k(0) \tag{3}$$

gives damped wavefunctions on each order of the interaction V in a series-expanded form. Here $V(t) = e^{itH}Ve^{-itH}$ is in the interaction representation. With regard to the lowest order, the diagonal part of the interaction Hamiltonian is $\Delta E_k - \frac{i}{\tau_k}$, of which $1/\tau_k$ was given above. We discard the energy shift term for simplicity.

Let us take the simplest δ-type interaction for V which assumes optic phonon-like scatterers and accordingly $\omega_q = \omega$. Considering that there are four scattering processes: *abs-f* (absorption of a "phonon" on forward scattering), *abs-b* (absorption of a "phonon" on backward scattering), *em-f* (emission of a "phonon" on forward scattering) and *em-b* (emission of a "phonon" on backward scattering), the expression finally reached for the damping rate is

$$\frac{1}{\tau_k} = \frac{\hbar}{2m} G \left(\frac{1}{K_+} + \frac{1}{K_-} \right), \tag{4}$$

where $G = (2m)^2 n_i g^2 / \hbar^4$ with the density of scatterers n_i and the strength of the interaction g. $K_\pm = \sqrt{k^2 \pm k_0^2}$ with $k_0^2 = 2m\omega/\hbar$.

Now writing down expressions for damped WP and generated wavelets is straightforward. In this study, we are interested in scattered WP after a finite time inside a finite region. We take up to the lowest order in perturbation for simplicity and disregard the scattering of the generated wavelet by the same scatterers, which will be given with higher order terms in perturbation.

The damped WP is

$$\varphi(x,t) = \varphi_0(x,t) - N^{-1/2} \sum_k A_k^p e^{-iE_k t/\hbar} \chi_k(x) \sum_{k'} \int_0^t dt_i \frac{e^{-t_i/\tau_k}}{\tau_{kk'}}. \tag{5}$$

Similarly, a generated wavelet $\tilde{\varphi}(x,t)$ is

$$\tilde{\varphi}(x,t) = N^{-1/2} \sum_k A_k^p \int_0^t dt_i e^{-iE_k t_i/\hbar} e^{-t_i/\tau_k} \sum_{k'} e^{-iE_{k'}(t-t_i)/\hbar} \frac{1}{\tau_{kk'}} \chi_{k'}(x - x_i) e^{ikx_i}, \tag{6}$$

with $x_i = \hbar k t_i / m$.

Exactly speaking, the energy-momentum conservation law in $1/\tau_{kk'}$ regarding these equations does not hold because the wavelet is considered within a finite time interval after scattering. Although the following calculation is valid only for $t - t_i \to \infty$, all results must be correct at least qualitatively, even if we assume the conservation relation. The energy deviation from the conservation condition dies away in proportion to $\hbar/(t - t_i)$.

2. CALCULATION OF GENERATED WAVELET

Calculation proceeds in parallel to the well-established treatment of the WP without scattering. We show only one example, namely, *abs-f*. In *abs-f* the central wavevector of the generated wavelet is $p_+ = \sqrt{p^2 + k_0^2}$. That is, the wavelet is accelerated by absorbing a "phonon" energy. The final expression is

$$\tilde{\varphi}(x,t)_{abs-f} \equiv e^{ipx - i\hbar p^2 t/2m} I_{abs-f}(t) \tag{7}$$

and

$$I_{abs-f} \approx N^{-1/2} \frac{\hbar G}{2m} \cdot \frac{1}{2p_+} \cdot \frac{1}{2\pi} \int dk e^{-\frac{(K_+ - p_+)^2}{4(\Delta k_+)^2}}$$

$$\times \int_0^t dt_i e^{-(\Delta k_+)^2 \left(x - x_i - \frac{\hbar p_+}{m}(t-t_i) \right)^2} e^{i(p_+ - p)(x - x_i) - i\frac{\hbar}{2m}(p_+^2 - p^2)(t - t_i) - \frac{\hbar G}{2m} t_i \left(\frac{1}{p_+} + \frac{1}{p_-} \right)},$$

with $\Delta k_+ = \Delta k (1 - k_0^2/2p^2)$ $(p \gg k_0)$. The steepest descent method, that is $k \approx p$, was applied to the mild part of the integrand and, consequently, $x_i = \hbar p t_i / m$.

When deriving this expression, we assumed that the time interval was too short for the WP to spread according to quantum diffusion, namely, $T = 2m/4\hbar(\Delta k)^2 \gg t$ and

$p \gg (\Delta k)^2 x$. Similar integrals for the other processes, $\tilde{\varphi}_{abs-b}$, $\tilde{\varphi}_{em-f}$, and $\tilde{\varphi}_{em-b}$ were easily obtained by the following substitution in $\tilde{\varphi}_{abs-f}$ given above except in $1/\tau_k$.

$$
\begin{aligned}
abs - b : K_+ &\rightarrow -K_+; p_+ \rightarrow -p_+; \Delta k_+ \rightarrow \Delta k_+, \\
em - f : K_+ &\rightarrow K_-; p_+ \rightarrow p_-; \Delta k_+ \rightarrow \Delta k_-, \\
em - b : K_+ &\rightarrow -K_-; p_+ \rightarrow -p_-; \Delta k_+ \rightarrow \Delta k_-.
\end{aligned}
\tag{8}
$$

.1. Coherence between the original and generated WPs

Since the original WP is given by

$$
\varphi(t) = e^{ipx - i\hbar p^2 t/2m} I_0(t) \tag{9}
$$

with

$$
I_0(t) \approx N^{-1/2} e^{-\frac{\hbar G}{2m} t(\frac{1}{p_+} + \frac{1}{p_-})} e^{-(\Delta k)^2 (x - \frac{\hbar p}{m} t)^2} \times \frac{1}{2\pi} \int dk e^{-\frac{(k-p)^2}{4(\Delta k)^2}},
$$

the interference between the original and generated WP (e.g., I_{abs-f}) is given by

$$
|I_0(t) + I_{abs-f}(t)|^2 = I_0^2(t) + 2I_0(t)\Re(I_{abs-f}(t)) + |I_{abs-f}(t)|^2. \tag{10}
$$

3. RESULTS AND DISCUSSION

(1) $|I_{abs-f}|^2$ is shown in Fig. 2. The assumed parameters are: $p/\Delta k = 20$, $k_0/\Delta k = 10$, $G/\Delta k^3 = 0.05$. The generated wavelet grows its width with time, since the width is mainly determined by the difference between $x = \hbar p_+ t/m$ and $x = \hbar p t/m$ for a fixed t, as suggested in Fig. 1. ($x = \hbar p t/m$ line is the diagonal through the origin O.) All generated components are coherent to one another since they all have the same energy. This is assured by integrating all components for $t = t_i$ in Eq.(6). This coherence results from the fact that all generated wavelets come from a single envelope specified by a coherence length $1/2\Delta k$. Its intensity obviously grows according that the original WP decays.

(2) Since the original and generated WPs have different energies according to inelastic scattering, it is considered that the two WPs are incoherent. However, we deal with WP with a finite spatial coherence length, so that there exists a partial coherence between the original and generated WP immediately after its generation, because the generated component is still inside the coherence length of the the original. The component generated long before and sufficiently apart from the original obviously decohered to the original. This is not shown here, but the interference term in Eq.(10) assures this behavior.

(3) This calculation was limited to the lowest Born approximation. If we take into account the possibility that the generated abs-f wavelet can emit "phonon" in a higher scattering process, then the very new wavelet generated after a sequence of absorption-emission (or emission-absorption) processes must have its central wavevector equal to $\sqrt{p_+^2 - k_0^2} = p$ which is coherent to the original WP. This means that coherence can be recovered after higher-order inelastic scattering.

(4) So far we discussed that coherence survives only between wavepackets having equal energy, as accepted by most physicists. Here we propose a quest. Does coherence really exist between waves with different energies? If it does exist, the condition of equal energy

342

is sufficient, but not necessary. As previously discussed [2], we need some specific physical device for superposition to occur between waves. The Si interferometer for neutron beam and X-ray is one example, because it can superpose waves only when both waves satisfy the same Bragg condition, i.e., the same energy.

Let us devise a Hanbury Brown=Twiss (HBT) type interferometer in order to measure the second-order coherence for electrons in a 2-source-2-detector system. Here interference occurs in the form of $\Re\{\psi_{k_1}^*(D1)\psi_{k_2}^*(D2)\psi_{k1}(D2)\psi_{k_2}(D1)\}$, where D1 and D2 refer to the detector 1 and 2 [3]. This kind of formulation implicitly implies that two waves whatever their energies are are superposed and simultaneously detected at the detectors. In this experiment, the detector detects the wave as a particle. Therefore, the detected point-like particle has its wavevectors spread over a wide spectrum. Accordingly, the characteristic of the detector must be such that is insensitive to energy.

There is so far no clear-cut experimental finding of the HBT interferometry or anti-bunching effect of electrons. If such effect should be observed, it can be said that electrons with different energies can be superposed at least in higher-order interfering phenomena.

For boson like photons, this is the case. It is well known that two photons with frequencies ω_1 and ω_2 can be superposed and give the other photon with $\omega = \omega_1 + \omega_2$ (parametric up-conversion process) when passing through a nonlinear optical crystal. In order for this process to occur, at least partial coherence should exist between photons with different energies.

We thank Prof. F. Hasselbach, Prof. N. Achiwa, and Prof. H. Ezawa for their valuable discussions.

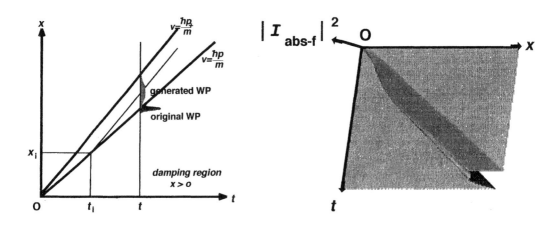

Fig. 1. Conceptual display of the WP dynamics.

Fig. 2. 3-D graphic display of a simulated $|I_{abs-f}|^2$ in the $x - t$-plane.

REFERENCES
1. Y. Murayama, in: *"Quantum Control and Measurement"*, ed. H. Ezawa and Y. Murayama, Elsevier, 1993, pp. 93-98.
2. Y. Murayama, Physica A**147** (1990) 334.
3. M. P. Silverman, IL Nuovo Cimento B**97** (1987) 200.

Quantum Coherence and Decoherence - ISQM - Tokyo '98
Y.A. Ono and K. Fujikawa (Editors)
© 1999 Elsevier Science B.V. All rights reserved.

Time-dependent Bell's inequality in two neutral kaon systems

Fumiyo Uchiyama
Institute of Applied Physics, University of Tsukuba,
Tsukuba 305, Japan (e-mail: uchiyama@kekvax.kek.jp)

Bell's inequality for two neutral kaon systems is extended to taking into account time-dependent situations where "mass" detectors are placed far from two kaon production points, which are more realistic "experimental" setups. In this gedanken experiment, two out of three quantum numbers, strangeness, CP, and mass are "measured". The amount of violation of Bell's inequalities depends on time of measurement of mass quantum numbers. Explicit discussions are given only for two kaons from Φ decays.

1 Introduction

Many tests on Einstein-Podolsky-Rosen nonlocality in the forms of Bell's inequalities have been carried out and thier measurments confirm the predictions from quantum mechanics. Most of these experimental tests however have been limitted in the systems which are interacting electro-magnetically. As for strongly interacting systems, the observation of the interference of two neutral kaons is the first triumphal results justifing kaon-interactions to be described by quantum mechanics. But no experiment has been performed on the direct questions of EPR itself so far eventhough there are several proposals[1] for testing it in strongly interacting systems.

In the gedanken experiment[2] using three quantum numbers, strangeness, CP, and " mass", the time-independent Bell-inequalities for correlated two neutral kaon systems are contradicted with quantum mechanics: A detector system is set up in such a way it could measure probabilities of two quantum numbers of two kaons, their strangeness quantum numbers could be measured at the set S_1 and S_2, CP at the set CP_1, CP_2 and their "mass" at the last set M_1 and M_2 as shown in Fig.1. Denoting the probability of finding kaons with quantum numbers a and b on the two trajectories 1 and 2 respectively by P(a,b), one of Bell's inequalities violating quantum mechanics is

$$P(+1, M_L) \leq P(CP-, M_L) + P(+1, CP-)$$

where the three probability measurements are performed at time t before kaon-deacys become significant, namely $t \ll \tau_S$ where τ_S is the life time of short lived kaons. Calculation of the ineqaulity using quantum mechanics gives

$$\frac{\frac{1}{2}(\varepsilon + \varepsilon^*) - |\varepsilon|^2}{2(1 + |\varepsilon|^2)} \leq 0 \tag{1}$$

which is not satisfied for ε , CP violating parameter, having absolute value and phase $\sim 10^{-3}$ and $\sim 45°$ respectively. There are five other similar ineqalities which is violated quantum mechanically by the same amount given in eq. (1).

In this paper, we introduce a finite time at which the observation of "mass" will be made. This enables us to separate K_L and K_S by utilizing their large life-time difference. Strangeness or CP quantum number of a kaon may be identified by the ordinary methods of observations of their decay products.

Fig. 1. Scheme of the detector system for the gedanken experiment for two correlated kaons.

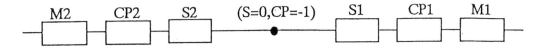

2   Time-dependent Bell's inequalities

For the time-independent case[2], we have used the fact that when one of the kaons belongs to some definite type, say (+1, CP+, m_L), then the other kaon belongs to a matching quantum number type (-1, CP-, m_S) because the same kinds of two kaons can not be detected at the same time when two kaons are produced from a state of CP=-1 and S=0. (+1, CP+, m_L) means that if strangeness is measured, we obtain S=+1 with certainty; if CP is measured we obtain CP=+1 with certainty; if mass is measured we obtain m_L with certainty. In construction of time-dependent Bell's ineqalities, the significant difference from the time-independent arguments is that the times involved may be greater than the life time of the short lived kaon τ_S. This allows the other kaon with a matching quantum numbers to disappear.

However this does not change the course of Bohm's type arguments. We denote the time for kaons to reach at the detectors, S, CP, and M by t_1, t_2, and t_3 where we set $t_1 = t_2 = 0$ for simplicity and $t_3 \gg \tau_S$ so that K_S decays out before they reach "mass" detectors and then set our gedanken experiment for selective events only; namely we use the events which registered arrivals of particles at either of the mass detectors M_1 or M_2. If M_1 register an arrival of a K_L, the kaon pair must belong to group 1 in table 1 and if M_2 register an arrival, the kaon pair must belong to group 2 and outcomes of S and CP counters on either trajectories classify the kaon pair uniquely in the eight mutually exclusive and disjoint types as listed in table 1. n_i (i=1,2,...8) in the forth column denotes the population of each type, which is positive definite. Suppose we make only two measurements, one on each different track at two different times. We denote the probability that the kaon on track 1 at time t_1 is detected having, say, S=+1, and the kaon on track 2 at time t_2 is detected having, say mass $CP = +1$, $P(+1, CP+; t_1, t_2)$. Namely $P(CP+, m_L; t_2, t_3)$ is the probability of finding a pair of kaons, one of which on track 1 is in a state with CP=+1 at time t_2 and the other on track 2, is in a state with m_L at time t_3.

Then reading from table 1, we have

$$P(+1, m_L; t_1, t_3) = \frac{(n_1 + n_2)}{\sum_1^8 n_i}, \quad P(CP+, m_L; t_1, t_3) = \frac{(n_1 + n_3)}{\sum_1^8 n_i}$$

$$P(+1, CP+; t_1, t_1) \geq \frac{(n_2 + n_6)}{\sum_1^8 n_i} \tag{2}$$

The inequality in the last equation is derived from the consideration that some of the kaon pairs in that category may decay before one of them reaches to the last sets of detectors, M_1 or M_2. Since all n_i are positive definite or 0, we get one of time-dependent Bell's inequalities;

$$P(+1, m_L; t_1, t_3) \leq P(CP+, m_L; t_2, t_3) + P(+1, CP+; t_1, t_2) \tag{3}$$

Group	Particle 1	Particle 2	Population
	$(CP+, +1, m_S)$	$(CP-, -1, m_L)$	n_1
1	$(CP-, +1, m_S)$	$(CP+, -1, m_L)$	n_2
	$(CP+, -1, m_S)$	$(CP-, +1, m_L)$	n_3
	$(CP-, -1, m_S)$	$(CP+, +1, m_L)$	n_4
	$(CP+, +1, m_L)$	$(CP-, -1, m_S)$	n_5
2	$(CP-, +1, m_L)$	$(CP+, -1, m_S)$	n_6
	$(CP+, -1, m_L)$	$(CP-, +1, m_S)$	n_7
	$(CP-, -1, m_L)$	$(CP+, +1, m_S)$	n_8

Table 1. Eight mutually exclusive and disjoint measurements at $t_1 = t_2$ and $t_3 \gg \tau_S$ with frequency n_i.

3 Quantum mechanical calculation

The two kaon system produced from hadronic decays of Φ, is characterized at time t by

$$\frac{1}{\sqrt{2}}(|K_0(t) >_1 |\bar{K}_0(t) >_2 - |K_0(t) >_2 |\bar{K}_0(t) >_1) \tag{4}$$

where 1 and 2 are trajectory numbers. Setting the origin of time at the time of decay of Φ,

$$|K_0(t) >_{1,2} = \frac{\sqrt{1+|\epsilon|^2}}{\sqrt{2}(1+\epsilon)} \left\{ e^{-i\lambda_S t}|K_S >_{1,2} + e^{-i\lambda_L t}|K_L >_{1,2} \right\}$$
$$= \frac{1}{\sqrt{2}(1+\epsilon)} \left\{ (e^{-i\lambda_S t} + \epsilon e^{-i\lambda_L t})|K_+ >_{1,2} + (\epsilon e^{-i\lambda_S t} + e^{-i\lambda_L t})|K_- >_{1,2} \right\} \tag{5}$$
$$= e^{-i\lambda_L t} \left\{ \tfrac{1}{2}(1 + e^{-i(\lambda_S - \lambda_L)t})|K^0 >_{1,2} - \tfrac{1}{(1-\epsilon)}(1 - e^{-i(\lambda_S - \lambda_L)t})|\bar{K}^0 >_{1,2} \right\}$$

Omitting subscripts for trajectory numbers,

$$|\bar{K}^0(t) >= \frac{\sqrt{1+|\epsilon|^2}}{\sqrt{2}(1-\epsilon)} \left\{ -e^{-i\lambda_S t}|K_S > + e^{-i\lambda_L t}|K_L > \right\}$$
$$= \frac{1}{\sqrt{2}(1+\epsilon)} \left\{ (-e^{-i\lambda_S t} + \epsilon e^{-i\lambda_L t})|K_+ > + (-\epsilon e^{-i\lambda_S t} + e^{-i\lambda_L t})|K_- > \right\} \tag{6}$$

where (K_S, K_L), $(K_0, \bar{K}_0)$, and $(K+, K_-)$ are eigenstates of mass, strangeness, and CP, respectively. The two mass eigenstates are described by

$$|K_S> = \tfrac{1}{N}[(1+\varepsilon)|K_0 > -(1-\varepsilon)|\bar{K}_0 >]$$
$$|K_L> = \tfrac{1}{N}[(1+\varepsilon)|K_0 > +(1-\varepsilon)|\bar{K}_0 >]$$

$$(7)$$

where $N^{-1} = \sqrt{2(1+|\varepsilon|^2)}$. The probabilities in the inequality can be calculated from these equations;

$$P(+1, m_L; t_1, t_3) = e^{-(\Gamma_L+\Gamma_S)t_1} \left\{ \tfrac{1}{4}(1 + 2Re\ \varepsilon + 4(Re\ \varepsilon)^2)e^{-\Gamma_L(t_3-t_1)} \right.$$
$$-(Re\ \varepsilon + 2(Re\ \varepsilon)^2)\cos\left[\Delta m(t_3 - t_1)\right]e^{-\frac{\Gamma_S+\Gamma_L}{2}(t_3-t_1)}$$
$$\left. +(Re\ \varepsilon)^2 e^{-\Gamma_S(t_3-t_1)} \right\}$$

$$(8)$$

$$P(CP+, m_L; t_1, t_3) = e^{-(\Gamma_L+\Gamma_S)t_1} \left\{ \tfrac{1}{2}|\varepsilon|^2 e^{-\Gamma_L(t_3-t_1)} - 2|\varepsilon|e^{-\frac{\Gamma_S+\Gamma_L}{2}(t_3-t_1)} \right.$$
$$\left. (Re\ \varepsilon)\cos\left[\Delta m(t_3 - t_1) - \theta\right] + 2(Re\ \varepsilon)^2 e^{-\Gamma_S(t_3-t_1)} \right\}$$

$$(9)$$

$$P(+1, CP+; t_1, t_2) = e^{-(\Gamma_L+\Gamma_S)t_1} \left\{ \tfrac{1}{4}(1 - 2Re\ \varepsilon + 4(Re\ \varepsilon)^2)e^{-\Gamma_L(t_2-t_1)} \right.$$
$$-\tfrac{1}{2}|\varepsilon|(1 + 2(Re\ \varepsilon))\cos\left[\Delta m(t_2 - t_1) + \theta\right]$$
$$\left. e^{-\frac{\Gamma_S+\Gamma_L}{2}(t_2-t_1)} + \tfrac{1}{4}|\varepsilon|^2 e^{-\Gamma_S(t_2-t_1)} \right\}$$

$$(10)$$

where $\tan\theta = \frac{Im\ \varepsilon}{Re\ \varepsilon}$. The quantum mechanical prediction for a time-dependent Bell-inequality for $t_1 = t_2$ can be obtained by substituting equations (8), (9) and (10) into Bell's inequality (3). It can be easily seen that the inequality of equation (3) is violated for the case that all t_i (i=1,2,3) are the same. However as soon as t_3 starts to increase keeping $t_1 = t_2$, the inequality is quickly satisfied due to the dominance of the term $P(+1, CP+; t_1, t_2)$ which remains constant. More complete description of analysis on the time-dependent Bell-inequalities will be reported elsewhere.

In this derivation of time-dependent Bell's inequalities, we have used the large life time differences between K_L and K_S in order to "measure" masses of kaons, which are not applicable for B-mesons.

In summary, we apply a logic similar to the one used for the derivation of the foolproof time-independent Bell-inequality, to the limited Φ-decay events in which one of kaons in pairs survives to one of mass detectors placed far from Φ-decay points and demonstrate that time-dependent Bell's inequality for systems of two correlated neutral kaons using only observables can be derived.

Acknowledgment: We thank professors J.Arafune and P.Eberhard for comments.

References

[1] P.H. Eberhard, Nucl. Phys. B 398 (1993) 115; A.Di. Domenico, Nucl. Phys. B450 (1995) 293 and the references therein.

[2] F. Uchiyama, Phys. Lett. A231 (1997) 295 and the references therein.

Quantum Coherence and Decoherence - ISQM - Tokyo '98
Y.A. Ono and K. Fujikawa (Editors)
© 1999 Elsevier Science B.V. All rights reserved.

Decoherence on Discretized Detectors

Ziro Maki[1]

Department of Physics, Faculty of Science and Technology, Kinki University, Higashi-Osaka, 577 Japan

A simple model of measurement apparatus is discussed, in which the decoherence among different pointer-states is demonstrated rigorously. The apparatus is discretized into a set of independent sub-detectors corresponding to each eigen-state of the observable of a system. The model satisfies all axioms of the first kind measurements, and purely quantum-mechanical and mathematically unambigious.

1.MEASURING A SYSTEM BY DISCRETIZED DETECTORS [1]

Let 'A' be an observable of a system S to be measured by the detector M. We assume, for simplicity, the corresponding operator A on the space $\mathcal{H}^S$ has a non-degenerate eigenvalue spectrum $\{a_k | k = 1, 2, \cdots, n\}$, with equations $A |a_k\rangle = a_k |a_k\rangle$ and $\langle a_k | a_\ell \rangle = \delta_{k\ell}$.

Suppose the apparatus M to be a set of sub-detectors $M^{(k)}$ $(k = 1, 2, \cdots, n)$ which are physically independent each other as long as they do not interact with S. Then, the space $\mathcal{H}^M$ for M is a tensor product of that for $M^{(k)}$'s ;

$$
\left.
\begin{aligned}
M &= \{M^{(1)}, M^{(2)}, \cdots, M^{(n)}\} \\
\mathcal{H}^M &= \mathcal{H}^{(1)} \otimes \mathcal{H}^{(2)} \otimes \cdots \otimes \mathcal{H}^{(n)}.
\end{aligned}
\right\}
\tag{1}
$$

Evidently, any observable $\mathcal{O}^k$ of $M^{(k)}$ is detected, upon reading-out, individually after switching off the interaction between S and M, $H_{SM}(t)$, which we introduce only during a short time-interval $\Delta\tau = \tau_2 - \tau_1, \tau_1 < t < \tau_2$. We further assume that, each detector acts digitally ('yes' or 'no') responding to the component eigen-state of A through the interaction. Denoting the 'yes'-state ('no'-state) of the k-th detector by $|1\rangle^{(k)}$ $(|0\rangle^{(k)})$, the whole states of M, i.e. $|\Phi\rangle$ in the space $\mathcal{H}^M$ are represented by the basic (complete and orthonormal) states

$$
\left.
\begin{aligned}
|0\rangle &\equiv |0\rangle^{(1)} \otimes |0\rangle^{(2)} \otimes \cdots \otimes |0\rangle^{(n)} \\
|k\rangle &\equiv |0\rangle^{(1)} \otimes \cdots |1\rangle^{(k)} \otimes \cdots \otimes |0\rangle^{(n)},
\end{aligned}
\right\}
\tag{2}
$$

[1]E-mail ziromaki@phys.kindai.ac.jp

where $k = 1, 2, \cdots, n$. (We shall drop the product symbol $\otimes$ hereafter, unless confused.)
The total state $\Psi(t)$ is subjected to the equation of motion

$$i\hbar \frac{\partial \Psi(t)}{\partial t} = H(t)\Psi(t) \tag{3}$$

with the total hamiltonian

$$H = H_S + H_M + H_{SM}(t), \tag{4}$$

where $H_{SM}(t)$ is the measurement interaction between S and M. The response of the
k-th detector is realized either as a change of the state from $|0\rangle^{(k)}$ to $|1\rangle^{(k)}$ ('yes') , or no
change from $|0\rangle^{(k)}$ ('no'). Employing (fictitous) spin variables σ_i ($i = 1, 2, 3$) one associates
the states $|1\rangle^{(k)}$ and $|0\rangle^{(k)}$ with the vectors $\begin{pmatrix} 1 \\ 0 \end{pmatrix}$ and $\begin{pmatrix} 0 \\ 1 \end{pmatrix}$, respectively in the matrix
representation $\sigma_3^{(k)} = \begin{pmatrix} 1 & 0 \\ 0 & -1 \end{pmatrix}$. Then, the change $|0\rangle^{(k)} \leftrightarrow |1\rangle^{(k)}$ is interpreted as a $180°$
rotation, generated by $\sigma_1^{(k)} = \begin{pmatrix} 0 & 1 \\ 1 & 0 \end{pmatrix}$ during the period $\Delta\tau$.

This kind of changing is brought about dynamically by choosing $H_{SM}(t)$ such that

$$H_{SM}(t) = g(t)\left(\frac{\pi}{2}\right)\hbar \sum_{k=1}^{n} |a_k\rangle \langle a_k| \sigma_1^{(k)}, \tag{5}$$

where $g(t)$ is a coupling parameter normalized as

$$\int_{\tau_1}^{\tau_2} g(t')dt' = 1, \quad g(t) = 0 \quad \text{outside the region} \quad (\tau_1, \tau_2). \tag{6}$$

Now we define t_I and t_F as the time of starting and of finishing the measurement,
respectively, and take the duration $\Delta\tau \equiv t_F - t_I$ to be finite, into which $\Delta\tau$ is embedded
($\Delta\tau < \Delta T$). However, we do not need to specify the time instants t_I and t_F very sharply,
except for imaging that t_I is close enough to τ_1 such that the evolution of S and M by free
motions in $t_I - \tau_1$ are safely negligible, and the same is the case for the interval τ_2 and t_F.

Our measurement processes go on as in usual way. At t_I, if the state of S is prepared
in an eigen-state $|a_k\rangle$ and that of M is in the state $|0\rangle$, the neutral position, the combined
initial state is $\Psi_k(t_I) = |a_k\rangle |0\rangle$. After the unitary time-evolution $U(t_F, t_I)$ by Eq.(3), $\Psi_k(t_I)$
developes to $|a_k\rangle |k\rangle$ (the bases defined at t_F),

$$|a_k\rangle |0\rangle \Big|_{t_I} \underset{\Delta T}{\longrightarrow} U(t_F, t_I) |a_k\rangle |0\rangle = |a_k\rangle |k\rangle \Big|_{t_F} \tag{7}$$

and, therefore, for a general state $|\psi\rangle = \sum_k c_k |a_k\rangle$ ($c_k = \langle a_k|\psi\rangle$) we have a well-known
formula

$$|\psi\rangle |0\rangle \Big|_{t_I} \underset{\Delta T}{\longrightarrow} \sum_k \langle a_k|\psi\rangle |k\rangle \Big|_{t_F}. \tag{8}$$

Or, in terms of the state-operator (or statistical operator) [3] $E^{S+M}(t) \equiv |\Psi(t)\rangle \langle \Psi(t)|$, we see

$$E^{S+M}(t_F) = \sum_{k,\ell} \langle a_k|\psi\rangle \langle \psi|a_\ell\rangle |a_k\rangle \langle a_\ell| \otimes |k\rangle \langle \ell| \bigg|_{t_F}. \tag{9}$$

showing that our discretized apparatus $\{M^{(1)}, M^{(2)}, \cdots, M^{(n)}\}$ behaves as a whole just as usual as conventional one where M is conceived of a single object whatsoever a huge number of degrees of freedom were involved.

In contrast, our sub-detectors have been specified merely by 'spin' variables $\{1, \sigma\}$ and moreover, σ_1 appears only in the time interval $\Delta\tau = \tau_2 - \tau_1$ when $H_{SM}(t)$ is operative. In the time later than t_F, $M^{(k)}$ loses certainly the variable σ_1 from its description, and hence remaining variables take generically the form $a + b\sigma_3$, with a, b being real constants. Therefore, after the measurement interaction went away we are obliged to work with the re-defined system involving no physical operators to shift their states from yes to no or *vice versa*. The exhibited states of M as a results of measurements are well readout, for instance, by the use of some external field couples to the 'charge' operator $(1 + \sigma_3)/2$ of individual detector.

2. DECOHERENCE, WHY AND WHEN ?

The state-operator $E^{S+M}(t)$ contains the interference term (the cross terms with respect to k and ℓ in Eq.(9)) and it persists even after the measurement ($t \geq t_F$). Notice that the state-operator is by no means an physically measurable quantity as well as the state-vector itself. What we can determine by experiments are the matrix elements of observables $\mathcal{O}$. For observables $\mathcal{O}^M$ of M, they are expressed as $tr(\mathcal{O}^M |k\rangle \langle \ell|) = \langle \ell| \mathcal{O}^M |k\rangle$. In the present model, however, $\mathcal{O}^M$ takes in general a form like $\mathcal{O}^M \sim (a^{(1)} + b^{(1)}\sigma_3^{(1)}) \otimes \cdots \otimes (a^{(n)} + b^{(n)}\sigma_3^{(n)})$ and therefore, for $k \neq \ell$, $\langle k| \mathcal{O}^M |\ell\rangle = 0$ identically, implying that the interference term in $E^{S+M}(t_F)$ does not contribute to $tr(\mathcal{O}^M E^{S+M}(t_F))$ for *all observables* $\mathcal{O}^M$ concerned. This amounts to saying that we are free to substitute $E^{S+M}(t_F)$ by

$$E'(t_F) = \sum_k |\langle a_k|\phi\rangle|^2 |a_k\rangle \langle a_k| \otimes |k\rangle \langle k| \bigg|_{t_F}, \tag{10}$$

which is just the expression for a mixture ensemble composed of 'decoherent' pure ensembles $E_k^S \otimes E_k^M \equiv |a_k\rangle \langle a_k| \otimes |k\rangle \langle k|$ with the frequency ratio $|\langle a_k|\psi\rangle|^2$. Note that $|\psi\rangle$ is a state-vector prepared at t_I and $|a_k\rangle$ is of the time t_F [2] and the probabilistic interpretation for $|\langle a_k|\psi\rangle|^2$ finds its logical ground only in conjunction with the state reduction which, as we see, is to occur immediately after $H_{SM}(t)$ has gone away before some irreversible interactions are introduced to make the results of 'reading-out' be an indelible recording.

Our final claim is the necessity of the use of $E'(t_F)$ instead of $E^{S+M}(t_F)$. Although both of them predict the same expectation value for *all the* $\mathcal{O}^M$ *involved*, we had better to adopt

$E'(t_F)$ for describing directly the experimental situation realized after the measurement, where the apparatus is to be well separated off without disturbing the measured system. Contrary to $E^{S+M}(t_F)$, $E'(t_F)$ is not attained by the unitary time evolution from the initial $E^{S+M}(t_I)$ as it should be, but understood as a re-definition of the state of $S + M$ for specifying an initial condition of the system to be developed later on.

In concluding, few comments are set in order.

a. Our model apparatus and its interaction are essentially time-dependent. In such a situation the quantal definition for the whole system neccessarily depends upon the time involved, and the apparatus, too, varies its individual-identity with time accordingly. [4]

b. The decoherence mechanism discussed here is different from that of other authors [5] in that we need not require the apparatus to be macroscopic, nor introduce any inter play with enviromental variables. Rather, we expect the present model, be it felt far from 'realistic', could be seen as a kind of the *logical skelton* for actual measurement situation.

REFERENCES

1. Z. Maki, *Soryushiron Kenkyu* (in Japanese) 96(1997), 79.

2. Z. Maki, "Von Neumann's Chain, Decoherence and Limit of Validity of the Probability Rule in Quantum Mechanics", in *Quantum Coherence and Decoherence*, K. Fujikawa and Y.A. Ono(edit.) (Elsvier Science B.V., Amsterdam 1996), p.327.

3. J. von Neumann, *Mathematische Grundlagen der Quantenmechanik* (Springer, 1932).

4. Z. Maki, "Concept of Individual-identity in Quantum Mechanical Objects", *Kagakukisoron Kenkyu* (in Japanese) 25 No.(1997), 33.

5. See for example
   R. Omnés, *The Interpretation of Quantum Mechanics* (Princeton University Press, Princeton, 1994); W.H. Zurek, Phys.Rev. D26(1982), 1862.

List of Participants

Achiwa, Norio
Dept. of Physics, Faculty of Science
Kyushu University
Hakozaki, Higashi-ku, Fukuoka 812-8581
Japan
  Phone: +81-92-642-2534
  Fax: +81-92-642-2553
  e-mail: achi8scp@mbox.nc.kyushu-u.ac.jp

Ajiki, Hiroshi
Dept. of Physical Science
Graduate School of Engineering Science
Osaka University
1-3 Machikaneyama
Toyonaka, Osaka 560-8531
Japan
  Phone: +81-6-850-6402
  Fax: +81-6-850-6401
  e-mail: ajiki@mp.es.osaka-u.ac.jp

Ando, Masato
Institute for Solid State Physics
University of Tokyo
7-22-1 Roppongi, Minato-ku, Tokyo 106-8666
Japan
  Phone: +81-3-3478-6811 (ext. 5602)
  Fax: +81-3-3478-0536
  e-mail: andom@tsubasa.issp.u-tokyo.ac.jp

Anglin, James R.
Institute for Theoretical Physics
University of Innsbruck
Technikerstrasse 25/4, A-6020 Innsbruck
Austria
  Phone: +43-512-507-6204
  Fax: +42-512-507-2919
  e-mail: James.Anglin@uibk.ac.at

Aoki, Takatoshi
Department of Physics
Faculty of Science and Technology
Science University of Tokyo
2641 Yamazaki, Noda-shi, Chiba 278-8510
Japan
  Phone: +81-471-24-1501 (ext. 3212)
  Fax: +81-471-23-9361
  e-mail: morinaga@ph.noda.sut.ac.jp

Aoyagi, Yoshinobu
Semiconductor Lab.
The Institute of Physical and Chemical
  Research (RIKEN)
2-1 Hirosawa, Wako-shi, Saitama 351-0166
Japan
  Phone: +81-48-467-9366
  Fax: +81-48-462-4659
  e-mail: aoyagi@postman.riken.go.jp

Asakura, Hiroshi
Dept. of Physics, Kyoto Sangyo University
Kita-ku, Kyoto 603-8555
Japan
  Phone: +81-75-701-2151 (ext. 3660)
  e-mail: asakura@cc.kyoto-su.ac.jp

Avishai, Yshai
NTT Basic Research Labs.
3-1 Morinosato-Wakamiya
Atsugi, Kanagawa 243-0198
Japan
  Phone: +81-462-40-3421
  Fax: +81-462-40-4723
  e-mail: yshai@bgh-mail.bgh.ac.il
Present Address
Department of Physics
Ber Gurion University
Beer Shera
Israel
  Phone : +972-7-6461763
  Fax: +972-7-6472904
  e-mail: yshai@bgh-mail.bgh.ac.il

Azuma, Hiroo
Mathematical Engineering Dept. 2
Canon Research Center
5-1 Morinosato-Wakamiya, Atsugi
Kanagawa 243-0193
Japan
  Phone: +81-462-47-2111
  Fax: +81-462-48-0306
  e-mail: hiroo@crc.canon.co.jp

352

Barbara, Bernard A.
Laboratoire de Magnétisme Louis Néel
CNRS
25 Av. des Martyrs, BP 166
38042 Grenoble, Cedex 9
France
    Phone: +33-4-76-88-11-92
    Fax: +33-4-76-88-11-91
    e-mail: barbara@labs.polycnrs-gre.fr

Chan, Moses H. W.
Dept. of Physics, 104 Davey Lab.
Pennsylvania State University
University Park, PA 16802
USA
    Phone: +1-814-863-2622
    Fax: +1-814-865-3604
    e-mail: chan@phys.psu.edu

Chatzidimitriou-Dreismann, C. Aris
I. N. Stranski-Institute, TU Berlin (Sekr:ER 1)
Strasse des 17 Juni 112, D-10623 Berlin
Germany
    Phone: +49-30-314-22692
    Fax: +49-30-314-26602
    e-mail: drei0533@w224zrz.zrz.tu-berlin.de

Chudnovsky, Eugene M.
Physics Dept., CUNY Lehman College
Bedford Park Boulevard West
Bronx, NY 10468-1589
USA
    Phone: +1-718-960-8770
    Fax: +1-718-960-8627
    e-mail: chudnov@lcvax.lehman.cuny.edu

Cosmelli, Carlo
Dept. of Physics, University of Rome
P.A. Moro 2, 00185, Roma
Italy
    Phone: +3906-4991-4216
    Fax: +3906-4957-697
    e-mail: carlo.cosmelli@roma1.infn.it

Dito, Giuseppe
Research Institute for Mathematical Sciences
Kyoto University
Kitashirakawa Oiwake-cho
Sakyo-ku, Kyoto 606-8502
Japan

    Phone: +81-75-753-7244
    Fax: +81-75-753-7272
    e-mail: dito@kurims.kyoto-u.ac.jp

DiVincenzo, David P.
T.J. Watson Research Center
IBM Research Division
PO Box 218, Yorktown Heights
New York 10598
USA
    Phone: +1-914-945-3076
    Fax: +1-914-945-4421
    e-mail: divince@watson.ibm.com

Ebisawa, Hiromichi
Graduate School of Information Sciences
Tohoku University
Aramaki, Aoba, Sendai 980-8579
Japan
    Phone: +81-22-217-5846
    Fax: +81-22-217-5851
    e-mail: ebi@cmt.is.tohoku.ac.jp

Ekert, Artur
Clarendon Lab., Dept. of Physics
University of Oxford
Parks Road, Oxford OX1 3PU
UK
    Phone: +44-1865-282202
    Fax: +44-1865-272375
    e-mail: ekert@physics.ox.ac.uk

Ezawa, Hiroshi
Dept. of Physics, Gakushuin University
1-5-1 Mejiro, Toshima-ku
Tokyo 171-8588
Japan
    Phone: +81-3-3986-0221 (ext. 6483)
    Fax: +81-3-5992-1029
    e-mail: hiroshi.ezawa@gakushuin.ac.jp

Fujikawa, Kazuo
Dept. of Physics, University of Tokyo
7-3-1 Hongo, Bunkyo-ku, Tokyo 113-0033
Japan
    Phone: +81-3-3812-2111 (ext. 4190)
    Fax: +81-3-5684-9642
    e-mail: fujikawa@phys.s.u-tokyo.ac.jp

Fukuyama, Hidetoshi
Dept. of Physics, University of Tokyo
7-3-1 Hongo, Bunkyo-ku, Tokyo 113-0033
Japan
    Phone: +81-3-3812-2111 (ext. 4184)
    Fax: +81-3-5600-6791
    e-mail: fukuyama@phys.s.u-tokyo.ac.jp
Present Address
Institute for Solid State Physics
University of Tokyo
7-22-1 Roppongi, Minato-ku
Tokyo 106-8666
Japan
    Phone: +81-3-3478-6987
    Fax: +81-3-3478-6987
    e-mail: fukuyama@issp.u-tokyo.ac.jp

Funahashi, Haruhiko
Dept. of Physics, Kyoto University
Kitashirakawa, Oiwake-cho
Sakyo-ku, Kyoto 606-8502
Japan
    Phone: +81-75-753-3861
    Fax: +81-75-753-3795
    e-mail: hal@pn.scphys.kyoto-u.ac.jp

Glauber, Roy J.
Lyman Laboratory of Physics
Harvard University
Cambridge, MA 02138
USA
    Phone: +1-617-495-2867
    Fax: +1-617-496-8396
    e-mail: glauber@huhepl.harvard.edu

Hall, David S.
JILA, CB 440, University of Colorado
Boulder, CO 80309-0440
USA
    Phone: +1-303-492-7784
    Fax: +1-303-492-5235
    e-mail: dshall@jilau1.colorado.edu

Harigaya, Kikuo
Physical Science Division
Electrotechnical Laboratory
1-1-4 Umezono, Tsukuba 305-8568
Japan

Phone: +81-298-54-5151
Fax: +81-298-54-5375
e-mail: harigaya@etl.go.jp

Haroche, Serge
Dept. de Physique, Ecole Normale Superieure
24, rue Lhomond, 75231 Paris Cedex 05
France
    Phone: +33-1-44 32 33 59
    Fax: +33-1-44 32 20 08
    e-mail: serge.haroche@physique.ens.fr

Hashizume, Tomihiro
Advanced Research Laboratory, Hitachi, Ltd.
Hatoyama, Saitama, 350-0395
Japan
    Phone: +81-492-96-6111 (ext. 6356)
    Fax: +81-492-96-6006
    e-mail: tomi@harl.hitachi.co.jp

Hasselbach, Franz
Institut für Angewandte Physik
Universität Tübingen
Auf der Morgenstelle 10
D-72076 Tübingen
Germany
    Phone: +49-7071-2976328
    Fax: +49-7071-295400
    e-mail: franz.hasselbach@uni-tuebingen.de

Hatano, Naomichi
MS-B262, Theoretical Division, T-11
Los Alamos National Laboratory
Los Alamos, NM 87545
USA
    Phone: +1-505-665-0478
    Fax: +1-505-665-4063
    e-mail: hatano@viking.lanl.gov
Present Address
Department of Physics
Aoyama Gakuin University
6-16-1 Chitosedai,
Setagaya-ku, Tokyo 157-8572
Japan
    Phone: +81-3-5384-2642
    Fax: +81-3-5384-6100
    e-mail: hatano@phys.aoyama.ac.jp

Hayashi, Masahiko
Graduate School of Information Sciences
Tohoku University
Aramaki, Aoba-ku, Sendai 980-8579
Japan
Phone: +81-22-217-5847
Fax: +81-22-217-5851
e-mail: hayashi@cmt.is.tohoku.ac.jp

Hess, Ortwin G.
Institute of Technical Physics, DLR
Pfaffenwaldring 38-40, D-70569 Stuttgart
Germany
Phone: +49-711-6862-214
Fax: +49-711-6862-788
e-mail: ortwin.hess@dlr.de

Hino, Masahiro
Research Reactor Institute, Kyoto University
Kumatori-cho, Sennan-gun, Osaka 590-0494
Japan
Phone: +81-724-51-2341
Fax: +81-724-51-2620
e-mail: hino@rri.kyoto-u.ac.jp

Hirano, Takuya
Dept. of Physics, Gakushuin University
1-5-1 Mejiro, Toshima-ku, Tokyo 171-8588
Japan
Phone: +81-3-3986-0221 (ext. 6466)
Fax: +81-3-5992-1029
e-mail: takuya.hirano@gakushuin.ac.jp

Hirasawa, Masahiko
Bio-Emission Laboratory
Division of Radiation Research
National Institute of Radiological Science
4-9-1 Anagawa, Inage-ku, Chiba 263-8555
Japan
Phone: +81-43-206-3066
Fax: +81-43-206-3069
e-mail: hirasawa@nirs.go.jp

Hitosugi, Taro
Dept. of Superconductivity
University of Tokyo
7-3-1 Hongo, Bunkyo-ku, Tokyo 113-8656
Japan
Phone: +81-3-3812-2111 (ext. 7204)
Fax: +81-3-3815-5632
e-mail: hitosugi@harl.hitachi.co.jp
Present Address
Gigabyte Research Laboratory
Sony Home-Network Company
6-7-35 Kitashinagawa, Shinagawa-ku
Tokyo 141-0001, Japan
Phone: +81-3-5448-6698
Fax: +81-3-5448-4717
e-mail: hitosugi@devm.crl.sony.co.jp

Hofmann, Holger F.
Institute of Technical Physics, DLR
Pfaffenwaldring 38-40, D-70569 Stuttgart
Germany
Phone: +49-711-6862-507
Fax: +49-711-6862-788
e-mail: holger.hofmann@dlr.de

Hosoya, Akio
Physics Department
Tokyo Institute of Technology
2-12-1 Oh-okayama, Meguro-ku
Tokyo 152-8550
Japan
Phone: +81-3-5734-2463
Fax: +81-3-5734-2463
e-mail: ahosoya@th.phys.titech.ac.jp

Imoto, Nobuyuki
NTT Basic Research Laboratories
3-1 Morinosato-Wakamiya, Atsugi-shi
Kanagawa 243-0198
Japan
Phone: +81-462-40-3405
Fax: +81-462-40-4726
e-mail: nobu@will.brl.ntt.co.jp
Present Address
School of Advanced Sciences
The Graduate University for Advanced
    Studies
Shonan-Village, Hayama
Kanagawa 240-0193
Japan
Phone: +81-468-58-1560
Fax: +81-468-58-1544
e-mail: imoto@soken.ac.jp

Imry, Yoseph
Dept. of Condensed Matter Physics
The Weizmann Institute of Science
Rehovot 76100
Israel
　　Phone: +972-8-9344064
　　Fax: +972-8-9344106
　　e-mail: fnimry@wicc.weizmann.ac.il

Inagaki, Satoru
Dept. of Physics
Faculty of Science and Technology
Meiji University
1-1-1 Higashi-Mita, Tama-ku
Kawasaki, Kanagawa 214-8571
Japan
　　Phone: +81-44-934-7432, 7171
　　Fax: +81-44-934-7911
　　e-mail: inagaki@isc.meiji.ac.jp

Inoue, Jun-ichi
Institute of Physics, University of Tokyo
3-8-1 Komaba, Meguro-ku, Tokyo 153-8902
Japan
　　Phone:+81-3-5454-4364
　　Fax: +81-3-5790-7229
　　e-mail: inoue@asone.c.u-tokyo.ac.jp

Ishibashi, Masayoshi
Advanced Research Laboratory, Hitachi, Ltd.
Hatoyama, Saitama 350-0395
Japan
　　Phone: +81-492-96-6111
　　Fax: +81-492-96-6006
　　e-mail: isibasi@harl.hitachi.co.jp

Ishii, Rikie
Dept. of Physics, Tohoku University
Aoba, Sendai 980-8578
Japan
　　Phone: +81-22-217-6446
　　Fax: +81-22-217-6447
　　e-mail: ishii@cmpt01.phys.tohoku.ac.jp

Iwabuchi, Shuichi
Dept. of Physics, Nara Women's University
Kitauoya-Nishimachi, Nara 630-8506
Japan
　　Phone: +81-742-20-3987
　　Fax: +81-742-20-3987

　　e-mail: iwabuchi@phys.nara-wu.ac.jp

Iwasawa, Hiroshi
Dept. of Electronics
Hiroshima-Denki Institute of Technology
Nakano 6-20-1, Aki-ku, Hiroshima 937-0321
Japan
　　Phone: +81-82-893-0381
　　Fax: +81-82-893-5014
　　e-mail: iwasawa@s.hiroshima-dit.ac.jp

Iwayanagi, Takao
Advanced Research Laboratory, Hitachi, Ltd.
Hatoyama, Saitama 350-0395
Japan
　　Phone: +81-492-96-6111
　　Fax: +81-492-96-6005
　　e-mail: takao@harl.hitachi.co.jp
Present Address
Central Research Laboratory, Hitachi, Ltd.
Kokubunji-shi, Tokyo 185-8601
Japan
　　Phone: +81-423-23-1111
　　Fax: +81-492-96-6005
　　e-mail: takao@harl.hitachi.co.jp

Iye, Yasuhiro
Institute for Solid State Physics
University of Tokyo
7-22-1 Roppongi, Minato-ku, Tokyo 106-8666
Japan
　　Phone: +81-3-3478-0536
　　Fax: +81-3-3478-0536
　　e-mail: iye@issp.u-tokyo.ac.jp

Kamei, Osamu
2-58-2 Takinogawa, Kita-ku, Tokyo 114-0023
Japan
　　Phone: +81-3-3917-3160

Kanda, Akinobu
Frontier Research Program
Laboratory for Nanoelectronics Materials
The Institute of Physical and Chemical
　　Research (RIKEN)
2-1 Hirosawa, Wako, Saitama 351-0198
Japan
　　Phone: +81-48-462-1111 (ext. 6312)
　　Fax: +81-48-467-5087
　　e-mail: akanda@postman.riken.go.jp

Karlsson, Anders
Department of Electronics
Laboratory of Photonics and Microwave
    Engineering
Royal Institute of Technology
S-164 40 Kista
Sweden
    Phone: +46-8-752-1272
    Fax: +46-8-752-1240
    e-mail: Andkar@ele.kth.se

Karlsson, Erik B.
Department of Physics, Uppsala Univ.
P.O.Box 530, S-75121 Uppsala
Sweden
    e-mail: erk@fysik.uu.se

Kato, Mayumi
Institute for Solid State Physics
University of Tokyo
7-22-1 Roppongi, Minato-ku, Tokyo 106-8666
Japan
    Phone: +81-3-3478-6811 (ext. 5602)
    Fax: +81-3-3478-0536
    e-mail: mayumi@kodama.issp.u-tokyo.ac.jp

Kato, Takeo
Institute for Solid State Physics
University of Tokyo
7-22-1 Roppongi, Minato-ku, Tokyo 106-8666
Japan
    Phone: +81-3-3478-6811 (ext. 5832)
    Fax: +81-3-3402-8174
    e-mail: kato@ginnan.issp.u-tokyo.ac.jp

Katsumoto, Shingo
Institute for Solid State Physics
University of Tokyo
7-22-1 Roppongi, Minato-ku, Tokyo 106-8666
Japan
    Phone: +81-3-3478-6811 (ext. 5631)
    Fax: +81-3-3478-6847
    e-mail: kats@issp.u-tokyo.ac.jp

Kawabata, Shiro
Physical Science Division
Electrotechnical Laboratory
1-1-4 Umezono, Tsukuba, Ibaraki 305-8568
Japan
    Phone: +81-298-54-5390
    Fax: +81-298-54-5375
    e-mail: shiro@etl.go.jp

Kawai, Takeshi
Neutron Science Department
Research Reactor Institute, Kyoto University
Kumatori-cho, Sennan-gun, Osaka 590-0494
Japan
    Phone: +81-724-51-2448
    Fax: +81-724-51-2600
    e-mail: kawai@rri.kyoto-u.ac.jp

Kawamura, Minoru
Institute for Solid State Physics
University of Tokyo
7-22-1 Roppongi, Minato-ku, Tokyo 106-8660
Japan
    Phone: +81-3-3478-6811 (ext. 5602)
    Fax: +81-3-3401-5169
    e-mail: minoru@kodama.issp.u-tokyo.ac.jp

Kawashima, Katsuhiro
Engineering Sciences Dept.
Tokyo Engineering University
1404-1 Katakura-machi, Hachioji-shi 192-8580
Japan
    Phone: +81-426-37-2111
    Fax: +81-426-37-2118
    e-mail: kawasima@cc.teu.ac.jp

Kim, Gwang-Hee
Dept. of Physics, Sejong University
Kunja-dong, Kwangjin-Gu, Seoul, 143-747
Republic of Korea
    Phone: +82-2-460-0211
    Fax: +82-2-460-0299
    e-mail: gkim@kunja.sejong.ac.kr

Kimble, H. Jeff
Norman Bridge Laboratory of Physics
California Institute of Technology
1200 E. California Blvd. (MC: 12-33)
Pasadena, CA 91106
USA
    Phone: +1-626-395-8342
    Fax: +1-626-793-9506
    e-mail: hjkimble@cco.caltech.edu

Kimura, Takashi
Dept. of Physical Electronics
Faculty of Engineering, Hiroshima University
1-4-1 Kagamiyama, Higashi-Hiroshima
Hiroshima 739-8527
Japan
    Phone: +81-824-24-7650
    Fax: +81-824-24-7650
    e-mail: kimura@qnt26.qp.hiroshima-u.ac.jp

Kitagawa, Masahiro
Graduate School of Engineering Science
Osaka University
1-3 Machikaneyama, Toyonaka,
Osaka 560-8531
Japan
    Phone: +81-6-850-6339
    Fax: +81-6-850-6338
    e-mail: kitagawa@ee.es.osaka-u.ac.jp

Koashi, Masato
NTT Basic Research Laboratories
3-1 Morinosato-Wakamiya, Atsugi-shi,
Kanagawa 243-0198
Japan
    Phone: +81-462-40-3498
    Fax: +81-462-40-4726
    e-mail: koashi@will.brl.ntt.co.jp

Kobayashi, Shun-ichi
The Institute of Physical and Chemical
    Research (RIKEN)
2-1 Hirosawa, Wako, Saitama 351-0198
Japan
    Phone: +81-48-462-1446
    Fax: +81-48-462-4604
    e-mail: skobaya@postman.riken.go.jp

Kohno, Yoshie
Fundamental Research Laboratories
NEC Corporation
34 Miyukigaoka, Tsukuba-shi, Ibaraki 305-8501
Japan
    Phone: +81-298-50-1582
    Fax: +81-298-56-6136
    e-mail: kohno@frl.cl.nec.co.jp

Koide, Koshi
AVC Research Laboratories
Sharp Corporation
1-9-2 Nakase, Mihama-ku, Chiba-shi
Chiba 261-8520
Japan
    Phone: +81-43-299-8711
    Fax: +81-43-299-8709
    e-mail: quan@iml.mkhar.sharp.co.jp

Kubo, Harunobu
Theory Department, KEK High Energy
    Accelerator Research Organization
1-1 Oh-ho, Tsukuba, Ibaraki 305-0801
Japan
    Phone: +81-298-64-5394
    Fax: +81-298-64-5755
    e-mail: kubo@theory.kek.jp

Kurosawa, Hitoshi
Dept. of Quantum Engineering and System
    Sciences, University of Tokyo
7-3-1 Hongo, Bunkyo-ku, Tokyo 113-0033
Japan
    Phone: +81-3-3812-2111 (ext. 6998)
    Fax: +81-3-3818-3455
    e-mail: kuroh@mail.dddd.ne.jp

Maki, Ziro
Dept. of Physics
Faculty of Science and Technology
Kinki University
3-4-1 Kowae, Higashi-Osaka, 577-8502
Japan
    Phone: +81-6-721-2332
    Fax: +81-6-727-4301
    e-mail: ziromaki@phys.kindai.ac.jp

Matsueda, Hideaki
Dept. of Information Science, Kochi University
2-5-1 Akebono-cho, Kochi 780-8072
Japan
    Phone: +81-888-44-8334
    Fax: +81-888-44-8345
    e-mail: matsueda@is.kochi-u.ac.jp

Matsuoka, Masahiro
Dept. of Physics, Kumamoto University
2-39-1 Kurokami, Kumamoto 860-8555
Japan
    Phone: +81-96-342-3359
    Fax: +81-96-342-3359
    e-mail: matsuoka@aster.sci.kumamoto-u.
        ac.jp

Matsuura, Shinobu
Dept. of Superconductivity
University of Tokyo
7-3-1 Hongo, Bunkyo-ku, Tokyo 113-8656
Japan
    Phone: +81-3-3812-2111 (ext. 7202)
    Fax: +81-3815-5632
    e-mail: matsuura@harl.hitachi.co.jp

Miyauchi, Katsuki
Advanced Research Laboratory, Hitachi, Ltd.
Hatoyama, Saitama 350-0395
Japan
    Phone: +81-492-96-6111
    Fax: +81-492-96-6005
    e-mail: katzmi@harl.hitachi.co.jp
Present Address
R&D Centre, Hitachi Europe Ltd.
Whitebrook Park, Lower Cookham Road
Maidenhead, Berkshire SL68YA
UK
    Phone: +44-1628-585000
    Fax: +44-1628-778322
    e-mail: Katsuki.Miyauchi@hitachi-eu.com

Morinaga, Atsuo
Dept. of Physics
Faculty of Science and Technology
Science University of Tokyo
2641 Yamazaki, Noda-shi, Chiba 278-8510
Japan
    Phone: +81-471-24-1501 (ex. 3212)
    Fax: +81-471-23-9361
    e-mail: morinaga@ph.noda.sut.ac.jp

Motoyoshi, Akio
Dept. of Physics, Faculty of Science
Kumamoto University
2-39-1 Kurokami, Kumamoto 860-8555
Japan

    Phone: +81-96-342-3363
    Fax: +81-96-342-3363
    e-mail: motoyosi@aster.sci.kumamoto-
        u.ac.jp

Murao, Mio
Optics Section, Blackett Laboratory
Imperial College
Prince Consort Road, London, SW7 2BZ
United Kingdom
    Phone: +44-171-594-7728
    Fax: +44-171-823-8376
    e-mail: m.murao@ic.ac.uk

Murayama, Yoshimasa
Dept. Materials Science and Technology
Faculty of Engineering, Niigata University
8050 Ikarashi-2-no-cho, Niigata-shi
Niigata 950-2181
Japan
    Phone: +81-25-262-6352
    Fax: +81-25-262-6352
    e-mail: murayama@geb.ge.niigata-u.ac.jp

Nakajima, Sadao
Superconductivity Research Laboratory
International Superconductivity Technology
    Center (ISTEC)
1-10-13 Shinonome, Koto-ku, Tokyo 135
Japan
    Phone: +81-3-3536-5700
    Fax: +81-3-3536-5714

Nakamura, Michiharu
Central Research Laboratory, Hitachi, Ltd.
1-280 Higashi-koigakubo, Kokubunji-shi,
Tokyo 185-8601
Japan
    Phone: +81-42-323-1111
    Fax: +81-42-327-7087
    e-mail: nakamura@crl.hitachi.co.jp
Present Address
Research & Development Group, Hitachi, Ltd.
New Marunouchi Bldg., 1-5-1 Marunouchi
Chiyoda-ku, Tokyo 100-8220
Japan
    Phone: +81-3-3212-1111
    Fax: +81-3-3212-3050
    e-mail: nakam@cm.head.hitachi.co.jp

Nakamura, Yasunobu
Fundamental Research Laboratories
NEC Corporation
34 Miyukigaoka, Tsukuba-shi, Ibaraki 305-8501
Japan
    Phone: +81-298-50-1148
    Fax: +81-298-56-6139
    e-mail: yasunobu@frl.cl.nec.co.jp

Nakano, Hayato
Materials Science Research Labs.
NTT Basic Research Laboratories
3-1 Morinosato-Wakamiya, Atsugi-shi
Kanagawa 243-0198
Japan
    Phone: +81-462-40-3528
    Fax: +81-462-40-4722
    e-mail: nakano@will.brl.ntt.co.jp

Nemoto, Kae
Centre of Laser Science, Dept. of Physics
University of Queensland
Qld 4072
Australia
    Phone: +61-7-3365-1344
    Fax: +61-7-3365-1242
    e-mail: nemoto@physics.uq.edu.au

Nikuni, Tetsuro
Dept. of Physics, Tokyo Institute of Technology
2-12-1 Oh-okayama, Meguro, Tokyo 152-8550
Japan
    Phone: +81-3-5734-2736
    Fax: +81-3-5734-2739
    e-mail: nick@stat.phys.titech.ac.jp

Obata, Shuji
Dept. of Natural Sciences
Tokyo Denki University
Hatoyama, Saitama 350-0394
Japan
    Phone: +81-492-96-2911 (ext. 8999)
    Fax: +81-492-96-2915
    e-mail: obata@u.dendai.ac.jp

Ogawa, Susumu
Advanced Research Laboratory, Hitachi, Ltd.
Hatoyama, Saitama 350-0395
Japan
    Phone: +81-492-96-6111
    Fax: +81-492-96-6006
    e-mail: ogawa@harl.hitachi.co.jp

Ogura, Tetsuya
Dept. of Physics, Faculty of Science
Kumamoto University
2-39-1 Kurokami, Kumamoto 860-8555
Japan
    Phone: +81-96-344-2111 (ext.3367)
    Fax: +81-96-342-3363
    e-mail: ogura@aster.sci.kumamoto-u.ac.jp

Ohno, Yuzo
Research Institute of Electrical Communication
Tohoku University
2-1-1 Katahira, Aoba-ku, Sendai 980-8577
Japan
    Phone: +81-22-217-5555
    Fax: +81-22-217-5555
    e-mail: oono@riec.tohoku.ac.jp

Oiwa, Akira
Institute for Solid State Physics
University of Tokyo
7-22-1 Roppongi, Minato-ku, Tokyo 106-8666
Japan
    Phone: +81-3-3478-6811 (ext. 5602)
    Fax: +81-3-3478-0536
    e-mail: oiwa@kodama.issp.u-tokyo.ac.jp

Ono, Yoshimasa A.
Advanced Research Laboratory, Hitachi, Ltd.
Hatoyama, Saitama 350-0395
Japan
    Phone: +81-492-96-6111
    Fax: +81-492-96-6005
    e-mail: yaono@harl.hitachi.co.jp

Onogi, Toshiyuki
Advanced Research Laboratory, Hitachi, Ltd.
Hatoyama, Saitama 350-0395
Japan
    Phone: +81-492-96-6111(ext.6275)
    Fax: +81-492-96-6006
    e-mail: onogi@harl.hitachi.co.jp

Otake, Yoshie
Coherent X-Ray Optics Laboratory
The Institute of Physical and Chemical
    Research (RIKEN)
Mihara, Mikazuki-cho, Sayo-gun
Hyogo 679-5143
Japan
    Phone: +81-7915-8-2806
    Fax: +81-7915-8-2810
    e-mail: yoshie@spring8.or.jp

Otani, Yoshichika
Dept. of Materials Science
Graduate School of Engineering
Tohoku University
Sendai 980-8579
Japan
    Phone: +81-22-217-7316
    Fax: +81-22-217-7316
    e-mail: chika@material.tohoku.ac.jp

Ponomarenko, Vadim V.
Dept. of Applied Physics, University of Tokyo
7-3-1 Hongo, Bunkyo-ku, Tokyo 113-0033
Japan
    Phone: +81-3-5689-7337
    Fax: +81-3-5689-7337
    e-mail: vadim@appi.t.u-tokyo.ac.jp

Roukes, Michael L.
Condensed Matter Physics 114-36
California Institute of Technology
Pasadena, CA 91125
USA
    Phone: +1-626-395-2916
    Fax: +1-626-683-9060
    e-mail: roukes@caltech.edu

Saikawa, Kazuhiko
Dept. of Physics, Faculty of Science
Shinshu University
3-1-1 Asahi, Matsumoto 390-8621
Japan
    Phone: +81-263-37-2459
    Fax: +81-263-37-2438
    e-mail: ksaikaw@gipac.shinshu-u.ac.jp

Saito, Hiroki
Dept. of Physical Electronics
Faculty of Engineering, Hiroshima University

1-4-1 Kagamiyama, Higashi-Hiroshima
Hiroshima 739-8527
Japan
    Phone: +81-824-24-7630
    Fax: +81-824-24-7650
    e-mail: hiroki@qp.hiroshima-u.ac.jp

Sakurai, Akio
Dept. of Physics, Kyoto Sangyo University
Kamigamo, Kita-ku, Kyoto 603-8555
Japan
    Phone: +81-75-705-1606
    Fax: +81-75-705-1640
    e-mail: sakurai@cc.kyoto-su.ac.jp

Sato, Hideki
Institute for Solid State Physics
University of Tokyo
7-22-1 Roppongi, Minato-ku. Tokyo 106-8666
Japan
    Phone: +81-3-3478-6811
    Fax: +81-3-3401-5169
    e-mail: hsatoh@kodama.issp.u-tokyo.ac.jp

Sawada, Anju
Dept. of Physics, Tohoku University
Aoba, Sendai 980-8578
Japan
    Phone: +81-22-217-6478
    Fax: +81-22-217-6477
    e-mail: sawada-a@mail.cc.tohoku.ac.jp

Sawada, Shoji
15-403 Momoyama-Jyutaku
1-77 Momoyama, Midori-ku, Nagoya 458-0002
Japan
    Phone: +81-52-876-0226
    Fax: +81-52-878-1193
    e-mail: sawadas@fb3.so-net.or.jp

Schön, Gerd
Inst. Theor. Festkörperphysik
University of Karlsruhe
D-76128 Karlsruhe
Germany
    Phone: +49-721-6083361
    Fax: +49-721-698150
    e-mail: schoen@tfp.physik.uni-karlsruhe.de

Shibata, Junya
Dept. of Physics, Tohoku University
Aoba, Sendai 980-8578
Japan
Phone: +81-22-217-6443
Fax: +81-22-217-6443
e-mail: shibata@cmpt01.phys.tohoku.ac.jp

Shimizu, Akira
Institute of Physics, University of Tokyo
3-8-1 Komaba, Meguro-ku, Tokyo 153-8902
Japan
Phone: +81-3-5454-6532
Fax: +81-3-5790-7229
e-mail: shmz@asone.c.u-tokyo.ac.jp

Shimizu, Fujio
Institute for Laser Science
University of Electro-Communications
1-5-1 Chofugaoka, Chofu, Tokyo 182-8585
Japan
Phone: +81-424-43-5701 or 5711
Fax: +81-424-85-8960
e-mail: fshimizu@ils.uec.ac.jp

Shimizu, Kaoru
Physical Science Research Laboratories
NTT Basic Research Laboratories
3-1 Morinosato-Wakamiya, Atsugi-shi
Kanagawa 243-0198
Japan
Phone: +81-462-40-3469
Fax: +81-462-40-4726
e-mail: shimizu@will.brl.ntt.co.jp

Shinba, Yutaka
Shizuoka Institute of Science and Technology
2200-2 Toyozawa, Fukuroi, Shizuoka 437-8555
Japan
Phone: +81-538-45-0190
e-mail: shinba@ns.sist.ac.jp

Sone, Jun'ichi
Fundamental Research Laboratories
NEC Corporation
34 Miyukigaoka, Tsukuba-shi, Ibaraki 305-8501
Japan

Phone: +81-298-50-1164
Fax: +81-298-56-6139
e-mail: sone@frl.cl.nec.co.jp

Sonnentag, Peter
Institut für Angewandte Physik
Universität Tübingen
Auf der Morgenstelle 10
D-72076 Tübingen
Germany
Phone: +49-7071-2976331
Fax: +49-7071-295400
e-mail: peter.sonnentag@uni-tuebingen.de

Sugano, Takuo
Frontier Materials Research
Institute of Physical and Chemical
    Research (RIKEN)
2-1 Hirosawa, Wako, Saitama 351-0198
Japan
Phone: +81-48-467-9597
Fax: +81-48-467-5087
e-mail: sugano@hakusrv.toyo.ac.jp

Takagi, Shin
Dept. of Physics, Tohoku University
Aoba, Sendai 980-8578
Japan
Phone: +81-22-217-6440
Fax: +81-22-217-6447
e-mail: takagi@cmpt01.phys.tohoku.ac.jp

Takayanagi, Hideaki
Material Science Research Labs.
NTT Basic Research Laboratories
3-1 Morinosato-Wakamiya, Atsugi-shi
Kanagawa 243-0198
Japan
Phone: +81-462-40-3530
Fax: +81-462-40-4722
e-mail: takayan@will.brl.ntt.co.jp

Tanaka, Atushi
Institute of Physics, University of Tsukuba
Tsukuba, Ibaraki 305-8571
Japan
Phone: +81-298-53-6118
Fax: +81-298-53-4492
e-mail: atanaka@am.ph.tsukuba.ac.jp

Tatara, Gen
Dept. of Earth and Space Science
Graduate School of Science, Osaka University
Machikaneyama, Toyonaka, Osaka 560-0043
Japan
    Phone: +81-6-850-5494
    Fax: +81-6-850-5494
    e-mail: tatara@ess.sci.osaka-u.ac.jp

Terashima, Hiroaki
Dept. of Physics, University of Tokyo
7-3-1 Hongo, Bunkyo-ku, Tokyo 113-0033
Japan
    Phone: +81-3-3812-2111 (ext. 4137)
    Fax: +81-3-5684-9642
    e-mail: terasima@hep-th.phys.s.
        u-tokyo.ac.jp

Tonomura, Akira
Advanced Research Laboratory, Hitachi, Ltd.
Hatoyama, Saitama 350-0395
Japan
    Phone: +81-492-96-6111
    Fax: +81-492-96-6006
    e-mail: tonomura@harl.hitachi.co.jp

Toyoshima, Kouichi
Faculty of Science and Engineering
University of Saga
1 Honjo, Saga 840-8502
Japan
    Phone: +81-952-28-8845
    Fax: +81-952-28-8845
    e-mail: toyo@cc.saga-u.ac.jp

Tsai, Jaw-Shen
Fundamental Research Laboratories
NEC Corporation
34 Miyukigaoka, Tsukuba, Ibaraki 305-8501
Japan
    Phone: +81-298-50-1161
    Fax: +81-298-56-6139
    e-mail: tsai@frl.cl.nec.co.jp

Tsai, S.Y.
Department of Physics
College of Science and Technology
Nihon University

Kanda-Surugadai, Chiyoda-ku
Tokyo 101-8308,
Japan
    Phone: +81-3-3259-0867
    Fax: +81-3-3293-8269
    e-mail: tsai@phys.cst.nihon-u.ac.jp

Tsutsui, Izumi
Institute of Particle and Nuclear Studies
High Energy Accelerator Research
   Organization (KEK)
Midori-cho, Tanashi-shi, Tokyo 188-8501
Japan
    Phone: +81-424-69-9522
    Fax: +81-424-62-0763
    e-mail: tsutsui@tanashi.kek.jp

Uchiyama, Fumiyo
Institute of Applied Physics
University of Tsukuba
1-1-1 Tennodai, Tsukuba, Ibaraki 305-8573
Japan
    Phone: +81-298-53-4993
    Fax: +81-298-53-5205
    e-mail: uchiyama@kekvax.kek.jp

Ueda, Masahito
Dept. of Physical Electronics
Faculty of Engineering, Hiroshima University
1-4-1 Kagamiyama
Higashi-Hiroshima 739-8527
Japan
    Phone: +81-824-24-7653
    Fax: +81-824-24-7650
    e-mail: mueda@ipc.hiroshima-u.ac.jp

Ulam-Orgikh, Duger
Dept. of Electrical Engineering
Osaka University
1-3 Machikaneyama
Toyonaka-shi, Osaka 560-8531
Japan
    Phone: +81-6-850-6336
    Fax: +81-6-850-6341
    e-mail: uka@laser.ee.es.osaka-u.ac.jp

Varoquaux, Eric
Laboratoire de Physique des Solides
Bât. 510 - Université Paris-Sud
91405 Orsay
France
    Phone: +33-1-69087333
    Fax: +33-1-69088786
    e-mail: evaroquaux@cea.fr
          varoquau@lps.u-psud.fr

Waki, Izumi
Advanced Research Laboratory, Hitachi, Ltd.
Hatoyama, Saitama 350-0395
Japan
    Phone: +81-492-96-6111(ext.6214)
    Fax: +81-492-96-6006
    e-mail: waki@harl.hitachi.co.jp

Webb, Richard A.
Center for Superconductivity Research
Dept. of Physics, University of Maryland
College Park, MD 20742-4111
USA
    Phone: +1-301-405-6175
    Fax: +1-301-405-3779
    e-mail: rawebb@squid.umd.edu

Weidenmüller, Hans A.
Max-Planck-Institut für Kernphysik
Postfach 103980, D-69029 Heidelberg
Germany
    Phone: +49-6221-516291
    Fax: +49-6221-516605
    e-mail: Hans.Weidenmueller@
          mpi-hd.mpg.de

Williams, David A.
Hitachi Cambridge Laboratory, Cavendish Lab.
Madingley Road, Cambridge CB3 0HE
United Kingdom
    Phone: +44-1223-467944
    Fax: +44-1223-467942
    e-mail: williams@phy.cam.ac.uk

Wineland, David J.
Ion Storage Group
National Institute of Standards and Technology
Div. 847.10
325 Broadway, Boulder, Colorado 80303
USA
    Phone: +1-303-497-5286
    Fax: +1-303-497-7375
    e-mail: dwineland@nist.gov

Yamada, Eizaburo
Faculty of Information Sciences
Meisei University
2-590 Nagabuchi, Ome-shi, Tokyo 198-8655
Japan
    Phone: +81-428-25-5267
    Fax: +81-428-25-5185

Yamamoto, Yoshihisa
E. L. Ginzton Laboratory, Stanford University
Stanford, CA 94305-4085
USA
    Phone: +1-650-725-3327
    Fax: +1-650-723-5320
    e-mail: yamamoto@loki.stanford.edu
NTT Basic Research Laboratories
3-1 Morinosato-Wakamiya, Atsugi-shi
Kanagawa 243-0198
Japan

Yamanaka, Yoshiya
Dept. of Physics
Waseda Univ. Senior High School
3-31-1 Kamishakujii, Nerima-ku
Tokyo 177-0044
Japan
    Phone: +81-3-5991-4182
    Fax: +81-3-3928-4110
    e-mail: yamanaka@mn.waseda.ac.jp

Yamashita, Makoto
NTT Basic Research Laboratories
3-1 Morinosato-Wakamiya, Atsugi-shi
Kanagawa 243-0198
Japan
    Phone: +81-462-40-3486
    Fax: +81-462-40-4726
    e-mail: makoto@will.brl.ntt.co.jp

Yoneda, Tetsuya
Dept. of Physics, Faculty of Science
Kumamoto University
2-39-1 Kurokami, Kumamoto 860-8555
Japan
    Phone: +81-96-373-5487
    Fax: +81-96-373-5487
    e-mail: tyoneda@aps.cms.kumamoto-u.ac.jp

Zanardi, Paolo
Inst. for Scientific Interchange Foundation
Villa Gualino, Viale Settimio Severo 65
IT-10133 Torino
Italy
    Phone: +39-11-6603090
    Fax: +39-11-6600049
    e-mail: zanardi@isiosf.isi.it

Zeilinger, Anton
Institut für Experimental Physik
Universität Innsbruck
Technikerstrasse 25/4
A-6020 Innsbruck
Austria
    Phone: +43-512-507-6300
    Fax: +43-512-507-2921
    e-mail: anton.zeilinger@uibk.ac.at

Zhang, Yi
Department of Applied Physics
Stanford University
Stanford CA 94305-4090
USA
    Phone: +1-650-725-8697
    e-mail: yzhang@stanford.edu

AUTHOR INDEX

Printed and bound by CPI Group (UK) Ltd, Croydon, CR0 4YY

03/10/2024

01040328-0009